TEXTS AND READINGS IN MATHEMATICS **84**

Dilations, Completely Positive Maps and Geometry

Dilations, Completely Positive Maps and Geometry

B V Rajarama Bhat
Indian Statistical Institute
Bengaluru

Tirthankar Bhattacharyya
Indian Institute of Science
Bengaluru

HINDUSTAN
BOOK AGENCY

Published by

Hindustan Book Agency (India)
P 19 Green Park Extension
New Delhi 110 016
India

Email: info@hindbook.com
www.hindbook.com

ISBN 978-81-957829-2-5

Raja dedicates this book to Vidya and Shikhara.

Tirtha dedicates this book to his mother and to the memory of his father.

Contents

Preface ix

Chapter 1. Dilation for One Operator 1
1.1. Positive kernels 1
1.2. Positive operators 11
1.3. Tensor product of bounded operators 14
1.4. Dilation of a single contraction 16
1.5. First proof of existence - an abstract method 20
1.6. Second proof of existence - Schäffer construction 24
1.7. The unilateral shift 25
1.8. The Hardy space 27
1.9. Pure contractions 29
1.10. The characteristic function and functional model 31
1.11. The characteristic function as a complete unitary invariant 37
1.12. Decomposition 39
1.13. The unitary dilation 40
1.14. The Berger-Coburn-Lebow Theorem 43

Chapter 2. C^*-Algebras and Completely Positive Maps 47
2.1. Banach algebras 47
2.2. C^*-algebras 58
2.3. Completely positive maps 78

Chapter 3. Dilation Theory in Two Variables - The Bidisc 91
3.1. Tuples of commuting contractions 91
3.2. Two applications 97
3.3. The realization formula in the unit disc 99
3.4. Contractions with finite dimensional defect spaces 102

Chapter 4. Dilation Theory in Several Variables - the Euclidean Ball 109
4.1. Contractive tuples 109
4.2. The full Fock space and the symmetric Fock space 110
4.3. The Toeplitz C^*-algebra 116
4.4. Standard commuting dilation 121

4.5. Standard non-commuting dilation 124
4.6. The limiting operator T_∞ 130
4.7. Maximality of the standard commuting dilation 132

Chapter 5. The Euclidean Ball - The Drury Arveson Space 141
5.1. The space 141
5.2. The characteristic function 151
5.3. Functional model of a pure commuting contractive tuple 154
5.4. The characteristic function as a complete unitary invariant 159

Chapter 6. Dilation Theory in Several Variables - The Symmetrized Bidisc 163
6.1. The joint spectrum 163
6.2. Normal operator tuples 166
6.3. Spectral sets 169
6.4. The symmetrized bidisc 171
6.5. Polynomial convexity 172
6.6. The Fejér-Riesz Theorem 173
6.7. The set Γ 180
6.8. The distinguished boundary of Γ 183
6.9. Structure theorems for $\Gamma-$isometries and $\Gamma-$unitaries 185
6.10. Γ-contractions - examples 195
6.11. Dilation 203

Chapter 7. An Abstract Dilation Theory 215
7.1. Set theoretic dilation 215
7.2. Single variable dilation theory 215
7.3. Multivariable dilation theory 226
7.4. Berger, Coburn and Lebow Theorem 230
7.5. Generalizations and variations 233

Bibliography 237

Index 245

Preface

The notion of dilation of a bounded operator on a Hilbert space was a seminal concept introduced by Sz.-Nagy in the mid-twentieth century. Sz.-Nagy provided abstract arguments for dilating a contraction to a unitary operator. Soon after, Schäffer produced an explicit construction. When Ando followed it up a dozen years later by dilating a commuting pair of contractions to a commuting pair of unitary operators, it started becoming clear that there is a deep relation between dilating a commuting operator d-tuple satisfying some natural contractivity condition and the geometry of the distinguished boundary (the Silov boundary with respect to a natural algebra of functions) of a domain in the complex Euclidean space $\mathbb{C}^d$ (to which the contractivity condition is attuned). Following this idea, Arveson proved a landmark theorem stating that a boundary normal dilation exists if and only if a certain contractive homomorphism is completely contractive. Thus, the topics of dilation, complex geometry and completely positive maps got related.

The aim of this book is to give concrete constructions of dilations of commuting operator tuples related to four domains where this theory has been beautifully successful so far. The first one is the classical unit disc. This is Sz.-Nagy's theory, later joined by Foias. The next is the Ando dilation of a commuting pair of contractions and related recent development. Indeed, there has been a renewed interest in the bidisc case and interesting new results about finding concrete dilations for a pair of commuting contractions with finite dimensional defect spaces have been incorporated in this book. We then move on to the case of the Euclidean unit ball. There, the nature of the dilation is different from that in the disc or in the bidisc. It is in this domain that the theory of completely positive maps plays a big role. That enables one not only to do an explicit construction of the dilation for a commuting contractive tuple, but also for a non-commuting contractive tuple. Finally, we discuss the case of a domain studied in great detail of late, viz., the symmetrized bidisc. This falls in the list of domains that admit the Arveson theme, viz., any pair of commuting bounded operators that has the domain as a spectral set also has it as a complete spectral set. This was proved abstractly in the end of twentieth century. Explicit constructions for the dilation which, then, is bound to exist, is done recently. The construction here is greatly interesting because of various inputs that are required like the Fejer-Riesz theorem for example. So, it connects to holomorphic function theory and geometry very well.

A lot of examples are discussed. The content of this book is fit for a two semester course to beginning Ph.D. students. The first semester may concentrate on Chapters 1 through 4. This portion assumes the knowledge of a standard M.Sc. in mathematics. For example, we assume the reader is comfortable with one variable complex analysis, measure theory and a first course in functional analysis. After these chapters, a couple of advanced themes have been used. These are certain elements of several variables complex analysis and joint spectral theory. We have started from the basics of holomorphic functions in several variables in Chapter 5. In Chapter 6, we have briefly described what the joint spectrum is. Here, we also need to use the Oka-Weil Theorem for a particular domain in $\mathbb{C}^2$. The material there could not be made completely self-contained because proving this landmark theorem in several variables complex analysis would have been too much of a detour for this book. For both of these advanced topics, appropriate references have been given. Thus, the second semester of the course may start with some discussion on these two classical topics followed by Chapters 5, 6 and 7. At the end, a student would be ready for research on current topics.

Some notational and linguistic comments are in order. The unit in a unital C^*-algebra is often denoted just by 1. Sometimes the constant function 1 on a set X is also denoted by 1. It will be clear from the context what 1 stands for. When X is a compact Hausdorff topological space, then of course the two 1's are same because the constant function 1 is the unit of the unital C^*-algebra $C(X)$. All the operators in this book are bounded operators on some Hilbert space. So, sometimes we drop the adjective bounded. It is understood that an operator is a bounded operator.

There are many who we would like to thank for their helpful comments on a preliminary version of the book. The valuable inputs given by Michael Dritschel, Jaydeb Sarkar and Orr Shalit have greatly improved the book. We are also thankful to Michael Dritschel and Jaydeb Sarkar for allowing us to reproduce parts of their work. Arghya Chongdar and Mainak Bhowmik did a great job to read and present the contents of the book to us which eliminated many typographical errors and led to rewriting some proofs for better presentation. Careful reading by Anindya Biswas, Astrid Swizell Dias, Mahesh Krishna K and Biswarup Saha caught some typos and inconsistencies.

We are very thankful to John Holbrook and two anonymous referees for careful, detailed and patient readings leading to innumerable suggestions which greatly improved the book. This book would not have been possible without the continuous pursuance of Mr. D. K. Jain. The overarching influence of Professor Rajendra Bhatia for giving encouragement and for being patient is gratefully acknowledged.

The works of both authors are supported by J. C. Bose fellowships of the Science and Engineering Research Board (SERB), Government of India. The grant numbers are JCB/2021/000041, JCB/2018/000020 and JBR/2021/000024.

Bangalore
February, 2023

CHAPTER 1

Dilation for One Operator

This book assumes a basic knowledge of Hilbert spaces and bounded operators on them. All our Hilbert spaces are over the complex field and are separable. What we do in this chapter is very classical. Appropriate references have been mentioned at the end of every section.

1.1. Positive kernels

To start with something simple and elegant, we choose *positive kernels*. These are natural generalizations of positive semi-definite matrices and appear at every corner in functional analysis. An $n \times n$ matrix A is called a positive semi-definite matrix if the associated quadratic form $\langle Ax, x\rangle$ is non-negative. Whenever we write an inner product $\langle \cdot, \cdot \rangle$ without any further qualification, it will denote the natural inner product of the ambient Hilbert space, in this case, $\mathbb{C}^n$, i.e., $\langle x, y\rangle = \sum x_k \overline{y_k}$. Positive semi-definite matrices have been studied greatly and several characterizations are known. In particular, these are self-adjoint matrices with non-negative eigenvalues. In the case when a positive semi-definite matrix is actually invertible, i.e., when the eigenvalues are all positive, it will be called a positive definite matrix. When the matrix A is positive definite, we can give a new inner product on $\mathbb{C}^n$, viz.,

$$\langle x, y\rangle_A = \langle Ax, y\rangle.$$

This new inner product on $\mathbb{C}^n$ can also be described as follows. Let e_j be the vector whose jth co-ordinate is 1 and all the rest are 0. Equip $\mathbb{C}^n$ with the new inner product by first declaring

$$\langle e_j, e_i\rangle_A = a_{ij}$$

and then extending by linearity. Note that here $\langle x, x\rangle_A = 0$ implies $x = 0$ due to invertibility of A. A new inner product $\langle \cdot, \cdot\rangle_A$ can also be constructed in case A is positive semi-definite and not positive definite, albeit on a smaller dimensional space, through a quotienting procedure. In what follows, we give a generalization of this concept to an arbitrary index set instead of the set $\{1, 2, \ldots, n\}$.

Taking a different view point we ask the question as to given a set X, when it can be embedded as a set of vectors in a Hilbert space. To do that, we need to know the inner product between these vectors. Recall that given any n vectors $v_1, v_2, \ldots, v_n$ in a Hilbert space, their *Gram matrix* is defined as the matrix of inner products:

$$[\langle v_j, v_i \rangle]_{1 \leq i,j \leq n}$$

and it is always positive semi-definite. This motivates the following definition.

Definition 1.1.1. *Let X be a non-empty set. A positive kernel k on X is function $k : X \times X \to \mathbb{C}$ such that, for any positive integer n, any $x_1, x_2, \ldots, x_n$ from X and any n complex numbers $c_1, c_2, \ldots, c_n$,*

$$\sum_{i=1}^{n} \sum_{j=1}^{n} c_i \overline{c_j} k(x_j, x_i) \geq 0.$$

Note that k is a positive kernel means that for every choice of the positive integer n and any $x_1, x_2, \ldots, x_n$, the $n \times n$ matrix

$$((k(x_j, x_i)))$$

is a positive semi-definite matrix. We are considering the matrix with (i, j)th entry as $k(x_j, x_i)$, instead of more natural looking $k(x_i, x_j)$, just so that later on our notation matches with standard kernels such as the Szegő kernel. Examples of positive kernels are easy to come by.

(1) Let X be the finite set $\{1, 2, \ldots, n\}$. Take an $n \times n$ positive semi-definite matrix $A = ((a_{ij}))$ and define $k(j, i) = a_{ij}$ for all $i, j = 1, 2, \ldots, n$.
(2) Let $X = H$, a Hilbert space. Let A be a positive operator, i.e., a bounded operator on H that satisfies $\langle Ah, h \rangle \geq 0$ for all $h \in H$. Let $k(x, y) = \langle Ax, y \rangle$.
(3) Let X be any non-empty set and let $g : X \to \mathbb{C}$ be a function. Take $k(x, y) = g(x)\overline{g(y)}$ $\forall x, y \in X$. Then k is a positive kernel.

Theorem 1.1.2. *[GNS Construction] Given a positive kernel k on a set X, there exists a Hilbert space $\mathcal{H}$ and a map $\lambda : X \to \mathcal{H}$ such that*

$$k(x, y) = \langle \lambda(y), \lambda(x) \rangle, \qquad x, y \in X. \tag{1.1.1}$$

Moreover, the space $\mathcal{H}$ can be chosen to be minimal in the following sense:

$$\mathcal{H} = \overline{span}\{\lambda(x) : x \in X\}. \tag{1.1.2}$$

If $(\mathcal{H}', \lambda')$ is another such minimal pair, then there exists a unitary $U : \mathcal{H} \to \mathcal{H}'$ satisfying $U(\lambda(x)) = \lambda'(x)$ for all $x \in X$.

This theorem will be a basic tool for us. As per Wikipedia, positive definite kernels first appeared in 1909 in a work on integral equations by James Mercer. The subject has a long history and has been re-discovered several times. It is intimately connected with the notion of *Reproducing Kernel Hilbert Spaces* (RKHS) . As we shall see, the Hilbert space of the theorem above can be naturally realized as a reproducing kernel Hilbert space and this is known as the Moore - Aronszajn theorem. The paper [12] of Aronszajn is a classic. M. G. Krein and A. N. Kolmogorov are some of the other names associated to this field. For an account of this history and further details, we refer to Szafraniec [114]. Paulsen and Raghupati provide a modern introduction to the theory of reproducing kernel Hilbert spaces in [90].

Before we start the proof, which is constructive, we set up some basic conventions. Given two Hilbert spaces $\mathcal{H}$ and $\mathcal{K}$, when we say $\mathcal{H}$ is contained in $\mathcal{K}$, usually denoted by $\mathcal{H} \subset \mathcal{K}$ (or $\mathcal{K} \supset \mathcal{H}$) we shall mean that $\mathcal{H}$ is physically contained in $\mathcal{K}$ as a closed subspace or that $\mathcal{H}$ is isometrically embedded into $\mathcal{K}$, i.e., there is a linear isometry V mapping $\mathcal{H}$ into $\mathcal{K}$. In the latter case, we shall identify $\mathcal{H}$ with the closed subspace Ran V of $\mathcal{K}$ and make no distinction between the elements h and Vh. In other words, V implements a Hilbert space isomorphism between $\mathcal{H}$ and Ran V. An isometry which is surjective is called a unitary map. If V is a unitary map, then we say that $\mathcal{H}$ and $\mathcal{K}$ are unitarily equivalent. Sometimes we say that V is a unitary isomorphism between the Hilbert spaces $\mathcal{H}$ and $\mathcal{K}$.

Proof: Given a positive kernel $k : X \times X \to \mathbb{C}$, one can construct a Hilbert space $\mathcal{H}$ and a function $\lambda : X \to \mathcal{H}$ as follows. Consider the vector space $\mathbb{C}X$ of all complex valued functions $\xi : X \to \mathbb{C}$ with the property that $\xi(x) = 0$ for all but a finite number of $x \in X$. We can define a sesquilinear form $\langle \cdot, \cdot \rangle$ on $\mathbb{C}X$ by

$$\langle \xi, \eta \rangle = \sum_{x,y \in X} k(x,y)\xi(y)\overline{\eta(x)}, \qquad \xi, \eta \in \mathbb{C}X.$$

Even if X is an infinite set, the sum here is a finite sum as ξ, η vanish except for finite sets. Indeed if for $x \in X$, δ_x is the indicator function of x, then $\{\delta_x : x \in X\}$ is a vector space basis for $\mathbb{C}X$ and if $\xi = \sum_j d_j \delta_{y_j}$ and $\eta = \sum_i c_i \delta_{x_i}$, then

$$\langle \xi, \eta \rangle = \sum_{i,j} d_j \overline{c_i} k(x_i, y_j).$$

It is then clear that $\langle \cdot, \cdot \rangle$ is a well-defined sesqui-linear form and it is positive semidefinite because of the positivity hypothesis on k. However, it may not be an inner product as $\langle \xi, \xi \rangle = 0$ may not mean that

$\xi = 0$. Now recall that the Cauchy-Schwarz inequality

$$|\langle \xi, \eta \rangle|^2 \leq \langle \xi, \xi \rangle \langle \eta, \eta \rangle,$$

is valid even for semi-inner products and this implies that the set

$$N := \{\xi \in \mathbb{C}X : \langle \xi, \xi \rangle = 0\}$$

is same as

$$\{\xi \in \mathbb{C}X : \langle \xi, \eta \rangle = 0, \ \forall \eta \in \mathbb{C}X\}.$$

Hence N is a linear subspace of $\mathbb{C}X$, and so we can quotient $\mathbb{C}X$ by it. This way the semi-inner product $\langle \cdot, \cdot \rangle$ can be promoted naturally to an inner product on the quotient $\mathbb{C}X/N$, i.e., if $\xi + N$ and $\eta + N$ are two cosets, then

$$\langle \xi + N, \eta + N \rangle = \langle \xi, \eta \rangle.$$

Let $\mathcal{H}$ be the Hilbert space obtained by completing the inner product space $\mathbb{C}X/N$. Define a function $\lambda : X \to \mathcal{H}$ as follows:

$$\lambda(x) = \delta_x + N, \qquad x \in X. \tag{1.1.3}$$

By construction, $k(x, y) = \langle \lambda(y), \lambda(x) \rangle$. Note too that the pair $(\mathcal{H}, \lambda)$ is *minimal* for k just by construction.

Now suppose $(\mathcal{H}', \lambda')$ is another such pair. We could try to define U by $U(\lambda(x)) = \lambda'(x)$ and extend it linearly to finite linear combinations of the $\lambda(x)$. Indeed, it then defines a linear operator on a dense subspace of $\mathcal{H}$ and we find that for points $x_1, x_2, \ldots, x_n$ and $y_1, y_2, \ldots, y_m$ from X, we have

$$\begin{aligned}
\langle U(\sum_{j=1}^{n} d_j \lambda(y_j)), U(\sum_{i=1}^{m} c_i \lambda(x_i)) \rangle_{\mathcal{H}'} &= \sum_{i,j} d_j \bar{c}_i \langle U(\lambda(y_j)), U(\lambda(x_i)) \rangle_{\mathcal{H}'} \\
&= \sum_{i,j} d_j \bar{c}_i \langle \lambda'(y_j), \lambda'(x_i) \rangle_{\mathcal{H}'} \\
&= \sum_{i,j} d_j \bar{c}_i k(x_i, y_j) \\
&= \sum_{i,j} d_j \bar{c}_i \langle \lambda(y_j), \lambda(x_i) \rangle_{\mathcal{H}} \\
&= \langle \sum_{j=1}^{n} d_j \lambda(y_j), \sum_{i=1}^{m} c_i \lambda(x_i) \rangle_{\mathcal{H}}.
\end{aligned}$$

Thus U is an isometry from a dense subspace of $\mathcal{H}$ to a dense subspace of $\mathcal{H}'$. As U is isometric, it is bounded and hence extends to a bounded linear map on the closure. Then it is clear that the extension remains an isometry and is in fact a unitary as the range was already dense. ■

Obtaining a Hilbert space $\mathcal{H}$ from a positive kernel k will be referred to as the Kolmogorov construction or the Gelfand-Naimark-Segal (GNS) construction . The space $\mathcal{H}$ is called the GNS space and the pair $(\mathcal{H}, \lambda)$ together is called the GNS pair . The only word of caution is that in general, the GNS space $\mathcal{H}$ need not be separable. The GNS pair $(\mathcal{H}, \lambda)$ is said to be *minimal* if (1.1.2) is also satisfied. The following proposition lists some basic properties of the GNS construction.

Proposition 1.1.3. *Let K be a positive kernel on a set X and let $(\mathcal{H}, \lambda)$ be a GNS pair associated with K.*

(i) Suppose X is a topological space and K is continuous with respect to product topology on $X \times X$. Then λ is continuous.

(ii) Suppose X is a separable topological space and $(\mathcal{H}, \lambda)$ is a minimal GNS pair. Then $\mathcal{H}$ is separable.

(iii) Suppose $k \in \mathbb{N}$ and $X = V_1 \times V_2 \times \cdots \times V_k$, where $V_1, V_2, \ldots, V_k$ are some complex vector spaces. Suppose K is multi-linear in the second variable. Then λ is multi-linear.

Proof. (i) For x, y in X, $\|\lambda(y)-\lambda(x)\|^2 = K(y,y)-K(y,x)-K(x,y)+K(x,x)$ Hence if a sequence or a net $\{x_n\}$ converges to x then $\lambda(x_n)$ converges to $\lambda(x)$. This proves (i).

(ii) If S is a countable dense set in X, by continuity of λ,

$\lambda(S)$ is dense in $\lambda(X)$. Now considering linear span of $\lambda(S)$ with complex coefficients whose real and imaginary parts are rational, we get a countable dense set of $\mathcal{H}$.

(iii) Observe that by positivity $K(y,x) = \overline{K(x,y)}$ and hence K is anti-multilinear in the first variable. Now by direct computation

$$\begin{aligned}&\|\lambda(x_1, x_2, \ldots, x_{i-1}, ax_i + by_i, x_{i+1}, \ldots, x_n)\\ &- a\lambda(x_1, \ldots, x_n) - b\lambda(x_1, \ldots, x_{i-1}, y_i, x_{i+1}, \ldots, x_n)\|^2 = 0.\end{aligned}$$

This shows multi-linearity of λ. ∎

Now we wish to give a different presentation of the GNS construction which takes us to the notion of reproducing kernel Hilbert spaces.

A different presentation (or a second proof): For $y \in X$, define a function k_y on X by setting

$$k_y(x) = k(x,y) \text{ for all } y \in X.$$

Now take $W = \text{span}\{k_x : x \in X\}$. On the vector space W, we wish to have an inner product satisfying

$$\langle \sum_j d_j k_{y_j}, \sum_i c_i k_{x_i} \rangle = \sum_{i,j} d_j \overline{c_i} k(x_i, y_j).$$

There is a subtle point here. It is possible that the same function in W can be expressed in several different ways as linear combinations of generating vectors $\{k_x : x \in X\}$. So we need to verify that this candidate for the inner product is well-defined. This can be seen as follows. For $g = \sum_j d_j k_{y_j}$ and $h = \sum_i c_i k_{x_i}$,

$$\sum_{i,j} d_j \overline{c_i} k(x_i, y_j) = \sum_i \overline{c_i} g(x_i) = \sum_j d_j \overline{h(y_j)}.$$

This shows that this value does not depend upon the expansion of g as $\sum_j d_j k_{y_j}$ or on the expansion of h as $\sum_i c_i k_{x_i}$. Therefore

$$\langle g, h \rangle := \sum_{i,j} d_j \overline{c_i} k(x_i, y_j) \tag{1.1.4}$$

is a well-defined sesqui-linear form on W. The hypothesis that k is a positive kernel shows that it is a semi-inner product.

Now if $\langle h, h \rangle = 0$, by the Cauchy-Schwarz inequality (it is valid for semi-inner products), $\langle g, h \rangle = 0$ for every $g \in W$. In particular, for every $y \in X$, $h = \sum_i c_i k_{x_i}$,

$$\langle k_y, h \rangle = \sum_i \overline{c_i} K(x_i, y) = \overline{h(y)} = 0.$$

Consequently $h = 0$. Therefore (1.1.4) defines a genuine inner-product on W. Now take $\mathcal{H}'$ as the completion of W with respect to the inner product defined above. Define $\lambda' : X \to \mathcal{H}'$ by $\lambda'(x) = k_x$. Clearly we have $\langle \lambda'(y), \lambda'(x) \rangle = k(x, y)$. This completes the alternative construction of GNS pair. ■

Definition 1.1.4. *A subset $\mathcal{T}$ of a Hilbert space $\mathcal{H}$ is called total if* $\overline{\textit{span}}\{h : h \in \mathcal{T}\} = \mathcal{H}$.

It is to be noted that in the construction above we don't need quotienting, but completion is still required. Prior to completion every element of the Hilbert space is a function. Does this still hold after completion? Surprisingly, it is the case. For $g \in \mathcal{H}'$ we set $g(x) = \langle g, k_x \rangle$. It is easy to verify that for $g, h \in \mathcal{H}'$, if $\langle g, k_x \rangle = \langle h, k_x \rangle$ for all $x \in X$ then $g = h$. Indeed, putting $g - h = f$, we get that $\langle f, k_x \rangle = 0$ for all $x \in X$. Recall that by definition of $\mathcal{H}'$, the vectors k_x form a total set in $\mathcal{H}'$. Hence,

f is 0. Hence the association $g(x) = \langle g, k_x \rangle$ is well-defined. This association respects linearity, that is, $(cg + dh)(x) = cg(x) + dh(x)$, for $c, d \in \mathbb{C}, g, h \in \mathcal{H}'$. Therefore $\mathcal{H}'$ can be considered as an inner product space of functions on X. Moreover, there is no change in the identification if g is already in W. Finally, for $g \in \mathcal{H}'$, if $\{f_n\}$ is a sequence in W such that $\lim f_n = g$, then by taking inner product with k_x, we get $\lim_{n\to\infty} f_n(x) = g(x)$. So we get pointwise convergence of functions. Here we have seen that evaluations become continuous linear functionals on the Hilbert space. This motivates the following notion.

Definition 1.1.5. *Given a non-empty set X, a Hilbert space $\mathcal{H}$ is said to be a reproducing kernel Hilbert space (RKHS) over X if $\mathcal{H}$ is a space of scalar valued functions on X and the evaluation maps $f \mapsto f(x)$ are continuous linear functionals on $\mathcal{H}$ for every $x \in X$.*

Clearly the space $\mathcal{H}'$ we got above in the second proof of Theorem 1.1.2 is a reproducing kernel Hilbert space over X.

Example 1.1.6. *Consider the space $X = C[0,1]$ of continuous functions on the unit interval. Define $k : X \times X \to \mathbb{C}$ by*

$$k(f,g) = \int_0^1 g(t)\overline{f(t)}dt.$$

Clearly k is a positive kernel. Indeed k is an inner product on X. However $C[0,1]$ is not complete with respect to the L^2 norm and on completion we get $L^2[0,1]$. It is to be noted that evaluations at points $t \in [0,1]$ are not well-defined in $L^2[0,1]$ as elements in it are not functions on $[0,1]$, but equivalence classes of functions. However $L^2[0,1]$ is a reproducing kernel Hilbert space on X.

Every Hilbert space $\mathcal{H}$ can be thought of as a reproducing kernel Hilbert space over any total set X of $\mathcal{H}$ as follows. We identify $g \in \mathcal{H}$, with a function X, defined by $g(x) = \langle g, x \rangle$, for all $x \in X$. Note that the totality of X guarantees that distinct vectors in $\mathcal{H}$ give rise to distinct functions. Since the evaluation functional $g \mapsto g(x)$ is an inner product with a vector, it is continuous.

The astute reader would have noticed by now that the fact that every Hilbert space $\mathcal{H}$ can be thought of as a reproducing kernel Hilbert space, is the Riesz representation theorem in disguise (the dual space of a Hilbert space is a Hilbert space of linear functions with bounded point evaluation being continuous).

We observed that given a positive kernel k on a set, we get a reproducing kernel Hilbert space through the GNS construction. Now we consider the converse direction. Suppose $\mathcal{H}$ is a reproducing kernel Hilbert space over a set X. Since evaluation maps are continuous linear functionals on $\mathcal{H}$, by Riesz's theorem, for every $x \in X$ we get a vector k_x in $\mathcal{H}$ such that $g(x) = \langle g, k_x \rangle$ for all $g \in \mathcal{H}$. Define $k : X \times X \to \mathbb{C}$ by

$$k(x, y) = \langle k_y, k_x \rangle \text{ for all } x, y \in X.$$

Then clearly k is a positive kernel as $[k(x_j, x_i)]_{1 \leq i,j \leq n}$ is a Gram matrix for every $x_1, x_2, \ldots x_n$ in X, $n \geq 1$. Define $\lambda : X \to \mathcal{H}$ by $\lambda(x) = k_x$. It is easy to see that $(\mathcal{H}, \lambda)$ is the GNS pair of k. It is also minimal, that is, $\mathcal{H} = \overline{\text{span}}\{\lambda(x) : x \in X\} = \overline{\text{span}}\{k_x : x \in X\}$, because if $g \in \mathcal{H}$, and $\langle g, k_x \rangle = 0$ for all x, then $g(x) = 0$ for all x and hence $g = 0$. So there is a one to one correspondence between reproducing kernel Hilbert spaces and positive kernels.

Thus, the set $\{k_x : x \in X\}$ considered above is a total set.

The kernel of an RKHS can be constructed through an ortho-normal basis as follows. Suppose $\mathcal{H}$ is an RKHS on a set X. Let B be an indexing set and $\{e_n : n \in B\}$ be an ortho-normal basis for $\mathcal{H}$. As before, by Riesz's theorem, we have vectors k_y in $\mathcal{H}$ such that $g(y) = \langle g, k_y \rangle$ for all $g \in \mathcal{H}$. Expanding k_y in the given basis, we get

$$k_y = \sum_{n \in B} \langle k_y, e_n \rangle e_n = \sum_{n \in B} \overline{e_n(y)} e_n.$$

Hence

$$k(x, y) = \langle k_y, k_x \rangle = \langle \sum_{m \in B} \overline{e_m(y)} e_m, \sum_{n \in B} \overline{e_n(x)} e_n \rangle = \sum_{n \in B} \overline{e_n(y)} e_n(x).$$

This is known as Zaremba formula for the reproducing kernel (See [114]).

Here is the best known RKHS. We will be revisiting this space frequently.

Example 1.1.7. *[Hardy space] Let H^2 be the space of all functions f on the open unit disc $D = \{z \in \mathbb{C} : |z| < 1\}$, having power series expansion:*

$$f(z) = \sum_{n=0}^{\infty} a_n z^n, \; z \in D,$$

where $\sum_{n=0}^{\infty} |a_n|^2 < \infty$, with inner product $\langle f, g \rangle := \sum_{n=0}^{\infty} a_n \overline{b_n}$, for $g(z) = \sum_{n=0}^{\infty} b_n z^n$. Then the functions $\{z^n : n \geq 0\}$ form an orthonormal basis. In other words we have an orthonormal basis $\{e_n : n \geq 0\}$ with $e_n(z) = z^n$, $z \in D$. Now by Zaremba formula, H^2 is an RKHS on D with kernel

$$k(z, w) = \sum_{n=0}^{\infty} e_n(z) \overline{e_n(w)} = \sum_{n=0}^{\infty} z^n \overline{w}^n = \frac{1}{1 - z\overline{w}} \text{ for all } z, w \in D.$$

This kernel k has a name. It is usually called the Szegő kernel.

Let us now see how a map between two sets with kernels preserving the kernel structures in a suitable way implies that the GNS space of the first kernel is contained in that of the other. Suppose there are two positive kernels k_1 and k_2 on two sets X_1 and X_2 respectively with their respective GNS pairs $(\mathcal{H}_i, \lambda_i)$ for $i = 1, 2$. Suppose $\varphi : X_1 \to X_2$ is a map that satisfies

$$k_2(\varphi(x), \varphi(y)) = k_1(x, y), \qquad x, y \in X.$$

Then, using the GNS pair, we immediately have

$$\begin{aligned}\langle \lambda_2(\varphi(y)), \lambda_2(\varphi(x)) \rangle_{H(k_2)} &= k_2(\varphi(x), \varphi(y)) \\ &= k_1(x, y) = \langle \lambda_1(y), \lambda_1(x) \rangle_{H(k_1)},\end{aligned}$$

holding for all $x, y \in X_1$. Define $U_\varphi : H(k_1) \to H(k_2)$ by defining it first on $\{\lambda_1(x) : x \in X\}$ by

$$U_\varphi \lambda_1(x) = \lambda_2(\varphi(x))$$

and then extending it to linear combinations. The equality above shows that U_φ is an inner product preserving linear map defined on a dense subset of $\mathcal{H}(k_1)$. Hence we get a unique linear isometry from $\mathcal{H}(k_1)$ into $\mathcal{H}(k_2)$ such that

$$U_\varphi(\lambda_1(x)) = \lambda_2(\varphi(x)), \qquad x \in X_1. \tag{1.1.5}$$

As expected, it is not difficult to verify that $U_{\varphi_1} U_{\varphi_2} = U_{\varphi_1 \circ \varphi_2}$ holds.

We close this section by applying the GNS construction of obtaining a Hilbert space from a positive kernel to construct tensor product of finitely many Hilbert spaces. There are two elementary ways in which we produce new Hilbert spaces from old ones. One of them is taking direct sum. Given countably infinitely many Hilbert spaces $\mathcal{H}_1, \mathcal{H}_2, \ldots$, their direct sum, denoted by $\mathcal{H}_1 \oplus \mathcal{H}_2 \oplus \cdots$ is defined as the set of all sequences

$$(h_1, h_2, \ldots); \text{ with } h_i \in \mathcal{H}_i \text{ and } \|h_1\|^2 + \|h_2\|^2 + \cdots < \infty.$$

A straightforward application of elementary Cauchy-Schwarz inequality shows that this is a vector space and the following defines an inner product on it.

$$\langle (h_1, h_2, \ldots), (k_1, k_2, \ldots) \rangle = \sum \langle h_i, k_i \rangle.$$

Completeness of $\mathcal{H}_1 \oplus \mathcal{H}_2 \oplus \cdots$ follows from completeness of each of the $\mathcal{H}_i$. An element of $\mathcal{H}_1 \oplus \mathcal{H}_2 \oplus \cdots$ will sometimes be written as the sequence $(h_1, h_2, \ldots)$ and sometimes as $h_1 \oplus h_2 \oplus \cdots$. At times, we shall also need to start with the index 0. For example, l^2 is the Hilbert space of all square summable sequences and it is the same as $\mathbb{C} \oplus \mathbb{C} \oplus \cdots$. An element of l^2 may be written as $(x_0, x_1, \ldots)$ where $x_0, x_1, \ldots$ are complex numbers such that $|x_0|^2 + |x_1|^2 + \cdots < \infty$.

The other way of producing new Hilbert spaces from old is by taking the tensor product. This may be viewed as an application of GNS construction. Before we embark on the construction of the tensor product, we need to observe that Schur product of two positive semi-definite matrices is positive semi-definite again. The *Schur product* of two $n \times n$ matrices A and B is defined to be the $n \times n$ matrix whose (i, j)th. entry is $a_{ij}b_{ij}$. Let A and B be positive semi-definite. Thus, they have their positive semi-definite square roots $A^{1/2}$ and $B^{1/2}$. For any n scalars $c_1, c_2, \ldots, c_n$, let C denote the $n \times n$ diagonal matrix whose diagonal entries are $c_1, c_2, \ldots, c_n$. Then,

$$\begin{aligned}
\sum_{i,j=1}^{n} c_i \overline{c_j} a_{ij} b_{ij} &= \operatorname{Tr} CA(BC^*)^t \\
&= \operatorname{Tr} CAC^* B^t \\
&= \operatorname{Tr} CA^{1/2}A^{1/2}C^*(B^{1/2})^t(B^{1/2})^t \\
&= \operatorname{Tr} (B^{1/2})^t CA^{1/2}A^{1/2}C^*(B^{1/2})^t \\
&= \operatorname{Tr} ((B^{1/2})^t CA^{1/2})((B^{1/2})^t CA^{1/2})^* \geq 0
\end{aligned}$$

because $((B^{1/2})^t CA^{1/2})((B^{1/2})^t CA^{1/2})^*$ is a positive semi-definite matrix. With this knowledge, we are ready to construct tensor product of two Hilbert spaces. More on the Schur product will be discussed in Chapter 3 where we show that it induces a completely positive map.

Consider the following positive kernel L on the set $\mathcal{H} \times \mathcal{K}$:

$$L((h_1, k_1), (h_2, k_2)) = \langle h_2, h_1 \rangle_{\mathcal{H}} \langle k_2, k_1 \rangle_{\mathcal{K}}, \text{ for } h_1, h_2 \in \mathcal{H} \text{ and } k_1, k_2 \in \mathcal{K}.$$

It is a positive kernel because given n vectors $h_1, h_2, \ldots, h_n \in \mathcal{H}$ and $k_1, k_2, \ldots, k_n \in \mathcal{K}$, the $n \times n$ matrices $((\langle h_j, h_i \rangle_{\mathcal{H}}))$ and $((\langle k_j, k_i \rangle_{\mathcal{K}}))$ are positive semi-definite matrices and the $n \times n$ matrix $((L((h_i, k_i), (h_j, k_j))$ is just the Schur product of these two positive semi-definite matrices. Thus, we have

$$\sum_{i,j} c_i \bar{c}_j L((h_i, k_i), (h_j, k_j)) \geq 0.$$

Consequently, by what we learnt from GNS construction, we know that there is a unique Hilbert space (up to isomorphism) which we denote by $\mathcal{H} \otimes \mathcal{K}$ and a minimal map $\lambda : \mathcal{H} \times \mathcal{K} \to \mathcal{H} \otimes \mathcal{K}$ such that

$$L((h_1, k_1), (h_2, k_2)) = \langle \lambda(h_2, k_2), \lambda(h_1, k_1) \rangle.$$

The element $\lambda(h,k)$ is denoted by $h \otimes k$. These are called elementary tensors. Because the map λ is minimal, the linear span of the elementary tensors span the whole space $\mathcal{H} \otimes \mathcal{K}$. And of course, from the definition of the kernel L, we have

$$\begin{aligned}\langle h_1 \otimes k_1, h_2 \otimes k_2 \rangle &= \langle \lambda(h_1,k_1), \lambda(h_2,k_2)\rangle \\ &= L((h_2,k_2),(h_1,k_1)) = \langle h_1, h_2\rangle\langle k_1, k_2\rangle.\end{aligned}$$

For $k = 1, 2, \ldots$, we denote the tensor product space $\mathcal{H} \otimes \mathcal{H} \otimes \ldots \otimes \mathcal{H}$ (the tensor product of $\mathcal{H}$ with itself k times) by $\mathcal{H}^{\otimes k}$.

Good references for the material presented above are the survey article [13] by Arveson and Section 15 of the book [87] by Parthasarathy.

1.2. Positive operators

This section is on positive operators on a Hilbert space. We shall revise here the existence of the unique square root of a positive operator.

Definition 1.2.1. *A bounded self-adjoint operator A on a Hilbert space $\mathcal{H}$ is called positive if $\langle Ax, x\rangle \geq 0$ for all x in $\mathcal{H}$. We write $A \geq 0$ to mean that A is positive.*

If $\mathcal{H}$ is a complex Hilbert space, then $\langle Ax, x\rangle \geq 0$ for all x in $\mathcal{H}$ is enough to imply that A is self-adjoint. If the field is real, then $\left(\begin{smallmatrix} 1 & 1 \\ -1 & 1\end{smallmatrix}\right)$ is a counter-example. The positive operators form a cone in the algebra of all bounded operators $\mathcal{B}(H)$. A *cone* is a set $\mathcal{C}$ which satisfies three properties:

(1) sum of two elements of $\mathcal{C}$ are in $\mathcal{C}$,
(2) product of a non-negative scalar and an element of $\mathcal{C}$ is in $\mathcal{C}$,
(3) $\mathcal{C} \cap (-\mathcal{C}) = \{0\}$.

The self-adjoint operators on H get a natural ordering by virtue of this cone: if A and B are self-adjoint operators on H, then we say that $A \leq B$ if $B - A$ is a positive operator.

For any A in $\mathcal{B}(\mathcal{H})$, clearly A^*A is a positive operator. The aim of this section is to quickly lead the reader to the fact that all positive operators are of this form.

We shall use the following well-known fact on the strong limit of a uniformly bounded monotone increasing sequence of bounded nonnegative selfadjoint operators. It is one of the fundamental limit results in the theory of linear operators in Hilbert spaces and is available in many text books, see for example [101].

Theorem 1.2.2. *Let $\{A_n\}_{n=1}^{\infty}$ be a nondecreasing sequence of self-adjoint operators in a Hilbert space $\mathcal{H}$ and assume that the sequence A_n is uniformly bounded from above, i.e., $A_n \le \alpha I$ for some positive constant α and all $n \in \mathbb{N}$. Then there exists a selfadjoint operator $A \in \mathcal{B}(\mathcal{H})$ with $A_n \le A \le \alpha I$ for all $n \in \mathbb{N}$ such that*

$$\lim_{n\to\infty} A_n h = Ah \tag{1.2.1}$$

for all h in $\mathcal{H}$.

The above is proved by an application of the Uniform Boundedness Principle. Using it, we prove the following theorem.

Theorem 1.2.3. *Let $A \in \mathcal{B}(\mathcal{H})$ and $A \ge 0$. Then there is a unique $B \in \mathcal{B}(\mathcal{H})$ such that $B \ge 0$ and $B^2 = A$.*

Proof. Assume without loss of generality that $A \le I$ (divide by $\|A\|$). This ensures that $I - A \ge 0$. Consider the sequence $\{X_n\}$ in $\mathcal{B}(\mathcal{H})$ defined by

$$X_0 = 0, X_{n+1} = \frac{I - A + X_n^2}{2}.$$

Each X_n is a polynomial in $I - A$ with positive coefficients. So, $X_n \ge 0$. It is easy to see that

$$X_1 \le X_2 \le \cdots \le X_n \le \cdots \le I.$$

Thus, by the Theorem quoted above, the sequence $\{X_n\}_{n=0}^{\infty}$ converges strongly to a positive operator X.

The sequence $\{X_n\}_{n=0}^{\infty}$ is norm-bounded by 1.

Since $X_n h \to Xh$ and since X is a bounded operator, $XX_n h \to X^2 h$. Since $\{X_n\}_{n=0}^{\infty}$ converges strongly, $X_n X h \to X^2 h$. So, $(X_n X - XX_n)h$ converges to 0. Now,

$$\begin{aligned}\|(X_n^2 - X^2)h\| &= \|X_n^2 + XX_n - X_nX - X^2)h + (X_nX - XX_n)h\| \\ &\le \|(X_n + X)(X_n - X)h\| + \|(X_nX - XX_n)h\| \\ &\le (1 + \|X\|)\|(X_n - X)h\| + \|(X_nX - XX_n)h\| \to 0.\end{aligned}$$

So, $X_n^2 \to X^2$ strongly. So, we get

$$X = \frac{I - A + X^2}{2}.$$

In other words, $X^2 - 2X + (I - A) = 0$. Thus, $A = I - 2X + X^2 = (I - X)^2$. Define $B = I - X$. Note that $B \geq 0$. Indeed, X is a positive operator and $\langle Xh, h\rangle = \lim_{n\to\infty}\langle X_n h, h\rangle \leq \langle h, h\rangle$ so that $X \leq I$. This proves the existence.

For uniqueness of B, first note that for a bounded self-adjoint operator T, if $T^m = 0$ for some positive integer m, then $T = 0$. Indeed, $\mathcal{H} = \text{Ker}\, T \oplus \overline{\text{Ran}T}$ so that given $T^m = 0$, we have $T^{m-1}|_{\text{Ker}T}$ is always 0 and $T^{m-1}|_{\text{Ran}T} = 0$. So, $T^{m-1} = 0$. Finally, $T = 0$.

If $A = C^2$ where C is another positive operator, $AC = C^3 = CA$ so that C commutes with A. Since B above is obtained as a strong limit of polynomials in $I - A$, hence C commutes with B.

Let $x \in \mathcal{H}$ and $y = (B - C)x$. Then,

$$\langle By, y\rangle + \langle Cy, y\rangle = \langle (B+C)y, y\rangle = \langle (B+C)(B-C)x, y\rangle = \langle (B^2 - C^2)x, y\rangle.$$

The last quantity is 0 because $B^2 = C^2 = A$. But $\langle By, y\rangle$ and $\langle Cy, y\rangle$ are non-negative. So, $\langle By, y\rangle = 0 = \langle Cy, y\rangle$. So,

$$0 = \langle (B - C)y, y\rangle = \langle (B - C)^2 x, (B - C)x\rangle = \langle (B - C)^3 x, x\rangle.$$

Since $B - C$ is a self-adjoint operator, this means that $(B - C)^3 = 0$. So, $B = C$. ■

The unique positive square root of a positive operator A is denoted by $A^{1/2}$ or $\sqrt{A}$. The astute reader would have noted in the proof above that what we did was to approximate the square root function on the interval $[0, 1]$ by a concrete sequence of polynomials and then use that sequence to form a sequence of polynomials in the operator A which approximates $A^{1/2}$. By Weirstrass's Theorem, it is always possible to approximate any continuous function on $[0, 1]$ by a sequence of polynomials. So, does this mean that, like the square root function of A, we can form other functions of A as well? This idea will be in full bloom in Chapter 2 and the answer to this question will be found when we study the continuous functional calculus for an element of a commutative C^*-algebra. We shall end this section with an observation that will be used to prove a crucial identity in Section 1.4.

Proposition 1.2.4. *Let A and B be two positive operators on $\mathcal{H}$. If $T \in \mathcal{B}(\mathcal{H})$ satisfies $TA = BT$, then $TA^{1/2} = B^{1/2}T$.*

Proof. The proof is immediate if we note that in the proof above, we found a sequence of polynomials $\{p_n\}_{n\geq 1}$ such that $\{p_n(A)\}_{n\geq 1}$ converges strongly to $A^{1/2}$. Since $TA = BT$, we have $Tp_n(A) = p_n(B)T$ and then taking strong limit, we have the conclusion. ■

Ending this section, let us note that what the above shows is that the following three statements about an operator A in $\mathcal{B}(\mathcal{H})$ are equivalent.

(1) A can be written as B^*B for some B in $\mathcal{B}(\mathcal{H})$.
(2) The quadratic form $\langle Ah, h\rangle$ is non-negative for all h in $\mathcal{H}$.
(3) A has a unique positive square root.

1.3. Tensor product of bounded operators

We saw the tensor product of Hilbert spaces. This brief section will introduce tensor product of operators between Hilbert spaces. Here is the basic theorem.

Theorem 1.3.1. *Let H_1, H_2, K_1, K_2 be Hilbert spaces. Let $S : H_1 \to K_1$, and $T : H_2 \to K_2$ be bounded operators. Then there exists a unique bounded linear operator $(S \otimes T) : H_1 \otimes H_2 \to K_1 \otimes K_2$ satisfying*

$$(S \otimes T)(x \otimes y) = Sx \otimes Ty, \ \forall x \in H_1, y \in H_2.$$

Moreover, $\|S \otimes T\| = \|S\|.\|T\|$.

Proof. We try to construct the operator $(S \otimes T)$ on the dense subset

$$\{\sum_{i=1}^{n} x_i \otimes y_i : x_i \in H_1, y_i \in H_2, 1 \leq i \leq n, n \in \mathbb{N}\}$$

of $H_1 \otimes H_2$. On this set we define,

$$(S \otimes T)(\sum_{i=1}^{n} x_i \otimes y_i) = \sum_{i=1}^{n} Sx_i \otimes Ty_i.$$

We need to show that $(S \otimes T)$ is well-defined and bounded on this set. To this end, let $\{v_1, v_2, \dots, v_m\}$ be an orthonormal basis for

$$\text{span}\ \{Ty_1, Ty_2, \dots, Ty_n\}.$$

Then taking $a_{ik} := \langle Ty_i, v_k\rangle, 1 \leq i \leq n, 1 \leq k \leq m$, we have $Ty_i = \sum_{k=1}^{m} a_{ik}v_k$ for every i, and

$$\langle Ty_i, Ty_j\rangle = \sum_{k=1}^{m} a_{ik}\bar{a}_{jk}.$$

Let $z_k := \sum_{i=1}^{n} a_{ik} x_i$ for $1 \leq k \leq m$. Then

$$\begin{aligned}
\|\sum_{i=1}^{n} Sx_i \otimes Ty_i\|^2 &= \langle Sx_i, Sx_j\rangle\langle Ty_i, Ty_j\rangle \\
&= \sum_{k=1}^{m}\sum_{i,j=1}^{n} \langle Sx_i, Sx_j\rangle a_{ik}\bar{a}_{jk} \\
&= \sum_{k=1}^{m} \langle Sz_k, Sz_k\rangle \\
&\leq \sum_{k=1}^{m} \|S\|^2\langle z_k, z_k\rangle \\
&= \|S\|^2.\sum_{k=1}^{m}\sum_{i,j=1}^{n} \langle a_{ik}x_i, a_{jk}x_j\rangle \\
&= \|S\|^2 \sum_{i,j=1}^{n} \langle x_i, x_j\rangle\langle Ty_i, Ty_j\rangle.
\end{aligned}$$

Now expanding x_j's using an orthonormal basis of

$$\text{span}\,\{x_1, x_2, \ldots, x_n\}$$

and repeating the same argument (with the role of S replaced by T), we get,

$$\begin{aligned}
\|\sum_{i=1}^{n} Sx_i \otimes Ty_i\|^2 &\leq \|S\|^2.\|T\|^2. \sum_{i,j=1}^{n} \langle x_i, x_j\rangle\langle y_i, y_j\rangle \\
&= \|S\|^2\|T\|^2\|\sum_{i=1}^{n} x_i \otimes y_i\|^2.
\end{aligned}$$

Hence $(S \otimes T)$ is well-defined and extends to a bounded operator (still denoted by $S \otimes T$) satisfying, $\|S \otimes T\| \leq \|S\|.\|T\|$. The uniqueness is obvious. Now note that,

$$\begin{aligned}
\|S \otimes T\| &\geq \sup_{\|x\|,\|y\|\leq 1} \|Sx \otimes Ty\| \\
&= \|S\|.\|T\|.
\end{aligned}$$

Therefore, $\|S \otimes T\| = \|S\|.\|T\|$. ■

We may observe that,

(1) $(S \otimes T)^* = S^* \otimes T^*$.
(2) $(S_1 \otimes T_1)(S_2 \otimes T_2) = S_1S_2 \otimes T_1T_2$.

(3) Multilinearity: For scalars a_1, a_2, $(a_1S_1 + a_2S_2) \otimes T = a_1(S_1 \otimes T) + a_2(S_2 \otimes T)$.

Consequently, we have that if S, T are self-adjoint (respectively normal, unitary, positive, projection) then $S \otimes T$ is self-adjoint (respectively normal, unitary, positive, projection).

1.4. Dilation of a single contraction

Suppose t is a complex number with $|t| \leq 1$. Then it is easily seen that

$$\begin{bmatrix} t & (1-|t|^2)^{\frac{1}{2}} \\ (1-|t|^2)^{\frac{1}{2}} & -\bar{t} \end{bmatrix}$$

is a 2×2 unitary matrix. This can be generalized to contractions on Hilbert spaces as follows. Let T be a contraction on a Hilbert space $\mathcal{H}$, that is $T \in \mathcal{B}(\mathcal{H})$ with $\|T\| \leq 1$. Then as $\|T^*T\| = \|T\|^2$, the operator T^*T is a positive contraction. Further more, as $\|T^*\| = \|T\|$, T^* is also a contraction.

Moreover, the operators $1 - TT^*$ and $1 - T^*T$ are positive because

$$\langle (1 - TT^*)h, h \rangle = \|h\|^2 - \|T^*h\|^2 \geq 0$$

and

$$\langle (1 - T^*T)h, h \rangle = \|h\|^2 - \|Th\|^2 \geq 0.$$

Every positive operator has a unique positive square root by the last section. Let $D_{T^*} = (1 - TT^*)^{1/2}$ and $D_T = (1 - T^*T)^{1/2}$. These are called the *defect operators*. Let $\mathcal{D}_{T^*} = \overline{\text{Ran}}D_{T^*}$ and $\mathcal{D}_T = \overline{\text{Ran}}D_T$. These are called the *defect spaces*.

Note that $T(I - T^*T) = (I - TT^*)T$. Hence, by Proposition 1.2.4, we have the intertwining relation: $TD_T = D_{T^*}T$.

It is now easy to see that $U \in \mathcal{B}(\mathcal{H} \oplus \mathcal{H})$ defined by

$$U = \begin{bmatrix} T & D_{T^*} \\ D_T & -T^* \end{bmatrix}$$

is a unitary operator. In other words, given a contraction T on a Hilbert space $\mathcal{H}$, we can find a bigger Hilbert space (in the present case, the bigger space is $\mathcal{H} \oplus \mathcal{H}$ with the identification of $\mathcal{H}$ with $\mathcal{H} \oplus \{0\}$, i.e., with the first component of $\mathcal{H} \oplus \mathcal{H}$) and a unitary U on the larger space such that $T = P_{\mathcal{H}}U|_{\mathcal{H}}$ where $P_{\mathcal{H}}$ is the projection on to $\mathcal{H}$. This is what we mean by dilation. This particular dilation is called Halmos dilation of T.

However, for the Halmos dilation, in general

$$T^n \neq P_{\mathcal{H}} U^n|_{\mathcal{H}}$$

for $n > 1$. To take care of this, we employ the following construction from [85]. It is called the 'N−dilation' in [75]. Fix any natural number $N \geq 2$. Let U_N be the operator on $\mathcal{H}^{N+1} := \mathcal{H} \oplus \mathcal{H} \oplus \cdots \oplus \mathcal{H}$ ($N+1$ times) defined by

$$U_N = \begin{bmatrix} T & 0 & 0 & \dots & 0 & 0 & D_{T^*} \\ D_T & 0 & 0 & \dots & 0 & 0 & -T^* \\ 0 & I & 0 & \dots & 0 & 0 & 0 \\ 0 & 0 & I & \dots & 0 & 0 & 0 \\ \vdots & \vdots & \vdots & \ddots & \vdots & \vdots & \vdots \\ 0 & 0 & 0 & \dots & 0 & I & 0 \end{bmatrix}$$

Then one may verify that U_N is a unitary and once again identifying $\mathcal{H}$ with the first component of $\mathcal{H}^{N+1}$,

$$T^n = P_{\mathcal{H}} U_N^n|_{\mathcal{H}}$$

for all $1 \leq n \leq N$. Let us call it the *N-step unitary dilation of T.*

The N-step unitary dilation of a contraction gives a quick proof of von Neumann's inequality. In its original proof [84], von Neumann first proved it for Mobius functions and then used the fact that the space of absolutely convergent sums of finite Blaschke products is isometrically isomorphic to the disc algebra, the algebra of all functions which are analytic in the interior and continuous on the closure of the unit disc. See Drury [60] or Pisier [92] for the details of this proof. The following proof using the dilation is due to Halmos [67].

Theorem 1.4.1. *[von Neumann's inequality] For every polynomial $p(z) = a_0 + a_1 z + \cdots + a_N z^N$, let*

$$\|p\| = \sup\{|p(z)| : |z| \leq 1\}.$$

If T is a contraction and p is a polynomial, then

$$\|p(T)\| \leq \|p\|.$$

Proof: First note that by spectral theory, if U is a unitary operator and p is a polynomial, then $\sigma(p(U)) = \{p(z) : z \in \sigma(U)\}$. Since U is unitary, $\sigma(U) \subset \mathbb{T}$ where $\mathbb{T}$ is the unit circle. Thus $\|p(U)\| = \sup\{|p(z)| : z \in \sigma(U)\} \leq \sup\{|p(z)| : z \in \mathbb{T}\} = \|p\|$.

Now take U to be the N-step unitary dilation of T. Then $p(T) = P_{\mathcal{H}} p(U)|_{\mathcal{H}}$ which gives

$$\|p(T)\| = \|P_{\mathcal{H}} p(U)|_{\mathcal{H}}\| \leq \|p(U)\| \leq \|p\|.$$

■

The survey article by Drury [60] is an excellent source for more discussions on von Neumann's inequality.

The N-step unitary dilation is a useful tool even for proving results in complex analysis. A classical result of Carathéodory states that a holomorphic self-map of the unit disc $\mathbb{D}$ can be approximated uniformly on compact subsets of $\mathbb{D}$ by finite Blaschke products. Rudin proved that any holomorphic map $f : \mathbb{D}^2 \to \mathbb{D}$ can be approximated uniformly on compact subsets of $\mathbb{D}^2$ by rational inner functions. Recently, a generalization of these results has been proved by using the N-step unitary dilation. More precisely, if $\Omega = \mathbb{D}^2$ or $\Omega = \mathbb{D}$, any holomorphic function $F : \Omega \to M_N(\mathbb{C})$ with $\|F(z)\| \leq 1$ for all $z \in \mathbb{D}$ can be approximated (uniformly on compact subsets) by matrix-valued rational inner functions. See [8]. Even for complex valued functions, this proof is new.

Now, one may get greedy and ask for a dilation which works for all $n \geq 0$. This is possible and that is the topic of this Chapter.

If there is an isometry $V : \mathcal{H} \to \widehat{\mathcal{H}}$, any bounded operator T on $\mathcal{H}$ is then identified with the bounded operator VTV^* on Ran V. The projection from $\widehat{\mathcal{H}}$ onto Ran V (equivalently $\mathcal{H}$) will be denoted by $P_{\mathcal{H}}$.

Definition 1.4.2. *Let $\mathcal{H}$ be a Hilbert space. Suppose T is bounded operator on $\mathcal{H}$. If $\widehat{\mathcal{H}}$ is a Hilbert space and A is a bounded operator on $\widehat{\mathcal{H}}$ satisfying $\mathcal{H} \subset \widehat{\mathcal{H}}$ and*

(1.4.1) $\quad T^n h = P_{\mathcal{H}} A^n h$ *for all $h \in \mathcal{H}$ and all non-negative integers n*

then $\widehat{\mathcal{H}}$ is called a dilation space and the operator A is called a dilation for the operator T. An isometric (respectively unitary) dilation of T is a dilation A which is an isometry (respectively unitary).

Let A_1 and A_2 be two dilations of the same operator T on $\mathcal{H}$. Let the dilation spaces be $\widehat{\mathcal{H}}_1$ and $\widehat{\mathcal{H}}_2$ respectively and let the embedding isometries be $V_1 : \mathcal{H} \to \widehat{\mathcal{H}}_1$ and $V_2 : \mathcal{H} \to \widehat{\mathcal{H}}_2$ respectively. Then the dilations $(\widehat{\mathcal{H}}_1, A_1)$ and $(\widehat{\mathcal{H}}_2, A_2)$ are called unitarily equivalent *if there is a unitary $U : \widehat{\mathcal{H}}_1 \to \widehat{\mathcal{H}}_2$ which satisfies $UA_1U^* = A_2$ and $UV_1 h = V_2 h$ for all $h \in \mathcal{H}$.*

If A is a bounded operator on $\widehat{\mathcal{H}} = \mathcal{H} \oplus (\widehat{\mathcal{H}} \ominus \mathcal{H})$ such that the operator matrix of A with respect to the decomposition above is

$$A = \begin{pmatrix} T & 0 \\ * & * \end{pmatrix}$$

then it is clear from matrix multiplication that

$$A^n = \begin{pmatrix} T^n & 0 \\ * & * \end{pmatrix}$$

for any positive integer n. Thus A is a dilation of T. Since A^* leaves $\mathcal{H}$ invariant, $\mathcal{H}$ is called a *is a co-invariant subspace for* A. Of course, this is equivalent to saying that $A^*|_{\mathcal{H}}$ has to be T^*.

Given an operator T on $\mathcal{H}$ and an isometric dilation A of T on a dilation space $\mathcal{K}$ that leaves $\mathcal{H}$ co-invariant, one can always consider the subspace $\widehat{\mathcal{H}} \stackrel{\text{def}}{=} \overline{span}\{A^n h : h \in \mathcal{H}, n = 0, 1, 2, \ldots\}$. This is a reducing subspace for A (because $\mathcal{H}$ is a co-invariant subspace of A) and it is a dilation space for T. Moreover, note that if $\mathcal{H}$ is separable, then $\widehat{\mathcal{H}}$ is the span closure of elements of the form $A^n h$ where n is a non-negative integer and h comes from a countable dense subset of $\mathcal{H}$. Thus $\widehat{\mathcal{H}}$ is separable.

Let $\mathcal{B}(\mathcal{H})$ denote the algebra of all bounded operators on the Hilbert space $\mathcal{H}$ and let $\|\cdot\|$ denote the operator norm on $\mathcal{B}(\mathcal{H})$. An element T of $\mathcal{B}(\mathcal{H})$ is called a contraction if $\|T\| \leq 1$. Clearly, if T has an isometric dilation, then T is a contraction. Conversely, given a contraction T acting on a Hilbert space $\mathcal{H}$, Sz.-Nagy showed that T always has a minimal isometric dilation. We shall give two proofs of the theorem in this chapter.

Theorem 1.4.3. *Given a contraction T on a Hilbert space $\mathcal{H}$, there is an isometric dilation A which is* minimal *in the sense that*

$$\overline{span}\{A^n h : h \in \mathcal{H}, n = 0, 1, 2, \ldots\} = \hat{\mathcal{H}}.$$

Any two such minimal isometric dilations are unique up to unitary equivalence.

Because of the uniqueness (up to a unitary), we shall call it *the minimal isometric dilation.* There are a few comments to make before we embark on the proof.

- First note that if a minimal isometric dilation A exists, then its uniqueness up to unitary equivalence is straightforward because $\langle A^n h, A^m h' \rangle$ does not depend on a particular minimal dilation A. Indeed,

$$\langle A^n h, A^m h' \rangle = \begin{cases} \langle A^{n-m}h, h' \rangle = \langle T^{n-m}h, h' \rangle \text{ if } n \geq m \geq 0, \\ \langle h, A^{m-n}h' \rangle = \langle h, T^{m-n}h' \rangle \text{ if } m \geq n \geq 0. \end{cases} \tag{1.4.2}$$

So given two minimal dilations A_1 and A_2 on the spaces $\widehat{\mathcal{H}}_1$ and $\widehat{\mathcal{H}}_2$ respectively, define U on $\operatorname{span}\{A_2^n h : h \in \mathcal{H}, n = 0, 1, 2, \ldots\}$ by

$$U(\sum_{n=0}^{N} A_2^n h_n) = \sum_{n=0}^{N} A_1^n h_n.$$

This is a well-defined isometric linear transformation from a dense subspace of $\widehat{\mathcal{H}}_2$ onto a dense subspace of $\widehat{\mathcal{H}}_1$. Thus U extends to a unitary operator from $\widehat{\mathcal{H}}_2$ to $\widehat{\mathcal{H}}_1$. Moreover, note that for $h \in \mathcal{H}$, we have $Uh = U(A_2^0)h = A_1^0 h = h$, thus the isometric embedding of $\mathcal{H}$ into the dilation spaces is respected by U. Of course, U has been so constructed that $UA_2U^* = A_1$.

- The minimal isometric dilation A that we shall construct has the property that $\mathcal{H}$ is a co-invariant subspace for A. Now note that if this is true for one minimal isometric dilation A, then it is true for any minimal isometric dilation. Indeed, let $(\widehat{\mathcal{H}}_1, A_1)$ and $(\widehat{\mathcal{H}}_2, A_2)$ be two minimal isometric dilations. Suppose that $\mathcal{H}$ is a co-invariant subspace for A_1. We know by minimality that there is a unitary operator $U : \widehat{\mathcal{H}}_1 \to \widehat{\mathcal{H}}_2$ such that $Uh = h$ for all $h \in \mathcal{H}$ and $A_2 = UA_1U^*$. With respect to the decompositions $\widehat{\mathcal{H}}_i = \mathcal{H} \oplus (\widehat{\mathcal{H}}_i \ominus \mathcal{H})$ for $i = 1, 2$, we have

$$A_1 = \begin{pmatrix} T & 0 \\ * & * \end{pmatrix} \text{ and } U = \begin{pmatrix} I & 0 \\ 0 & * \end{pmatrix}.$$

Obviously, then $A_2 = UA_1U^*$ also has the operator matrix form

$$\begin{pmatrix} T & 0 \\ * & * \end{pmatrix}$$

which means that $\mathcal{H}$ is a co-invariant subspace for A_2. Thus any minimal isometric dilation leaves $\mathcal{H}$ co-invariant.

1.5. First proof of existence - an abstract method

This method, although abstract (we shall give a concrete construction in the next section), is a beautiful application of the theory of positive kernels and their associated GNS spaces. Particularly noteworthy is the point at which the contractivity of T enters the proof. We begin with a very useful lemma, a special case of which provides the intuition for the proof.

Lemma 1.5.1. *Let $\mathcal{K}_1, \mathcal{K}_2$ be two Hilbert spaces. Consider a block operator*

$$R = \begin{bmatrix} A & B \\ B^* & C \end{bmatrix}$$

on $\mathcal{K}_1 \oplus \mathcal{K}_2$. Then the following are equivalent:

(i) R is positive;

(ii) There exist operators $D_i \in \mathcal{B}(\mathcal{K}_i, \mathcal{K}_1 \oplus \mathcal{K}_2)$ for $i = 1, 2$ such that $A = D_1^* D_1, C = D_2^* D_2$ and $B = D_1^* D_2$.

(iii) A, C are positive and there exists a contraction K in $\mathcal{B}(\mathcal{K}_2, \mathcal{K}_1)$ such that $B = A^{\frac{1}{2}} K C^{\frac{1}{2}}$.

Proof $(i) \Rightarrow (ii)$. As R is positive,

$$R = \begin{bmatrix} A & B \\ B^* & C \end{bmatrix} = \begin{bmatrix} D_{11} & D_{12} \\ D_{21} & D_{22} \end{bmatrix}^* \begin{bmatrix} D_{11} & D_{12} \\ D_{21} & D_{22} \end{bmatrix}$$

for some block operator

$$\begin{bmatrix} D_{11} & D_{12} \\ D_{21} & D_{22} \end{bmatrix}$$

on $\mathcal{K}_1 \oplus \mathcal{K}_2$. Taking

$$D_1 = \begin{bmatrix} D_{11} \\ D_{21} \end{bmatrix}, \; D_2 = \begin{bmatrix} D_{12} \\ D_{22} \end{bmatrix}$$

the result follows.

$(ii) \Rightarrow (iii)$. Let $D_i = V_i |D_i|$ be the polar decompositions of D_i, with partial isometry $V_i \in \mathcal{B}(\mathcal{K}_i, \mathcal{K}_1 \oplus \mathcal{K}_2)$ and $|D_i| = (D_i^* D_i)^{\frac{1}{2}}$ for $i = 1, 2$. Then from (ii), $A = |D_1|^2, C = |D_2|^2$ and

$$B = |D_1| V_1^* V_2 |D_2 = A^{\frac{1}{2}} K B^{\frac{1}{2}}$$

where $K = V_1^* V_2$. Clearly K is a contraction as V_1, V_2 are partial isometries and hence are contractions.

$(iii) \Rightarrow (ii)$. As K is a contraction, $K^* K \leq I$ and hence

$$\begin{bmatrix} I & K \\ K^* & I \end{bmatrix} = \begin{pmatrix} I \\ K^* \end{pmatrix} \begin{pmatrix} I & K \end{pmatrix} + \begin{bmatrix} 0 & 0 \\ 0 & (I - K^* K) \end{bmatrix}$$

is positive. Now the positivity of R is clear from the factorization:

$$\begin{bmatrix} A & B \\ B^* & C \end{bmatrix} = \begin{bmatrix} A^{\frac{1}{2}} & 0 \\ 0 & C^{\frac{1}{2}} \end{bmatrix} \begin{bmatrix} I & K \\ K^* & I \end{bmatrix} \begin{bmatrix} A^{\frac{1}{2}} & 0 \\ 0 & C^{\frac{1}{2}} \end{bmatrix}.$$

■

Some more information on this circle of ideas can be found in [81]. Currently we need only the following special case.

Lemma 1.5.2. *Let $\mathcal{K}_1$ and $\mathcal{K}_2$ be two Hilbert spaces, let $B \in \mathcal{B}(\mathcal{K}_1)$ be a positive operator and let $A \in \mathcal{B}(\mathcal{K}_1, \mathcal{K}_2)$. Then*

$$\begin{pmatrix} B & A^* \\ A & I \end{pmatrix} : \mathcal{K}_1 \oplus \mathcal{K}_2 \to \mathcal{K}_1 \oplus \mathcal{K}_2 \tag{1.5.1}$$

is a positive operator if and only if $A^*A \leq B$.

Corollary 1.5.3. *Given T in $\mathcal{B}(\mathcal{H})$, we have $\begin{pmatrix} I & T \\ T^* & I \end{pmatrix}$ to be a positive operator if and only if T is contraction.*

This leads us to ask whether the operator matrix

$$X_n := \begin{pmatrix} I & T & T^2 & \cdots & T^n \\ T^* & I & T & \cdots & T^{n-1} \\ (T^*)^2 & T^* & I & \cdots & T^{n-2} \\ \vdots & \vdots & \vdots & \cdots & \vdots \\ (T^*)^n & (T^*)^{n-1} & (T^*)^{n-2} & \cdots & I \end{pmatrix} \tag{1.5.2}$$

is a positive operator on $\mathcal{H}^{n+1}$ (the direct sum of $n+1$ copies of $\mathcal{H}$) for every n. It is indeed so because X_n can be factored. It is a product of three matrices. These three matrices are

$$\begin{pmatrix} I & 0 & 0 & \cdots & 0 \\ T^* & I & 0 & \cdots & 0 \\ (T^*)^2 & T^* & I & \cdots & 0 \\ \vdots & \vdots & \vdots & \cdots & \vdots \\ (T^*)^n & (T^*)^{n-1} & (T^*)^{n-2} & \cdots & I \end{pmatrix},$$

$$\begin{pmatrix} I & 0 & 0 & \cdots & 0 \\ 0 & I - T^*T & 0 & \cdots & 0 \\ 0 & 0 & I - T^*T & \cdots & 0 \\ \vdots & \vdots & \vdots & \cdots & \vdots \\ 0 & 0 & 0 & \cdots & I - T^*T \end{pmatrix}$$

and

$$\begin{pmatrix} I & T & T^2 & \cdots & T^n \\ 0 & I & T & \cdots & T^{n-1} \\ 0 & 0 & I & \cdots & T^{n-2} \\ \vdots & \vdots & \vdots & \cdots & \vdots \\ 0 & 0 & 0 & \cdots & I \end{pmatrix}$$

in that order. Thus X_n is positive if and only if T is a contraction. This fact, interestingly, can be rephrased to give us a positive kernel. Define the set

$$X = \{(n, g) : n \in \mathbb{Z}_+, g \in \mathcal{H}\}$$

and $k : X \times X \to \mathbb{C}$ by

$$k((n,h),(m,g)) = \begin{cases} \langle g, (T^*)^{m-n}h\rangle & \text{if } m \geq n, \\ \langle g, T^{n-m}h\rangle & \text{if } n \geq m. \end{cases}$$

A moment's thought shows that k is a positive kernel if and only if X_n is a positive operator for all n. Thus, we have shown that k is indeed a positive kernel.

Let $(\widehat{\mathcal{H}}, \lambda)$ be the GNS pair for the positive kernel k. Since for any g and h in $\mathcal{H}$, we have

$$\langle \lambda(0,g), \lambda(0,h)\rangle_{\widehat{\mathcal{H}}} = k((0,h)),(0,g)) = \langle g, h\rangle_{\mathcal{H}},$$

we can embed $\mathcal{H}$ isometrically in $\widehat{\mathcal{H}}$ by identifying h in $\mathcal{H}$ with $\lambda(0,h)$ in $\widehat{\mathcal{H}}$. Define A on the total set of vectors by

$$A\lambda(m,h) = \lambda(m+1,h), \text{ where } (m,h) \in X$$

and then extend it linearly to finite linear combinations of these vectors. Let $m_1 \leq m_2 \leq \cdots \leq m_k$ be non-negative integers and let $h_1, h_2, \ldots, h_k$ be vectors in $\mathcal{H}$. Then using the definition of the inner product in terms of the kernel we get

$$\begin{aligned}
&\|\sum_{i=1}^k c_i\lambda(m_i,h_i)\|^2 \\
= \;& \sum_{i=1}^k\sum_{j=1}^k \overline{c_i}c_j\langle\lambda(m_i,h_i),\lambda(m_j,h_j)\rangle \\
= \;& \sum_{i>j}\overline{c_i}c_j\langle h_i,(T^*)^{m_i-m_j}h_j\rangle + \sum_{i=1}^k |c_i|^2\|h_i\|^2 + \sum_{i<j}\overline{c_i}c_j\langle h_i, T^{m_j-m_i}h_j\rangle \\
= \;& \sum_{i=1}^k\sum_{j=1}^k \overline{c_i}c_j\langle\lambda(m_i+1,h_i),\lambda(m_j+1,h_j)\rangle \\
= \;& \|\sum_{i=1}^k c_i\lambda(m_i+1,h_i)\|^2 = \|A(\sum_{i=1}^k c_i\lambda(m_i,h_i))\|^2.
\end{aligned}$$

Thus A is an isometry on this dense subspace and hence extends to the whole space as an isometry. It is a straightforward computation to see that $A^*|_{\mathcal{H}} = T^*$.

1.6. Second proof of existence - Schäffer construction

In this section, we give a concrete construction for the minimal isometric dilation. This elegant construction appears in a one page paper by Schäffer [109]. The dilation space is constructed by setting

$$\widehat{\mathcal{H}} = \mathcal{H} \oplus \mathcal{D}_T \oplus \mathcal{D}_T \cdots .$$

The space $\mathcal{H}$ is identified with the subspace of $\widehat{\mathcal{H}}$ consisting of elements of the form $(h, 0, 0, \ldots)$ where $h \in \mathcal{H}$. This is the way $\mathcal{H}$ is isometrically embedded into $\widehat{\mathcal{H}}$. Define the operator A on $\widehat{\mathcal{H}}$ by

$$A(h_0, h_1, h_2, \ldots) = (Th_0, D_T h_0, h_1, h_2, \ldots).$$

For every $h \in \mathcal{H}$, we have $\|Th\|^2 + \|D_T h\|^2 = \|h\|^2$, so that the operator A defined above is an isometry. Now for $h_0 \in \mathcal{H}$ and $(k_0, k_1, k_2, \ldots) \in \widehat{\mathcal{H}}$, we have

$$\begin{aligned} \langle A^* h_0, (k_0, k_1, k_2, \ldots)\rangle &= \langle h_0, A(k_0, k_1, k_2, \ldots)\rangle \\ &= \langle (h_0, 0, 0, \ldots), (Tk_0, D_T k_0, k_1, k_2, \ldots)\rangle \\ &= \langle h_0, Tk_0\rangle \\ &= \langle T^* h_0, k_0\rangle = \langle T^* h_0, (k_0, k_1, k_2, \ldots)\rangle. \end{aligned}$$

Thus $A^* h = T^* h$ for all $h \in \mathcal{H}$. The isometry A on $\widehat{\mathcal{H}}$ has the property that $\mathcal{H}$ is left invariant by A^*. So A is an isometric dilation of T. Minimality becomes clear by observing that

$$A^n(h_0, h_1, h_2, \ldots) = (T^n h_0, D_T T^{n-1} h_0, \ldots, D_T h_0, h_1, h_2, \ldots)$$

from which it follows that

$$A^n(h_0, 0, 0, \ldots) = (T^n h_0, D_T T^{n-1} h_0, \ldots, D_T h_0, 0, 0, \ldots). \tag{1.6.1}$$

Thus

$$\overline{span}\{A^n(h, 0, 0, \ldots) : h \in \mathcal{H}, n = 0, 1, 2, \ldots\} = \mathcal{H} \oplus \mathcal{D}_T \oplus \mathcal{D}_T \cdots = \widehat{\mathcal{H}}.$$

The Schäffer construction, apart from being direct and explicit, emphasizes a very basic ingredient in all of operator theory. To detect it, note that with respect to the decomposition

$$\widehat{\mathcal{H}} = \mathcal{H} \oplus (\mathcal{D}_T \oplus \mathcal{D}_T \oplus \mathcal{D}_T \oplus \mathcal{D}_T \oplus \cdots)$$

of the dilation space, the block matrix of the dilation operator is

$$\left(\begin{array}{c|ccccc} T & 0 & 0 & 0 & 0 & \dots \\ \hline D_T & 0 & 0 & 0 & 0 & \dots \\ 0 & 1 & 0 & 0 & 0 & \dots \\ 0 & 0 & 1 & 0 & 0 & \dots \\ 0 & 0 & 0 & 1 & 0 & \dots \\ \vdots & \vdots & \vdots & \vdots & \vdots & \vdots \end{array}\right). \tag{1.6.2}$$

The bottom right hand corner of the block matrix above is an example of a *unilateral shift*. This is a class of operators which plays a fundamental role.

1.7. The unilateral shift

Definition 1.7.1. *Let $\mathcal{E}_n$ be a Hilbert space of dimension n where $n \in \mathbb{N} \cup \{\infty\}$. A unilateral shift of multiplicity n is a bounded operator S_n defined on the Hilbert space $\mathcal{E}_n \oplus \mathcal{E}_n \oplus \cdots$ such that*

$$S_n(h_0, h_1, h_2, \dots) = (0, h_0, h_1, \dots).$$

The space $\mathcal{E}_n$ identified with $\mathcal{E}_n \oplus 0 \oplus 0 \oplus \cdots$ is called the initial space or the generating subspace of the unilateral shift S_n.

If $\mathcal{F}_n$ is another Hilbert space of the same dimension n and if T_n is the unilateral shift defined on $\mathcal{F}_n \oplus \mathcal{F}_n \oplus \cdots$ in the above fashion, then S_n and T_n are unitarily equivalent. Indeed, start with any unitary $U_0 : \mathcal{E}_n \to \mathcal{F}_n$. This can be obtained because $\mathcal{E}_n$ and $\mathcal{F}_n$ have the same dimension. Now define $U : \mathcal{E}_n \oplus \mathcal{E}_n \oplus \cdots \to \mathcal{F}_n \oplus \mathcal{F}_n \oplus \cdots$ by

$$h_0 \oplus h_1 \oplus \cdots \to U_0 h_0 \oplus U_0 h_1 \oplus \cdots .$$

This U intertwines S_n and T_n. Indeed,

$$\begin{aligned} US_n(h_0 \oplus h_1 \oplus h_2 \oplus \cdots) &= U(0 \oplus h_0 \oplus h_1 \oplus \cdots) \\ &= (0 \oplus U_0 h_0 \oplus U_0 h_1 \oplus \cdots) \\ &= T_n(U_0 h_0 \oplus U_0 h_1 \oplus \cdots) \\ &= T_n U(h_0 \oplus h_1 \oplus \cdots). \end{aligned}$$

Thus there is only one unilateral shift (up to unitary isomorphism) of a given multiplicity.

Let S_n be a unilateral shift as defined above. Since it is an isometry, its range is closed and $S_n S_n^*$ is the projection (because it is self-adjoint and $(S_n S_n^*)^2 = S_n S_n^* S_n S_n^* = S_n S_n^*$) onto the range of S_n. In this case, range of S_n is $0 \oplus \mathcal{E}_n \oplus \mathcal{E}_n \oplus \cdots$. Also, $I - S_n S_n^*$ is the projection onto the kernel of S_n^* which is $\mathcal{E}_n \oplus 0 \oplus 0 \cdots$.

Next, suppose $n \neq m$ and $\mathcal{E}_n$ and $\mathcal{E}_m$ are Hilbert spaces of dimensions n and m respectively. If S_n and S_m were unitarily equivalent, so would be the projections $1 - S_n S_n^*$ and $1 - S_m S_m^*$. But their ranges have different dimensions. So, two unilateral shifts of different multiplicities cannot be unitarily equivalent.

The bottom right corner of (1.6.2) has a unilateral shift. Indeed, consider the space $\mathcal{D}_T \oplus \mathcal{D}_T \oplus \mathcal{D}_T \oplus \cdots$ and the generating subspace $\mathcal{E} = \mathcal{D}_T \oplus 0 \oplus 0 \oplus \cdots$ and the unilateral shift S_n defined as above where dim $\mathcal{D}_T = n \in \mathbb{N} \cup \{\infty\}$. Thus the block operator matrix of the dilation A is

$$\left(\begin{array}{c|ccc} T & 0 & 0 & 0 \\ \hline D_T & & & \\ 0 & & S_n & \\ 0 & & & \end{array}\right).$$

Lemma 1.7.2. *Let $n \in \mathbb{N} \cup \{\infty\}$. The unilateral shift S_n of multiplicity n is unitarily equivalent to $S \otimes I$ on $l^2 \otimes \mathcal{E}_n$.*

Proof. Let $\mathcal{T} = \{(x_0, x_1, x_2, \ldots) \otimes e : (x_0, x_1, x_2, \ldots) \in l^2$ and $e \in \mathcal{E}_n\}$ and $\mathcal{M} =$ span $\mathcal{T}$. Consider the map U, first defined on elements $(x_0, x_1, x_2, \ldots) \otimes e \in l^2 \otimes \mathcal{E}_n$ as $U((x_0, x_1, x_2, \ldots) \otimes e) = x_0 e \oplus x_1 e \oplus x_2 e \oplus \cdots \in \mathcal{E}_n \oplus \mathcal{E}_n \oplus \cdots$ and then extended linearly to $\mathcal{M}$. A straightforward computation shows that U is an isometry on $\mathcal{M}$ which is dense in $l^2 \otimes \mathcal{E}_n$. Hence U extends to the whole space $l^2 \otimes \mathcal{E}_n$ as an isometry. It is onto because every vector of the form $e_0 \oplus e_1 \oplus e_2 \oplus \cdots \oplus e_k \oplus 0 \oplus \cdots$, where $e_0, e_1, e_2, \ldots, e_k \in \mathcal{E}_n$, is in the range of U (in fact in $U\mathcal{M}$) and such vectors are dense in $\mathcal{E}_n \oplus \mathcal{E}_n \oplus \cdots$. To verify that $S_n U = U(S \otimes I)$, it is enough to verify on the total set $\mathcal{T}$. On $\mathcal{T}$,

$$\begin{aligned} S_n U((x_0, x_1, x_2, \ldots) \otimes e) &= S_n(x_0 e \oplus x_1 e \oplus x_2 e \oplus \cdots) \\ &= (0 \oplus x_0 e \oplus x_1 e \oplus x_2 e \oplus \cdots) \\ &= U((0, x_0, x_1, x_2, \ldots) \otimes e) \\ &= U(S(x_0, x_1, x_2, \ldots) \otimes e) \\ &= U(S \otimes I)((x_0, x_1, x_2, \ldots) \otimes e) \end{aligned}$$

That completes the proof. ∎

The unilateral shift S_1 of multiplicity 1 acts on the space $\mathbb{C} \oplus \mathbb{C} \oplus \cdots$ which is l^2. We shall denote S_1 by S.

1.8. The Hardy space

We have seen the Hardy space already as a reproducing kernel Hilbert space. This section studies it in more detail.

Definition 1.8.1. *The Hardy space $H^2_{\mathbb{D}}$ is the set of all functions analytic in an open set around 0 and having a Taylor series expansion $f(z) = \sum_{n=0}^{\infty} a_n z^n$ such that $\sum_{n=0}^{\infty} |a_n|^2 < \infty$.*

It is easy to see that $H^2_{\mathbb{D}}$ is a vector space.

Lemma 1.8.2. *Every function in $H^2_{\mathbb{D}}$ is analytic at all points of $\mathbb{D} = \{z \in \mathbb{C} : |z| < 1\}$.*

Proof. Let $f(z) = \sum_{n=0}^{\infty} a_n z^n$ be a function in $H^2_{\mathbb{D}}$ and let z_0 be any point in the open unit disc $\mathbb{D}$. Since the sequence $\{a_n\}$ is square summable, it is bounded, i.e., there is an M such that $|a_n| \leq M$ for all $n = 1, 2, \ldots$. Thus,

$$\sum_{n=0}^{\infty} |a_n z_0^n| \leq M \sum_{n=0}^{\infty} |z_0|^n < \infty$$

because $|z_0| < 1$. So, the power series is convergent in the whole of open unit disc. Hence f is analytic in $\mathbb{D}$. ■

On this vector space, give the inner product

$$\langle f, g \rangle = \sum_{n=0}^{\infty} a_n \overline{b_n}$$

where $f(z) = \sum_{n=0}^{\infty} a_n z^n$ and $g(z) = \sum_{n=0}^{\infty} b_n z^n$. This makes it a Hilbert space because this inner product space is isomorphic to l^2 by the unitary $U : H^2_{\mathbb{D}} \to l^2$ sending $f(z) = \sum_{n=0}^{\infty} a_n z^n$ to the sequence $\{a_0, a_1, \ldots\}$ in l^2 and l^2 is complete.

It is straightforward that under the isomorphism between l^2 and the Hardy space, the unilateral shift of multiplicity 1 on l^2 is unitarily equivalent to M_z, the bounded operator of multiplication by z on $H^2_{\mathbb{D}}$.

Lemma 1.8.3. *If f is analytic in the open unit disc, then $f \in H^2_{\mathbb{D}}$ if and only if*

$$\sup_{0<r<1} \frac{1}{2\pi} \int_0^{2\pi} |f(re^{i\theta})|^2 d\theta < \infty.$$

Proof. Since f is analytic in the unit disc, let $f(z) = \sum_{n=0}^{\infty} a_n z^n$ be its Taylor series expansion around 0. Then,

$$\int_0^{2\pi} |f(re^{i\theta})|^2 d\theta = \int_0^{2\pi} \sum_{n=0}^{\infty}\sum_{m=0}^{\infty} a_n \bar{a}_m r^{n+m} e^{i(n-m)\theta} d\theta = 2\pi \sum_{n=0}^{\infty} r^{2n} |a_n|^2.$$

Hence the result follows by appealing to the definition of Hardy space functions. ■

In view of this Lemma, we have a different expression for the Hardy space norm. If $f \in H^2_{\mathbb{D}}$, then

$$\|f\|^2 = \sup_{0<r<1} \frac{1}{2\pi} \int_0^{2\pi} |f(re^{i\theta})|^2 d\theta.$$

This immediately leads us to the fact that if φ is in $H^\infty_{\mathbb{D}}$, i.e., if φ is a bounded analytic function on the open unit disc, then M_φ defined by $M_\varphi(f) = \varphi f$ is a bounded linear operator on $H^2_{\mathbb{D}}$. Indeed, let $f \in H^2_{\mathbb{D}}$ and let $\varphi \in H^\infty_{\mathbb{D}}$ with $\|\varphi\|_\infty = \sup\{|\varphi(z)| : z \in \mathbb{D}\}$. Then we have

$$\begin{aligned} &\sup_{0<r<1} \frac{1}{2\pi} \int_0^{2\pi} |(\varphi f)(re^{i\theta})|^2 d\theta \\ =& \sup_{0<r<1} \frac{1}{2\pi} \int_0^{2\pi} |\varphi(re^{i\theta}) f(re^{i\theta})|^2 d\theta \\ \le& \|\varphi\|^2_\infty \sup_{0<r<1} \frac{1}{2\pi} \int_0^{2\pi} |f(re^{i\theta})|^2 d\theta = \|\varphi\|^2_\infty \|f\|^2 \end{aligned}$$

showing that M_φ maps $H^2_{\mathbb{D}}$ into $H^2_{\mathbb{D}}$ and $\|M_\varphi\| \le \|\varphi\|_\infty$. We shall soon see that this inequality is actually an equality.

Definition 1.8.4. *Given a holomorphic function φ on $\mathbb{D}$, we say that φ is a multiplier of the Hardy space if the linear transformation $M_\varphi(f) = \varphi f$ is a bounded linear operator on $H^2_{\mathbb{D}}$.*

The definition could be relaxed a bit, that is, one could demand just that φf belongs to $H^2_{\mathbb{D}}$ for every f in $H^2_{\mathbb{D}}$. Then, the Closed Graph Theorem would take care of the rest and make M_φ a bounded operator. Note that the multipliers form an algebra. It is called the *multiplier algebra*. We saw above that every φ in $H^\infty_{\mathbb{D}}$ is a multiplier. The converse is also true. We see it below.

Definition 1.8.5. *For every $w \in \mathbb{D}$, define the function $k_w(z) = \sum_{n=0}^{\infty} \overline{w}^n z^n$.*

(The reader may like to go back to the place where *A different presentation* of the proof of the Gelfand-Naimark-Segal construction was given and check that this notation is consistent with the k_x there.)

It is easy to see that the radius of convergence of this series is $1/|w|$ which is bigger than 1 since $w \in \mathbb{D}$. Thus, k_w is analytic in $\mathbb{D}$ and sums to $1/(1 - z\overline{w})$. Also, it is in $H^2_{\mathbb{D}}$ because the coefficients are square summable. Indeed, they form a geometric series in powers of w and $|w|$ is less than 1. If $f \in H^2_{\mathbb{D}}$ has the Taylor series expansion $\sum_{n=0}^{\infty} a_n z^n$, then

$$\langle f, k_w \rangle = \sum_{n=0}^{\infty} a_n w^n = f(w).$$

Thus, the family of functions $\{k_w : w \in \mathbb{D}\}$ reproduces the values of functions in the Hardy space.

The action of adjoint of a multiplication operator on k_w can be found in the following way.

$$\langle M_\varphi^* k_w, f \rangle = \langle k_w, \varphi f \rangle = \overline{\langle \varphi f, k_w \rangle} = \overline{\varphi(w) f(w)} = \overline{\varphi(w)} \langle k_w, f \rangle.$$

Thus, $M_\varphi^* k_w = \overline{\varphi(w)} k_w$. So, $|\varphi(w)| \leq \|M_\varphi\|$ for every $w \in \mathbb{D}$. Consequently, $\|\varphi\|_\infty \leq \|M_\varphi\|$. Thus, if φ is a multiplier, then it is automatically in $H^\infty_{\mathbb{D}}$ and $\|M_\varphi\| = \|\varphi\|_\infty$. Thus we have proved

Theorem 1.8.6. *The multiplier algebra of $H^2_{\mathbb{D}}$ is $H^\infty_{\mathbb{D}}$.*

1.9. Pure contractions

There is a special class of contractions for which a different proof of the dilation theorem can be given. These are called purecontractions.

Definition 1.9.1. *A contraction T on a Hilbert space H is said to be pure (or to belong to the class $C_{\cdot 0}$) if T^{*n} converges to the zero operator in the strong topology of $\mathcal{B}(H)$, i.e., $T^{*n}x \to 0$ as $n \to \infty$ for every $x \in H$.*

This terminology $C_{\cdot 0}$ is due to Sz.-Nagy and Foias (see [113], Chapter III, Section 4). The term *pure* is more recent (see [17], Remark 4.4).

It is easy to check that M_z on $H^2_{\mathbb{D}}$ is pure. Indeed, if $f(z) = \sum_{k=0}^{\infty} a_k z^k$ is in $H^2_{\mathbb{D}}$, then

$$(M_z^{*n} f)(z) = \sum_{k=n}^{\infty} a_k z^{k-n}$$

has norm $\sum_{k=n}^{\infty} |a_k|^2$ which, being a tail of a convergent series, converges to 0 as $n \to \infty$.

Moreover, if $\mathcal{E}$ is a Hilbert space, let $\mathcal{O}(\mathbb{D}, \mathcal{E})$ be the class of all holomorphic $\mathcal{E}$-valued functions on the unit disc. Consider the *vector valued*

Hardy space

$$H^2_{\mathbb{D}}(\mathcal{E}) = \{f \in \mathcal{O}(\mathbb{D}, \mathcal{E}) \\ : f(z) = \sum_{n=0}^{\infty} a_k z^k \text{ where } a_k \in \mathcal{E} \text{ and } \sum_{n=0}^{\infty} \|a_k\|^2 < \infty\}.$$

On this space too, we can define the operator of multiplication by z. Let us denote it by $M_z^{\mathcal{E}}$. Thus,

$$(M_z^{\mathcal{E}} f)(z) = zf(z) \text{ for all } f \in H^2_{\mathbb{D}}(\mathcal{E}).$$

This operator $M_z^{\mathcal{E}}$ is also pure for the same reason as why M_z is pure on $H^2_{\mathbb{D}}$. Now, if $\mathcal{M}$ is a closed subspace of $H^2_{\mathbb{D}}(\mathcal{E})$ that is invariant under $(M_z^{\mathcal{E}})^*$, then define a bounded linear operator T on $\mathcal{M}$ by

$$T^* = (M_z^{\mathcal{E}})^*|_{\mathcal{M}}.$$

In other words, $T = P_{\mathcal{M}} M_z^{\mathcal{E}}|_{\mathcal{M}}$. Then T is pure for the same reason as above. The following theorem says that the converse is true. Before we state the theorem, we would like to point out, as an astute reader must have noticed by now, that $H^2_{\mathbb{D}}(\mathcal{E})$, as a Hilbert space, is the same as $H^2_{\mathbb{D}} \otimes \mathcal{E}$, the identification being done by a unitary operator U taking

$$\sum_{n=0}^{\infty} a_n z^n \to \sum_{n=0}^{\infty} z^n \otimes a_n \text{ where } a_n \in \mathcal{E}.$$

In other words, if n is a non-negative integer and e is an element of $\mathcal{E}$, then the unitary map takes the element $z^n e$ to $z^n \otimes e$. With this identification, the operator $M_z^{\mathcal{E}}$ on $H^2_{\mathbb{D}}(\mathcal{E})$ is identified with $M_z \otimes I$ on $H^2_{\mathbb{D}} \otimes \mathcal{E}$. This extends linearly. Because of this identification of $H^2_{\mathbb{D}}(\mathcal{E})$ with $H^2_{\mathbb{D}} \otimes \mathcal{E}$, we shall sometimes consider M_φ as an operator from $H^2_{\mathbb{D}} \otimes \mathcal{E}$ to $H^2_{\mathbb{D}} \otimes \mathcal{F}$, remembering the underlying identification, where φ is a multiplier as above.

Theorem 1.9.2. *Let T be a contraction on H. Then there is a contraction $W : H \to H^2_{\mathbb{D}} \otimes \mathcal{D}_{T^*}$ such that*

$$TW^* = W^*(M_z \otimes I).$$

This W is an isometry if and only if T is $C_{\cdot 0}$. Consequently, in the case of a $C_{\cdot 0}$ contraction T, the closed subspace Ran W is invariant under $M_z^* \otimes I$ and

$$(M_z^* \otimes I)|_{\mathrm{Ran W}} = WT^*W^*|_{\mathrm{Ran W}}.$$

Thus the dilation of T (identified with WTW^* on Ran W) is $M_z \otimes I$.

Proof. Define W by

$$Wh = \sum_{n=0}^{\infty} z^n \otimes D_{T^*} T^{*n} h.$$

Then

$$\|Wh\|^2 = \sum_{n=0}^{\infty} \|D_{T^*}T^{*n}h\|^2 = \sum_{n=0}^{\infty} \langle T^n D_{T^*}^2 T^{*n}h, h\rangle.$$

This is a telescopic sum converging to $\|h\|^2 - \lim_{n\to\infty} \|T^{*n}h\|^2$. The limit exists because

$$\|h\| \geq \|T^*h\| \geq \|T^{*2}h\| \geq \cdots$$

by virtue of T being a contraction. Clearly, W is a contraction which is an isometry if and only if T is given to be pure (because then the limit is 0). Let us compute W^*. For any $h \in H$ and $\xi \in \mathcal{D}_{T^*}$, we have

$$\begin{aligned}
\langle W^*(z^n \otimes \xi), h\rangle &= \langle z^n \otimes \xi, Wh\rangle \\
&= \langle z^n \otimes \xi, \sum_{k=0}^{\infty} z^k \otimes D_{T^*}T^{*k}h\rangle \\
&= \langle \xi, D_{T^*}T^{*n}h\rangle \\
&= \langle T^n D_{T^*}\xi, h\rangle.
\end{aligned}$$

So, $W^*(z^n \otimes \xi) = T^n D_{T^*}\xi$. Hence,

$$W^*(M_z \otimes I)(z^n \otimes \xi) = W^*(z^{n+1} \otimes \xi) = T^{n+1}D_{T^*}\xi = TW^*(z^n \otimes \xi).$$

Thus, the equality $W^*(M_z \otimes I) = TW^*$ holds on the total set

$$\{z^n \otimes \xi : n = 0, 1, 2, \ldots \text{ and } \xi \in \mathcal{D}_{T^*}\}$$

and hence everywhere. ∎

This simple construction leads us to a functional model.

1.10. The characteristic function and functional model

One of the most well-studied class of functions in complex analysis is the class of Möbius transformations, viz., the class of functions φ_a, indexed by points a of the open unit disc and defined by

$$\varphi_a(z) = \frac{z - a}{1 - \bar{a}z}.$$

The function φ_a immediately lends itself to a generalization. If T is a contraction, then $I - zT^*$ is invertible for every z in the open unit disc. So, it is natural to consider the operator valued holomorphic function $(I - zT^*)^{-1}(z - T)$. A simple and quick computation shows that this function is the same as $-T + z(I - zT^*)^{-1}D_T^2$. Indeed,

$$
\begin{aligned}
(I - zT^*)^{-1}(z - T) &= (I - zT^*)^{-1}z - (I - zT^*)^{-1}T \\
&= (I - zT^*)^{-1}z - (I - zT^*)^{-1}(I - zT^* + zT^*)T \\
&= (I - zT^*)^{-1}z - T - (I - zT^*)^{-1}zT^*T \\
&= -T + z(I - zT^*)^{-1}D_T^2.
\end{aligned}
$$

A little modification of the above leads to a holomorphic function from $\mathbb{D}$ to $\mathcal{B}(\mathcal{D}_T, \mathcal{D}_{T^*})$. For that we need to use the formula

$$
TD_T = D_{T^*}T \tag{1.10.1}
$$

for a contraction T. This was deduced in section 1.4. Thus, if we consider the function $-T + zD_{T^*}(I - zT^*)^{-1}D_T$, then this is a function from $\mathcal{D}_T$ to $\mathcal{D}_{T^*}$.

Definition 1.10.1. *Let $\mathcal{H}$ be a Hilbert space and let T be a contraction on $\mathcal{H}$. The $\mathcal{B}(\mathcal{D}_T, \mathcal{D}_{T^*})$-valued holomorphic function on $\mathbb{D}$ defined by*

$$
\theta_T(z) = \left(-T + zD_{T^*}(I - zT^*)^{-1}D_T\right)|_{\mathcal{D}_T}, \quad z \in \mathbb{D}
$$

is called the characteristic function of T.

We strongly recommend that the reader notes the structure of the function and compares it with the realization formula in chapter 3.

Lemma 1.10.2. *The function θ_T is contraction valued, i.e., for every z in $\mathbb{D}$, the operator $\theta_T(z)$ has norm no greater than 1.*

Proof. We want to show that

$$
I - \theta_T(z)\theta_T(w)^* = (1 - z\overline{w})D_{T^*}(I - zT^*)^{-1}(I - \overline{w}T)^{-1}D_{T^*} \tag{1.10.2}
$$

for all z and w in $\mathbb{D}$. To that end,

$$
\begin{aligned}
&\theta_T(z)\theta_T(w)^* \\
=&(-T + zD_{T^*}(I - zT^*)^{-1}D_T)(-T^* + \overline{w}D_T(I - \overline{w}T)^{-1}D_{T^*}) \\
=&TT^* - \overline{w}TD_T(I - \overline{w}T)^{-1}D_{T^*} - zD_{T^*}(I - zT^*)^{-1}D_TT^* \\
&+z\overline{w}D_{T^*}(I - zT^*)^{-1}(I - T^*T)(I - \overline{w}T)^{-1}D_{T^*}
\end{aligned}
$$

leading to

$$\begin{aligned}
&I-\theta_T(z)\theta_T(w)^* \\
=&D_{T^*}^2+\overline{w}D_{T^*}T(I-\overline{w}T)^{-1}D_{T^*}+zD_{T^*}(I-zT^*)^{-1}D_TT^* \\
-&z\overline{w}D_{T^*}(I-zT^*)^{-1}(I-\overline{w}T)^{-1}D_{T^*} \\
+&z\overline{w}D_{T^*}(I-zT^*)^{-1}T^*T(I-\overline{w}T)^{-1}D_{T^*} \\
=&D_{T^*}(I+\overline{w}T(I-\overline{w}T)^{-1}+(I-zT^*)^{-1}zT^* \\
-&z\overline{w}(I-zT^*)^{-1}(I-\overline{w}T)^{-1} \\
+&(I-zT^*)^{-1}\big((zT^*-I)+I\big)\big((\overline{w}T-I)+I\big)(I-\overline{w}T)^{-1}\big)D_{T^*} \\
=&D_{T^*}(I+\overline{w}T(I-\overline{w}T)^{-1}+(I-zT^*)^{-1}zT^* \\
-&z\overline{w}(I-zT^*)^{-1}(I-\overline{w}T)^{-1} \\
+&(-I+(I-zT^*)^{-1})(-I+(I-\overline{w}T)^{-1}))D_{T^*} \\
=&D_{T^*}(I+\overline{w}T(I-\overline{w}T)^{-1}+(I-zT^*)^{-1}zT^* \\
-&z\overline{w}(I-zT^*)^{-1}(I-\overline{w}T)^{-1} \\
+&I-(I-zT^*)^{-1}-(I-\overline{w}T)^{-1}+(I-zT^*)^{-1}(I-\overline{w}T)^{-1}D_{T^*} \\
=&(1-z\overline{w})D_{T^*}(I-zT^*)^{-1}(I-\overline{w}T)^{-1}D_{T^*}.
\end{aligned}$$

Putting $z=w$, we get that the right hand side is a positive operator and hence the conclusion. ■

Recall the vector valued Hardy space introduced in the last section. For any Hilbert space $\mathcal{E}$, one can consider $H^2_{\mathbb{D}}(\mathcal{E})$ which is the Hardy space of $\mathcal{E}$-valued functions. Consider two spaces $H^2_{\mathbb{D}}(\mathcal{E})$ and $H^2_{\mathbb{D}}(\mathcal{F})$. A $\mathcal{B}(\mathcal{E},\mathcal{F})$-valued holomorphic function φ defined on the open unit disc $\mathbb{D}$ is called a multiplier if φf is in $H^2_{\mathbb{D}}(\mathcal{F})$ for every f in $H^2_{\mathbb{D}}(\mathcal{E})$. A standard Closed Graph Theorem argument then shows that the *operator of multiplication* M_φ defined by $M_\varphi f=\varphi f$ is a bounded operator from $H^2_{\mathbb{D}}(\mathcal{E})$ to $H^2_{\mathbb{D}}(\mathcal{F})$.

Just as in Lemma 1.8.3, we have an alternative description of $H^2_{\mathbb{D}}(\mathcal{E})$. A function f in $\mathcal{O}(\mathbb{D},\mathcal{E})$ is in $H^2_{\mathbb{D}}(\mathcal{E})$ if and only if

$$\sup_{0<r<1}\frac{1}{2\pi}\int_0^{2\pi}\|f(re^{i\theta})\|^2d\theta<\infty.$$

and in fact

$$\|f\|^2=\sup_{0<r<1}\frac{1}{2\pi}\int_0^{2\pi}\|f(re^{i\theta})\|^2d\theta.$$

We omit the proofs of these two assertions because they are exactly the same as the proof of Lemma 1.8.3 and the discussion following it.

Let φ be in $H^\infty_{\mathbb{D}}(\mathcal{B}(\mathcal{E},\mathcal{F}))$, i.e., φ is a holomorphic function on the open unit disc with values in $\mathcal{B}(\mathcal{E},\mathcal{F})$ and satisfying $\sup\{\|\varphi(z)\| : z \in \mathbb{D}\} < \infty$. Now, M_φ is a bounded linear operator from $H^2_{\mathbb{D}}(\mathcal{E})$ into $H^2_{\mathbb{D}}(\mathcal{F})$ because for any $f \in H^2_{\mathbb{D}}(\mathcal{E})$, we have

$$\begin{aligned}
&\sup_{0<r<1} \frac{1}{2\pi}\int_0^{2\pi} \|(\varphi f)(re^{i\theta})\|^2 d\theta \\
=&\sup_{0<r<1} \frac{1}{2\pi}\int_0^{2\pi} \|\varphi(re^{i\theta})f(re^{i\theta})\|^2 d\theta \\
\leq&\|\varphi\|^2_\infty \sup_{0<r<1} \frac{1}{2\pi}\int_0^{2\pi} \|f(re^{i\theta})\|^2 d\theta = \|\varphi\|^2_\infty \|f\|^2
\end{aligned}$$

where $\|\varphi\|_\infty = \sup\{\|\varphi(z)\| : z \in \mathbb{D}\}$. Along with showing that M_φ is a bounded operator from $H^2_{\mathbb{D}}(\mathcal{E})$ into $H^2_{\mathbb{D}}(\mathcal{F})$, this also shows that $\|M_\varphi\| \leq \|\varphi\|_\infty$.

Lemma 1.10.3. *Let φ be in $H^\infty_{\mathbb{D}}(\mathcal{B}(\mathcal{E},\mathcal{F}))$ and let θ be in $H^\infty_{\mathbb{D}}$. Then for any f in $H^2_{\mathbb{D}}(\mathcal{E})$, we have*

$$M_\varphi M_\theta f = M_\theta M_\varphi f.$$

Note that on the left hand side M_θ acts on $H^2_{\mathbb{D}}(\mathcal{E})$ whereas on the right hand side M_θ acts on $H^2_{\mathbb{D}}(\mathcal{F})$.

Proof. Let z be in $\mathbb{D}$. Then, since $\theta(z)$ is scalar valued, we have $M_\varphi M_\theta f(z) = \varphi(z)\theta(z)f(z) = \theta(z)\varphi(z)f(z) = M_\theta M_\varphi f(z)$. ■

The importance of Lemma 1.10.2 is that it tells us that θ_T is in the space $H^\infty_{\mathbb{D}}(\mathcal{B}(\mathcal{D}_T,\mathcal{D}_{T^*}))$ with $\|\theta_T\| \leq 1$. Thus, we now know, because of the discussion in the preceeding paragraph that M_{θ_T} is a contraction from $H^2_{\mathbb{D}}(\mathcal{D}_T)$ into $H^2_{\mathbb{D}}(\mathcal{D}_{T^*})$.

The identification of $H^2_{\mathbb{D}}(\mathcal{E})$ with $H^2_{\mathbb{D}} \otimes \mathcal{E}$ explained in the last section makes it easy to describe action of the adjoint M^*_φ.

Lemma 1.10.4. *Let φ be in $H^\infty_{\mathbb{D}}(\mathcal{B}(\mathcal{E},\mathcal{F}))$. Let $\xi \in \mathcal{F}$. Then, $M^*_\varphi(k_w \otimes \xi) = k_w \otimes \varphi(w)^*\xi$ for any w in $\mathbb{D}$ where $k_w(z) = (1-\overline{w}z)^{-1}$.*

Proof. Let η be in $\mathcal{E}$. First note that the action of M_φ on the vector $1 \otimes \eta$, which according to the identification above is the constant function whose value is η at every z in $\mathbb{D}$, is $(M_\varphi(1\otimes\eta))(z) = \varphi(z)\eta$ for every z in $\mathbb{D}$. Now, let p be any polynomial. Then,

$$\begin{aligned}
(M_\varphi(p\otimes\eta))(z) = (M_\varphi(p.\eta))(z) &= \varphi(z)p(z)\eta \\
&= p(z)\varphi(z)\eta = (M_p M_\varphi(1\otimes\eta))(z)
\end{aligned}$$

for every z in $\mathbb{D}$ showing that $M_\varphi(p\otimes\eta) = M_p M_\varphi(1\otimes\eta)$ for every polynomial p. Now, for any w in $\mathbb{D}$, we have

$$\begin{aligned}\langle p\otimes\eta, M_\varphi^*(k_w\otimes\xi)\rangle &= \langle M_\varphi(p\otimes\eta), k_w\otimes\xi\rangle \\ &= \langle M_p M_\varphi(1\otimes\eta), k_w\otimes\xi\rangle \\ &= \langle p(w)\varphi(w)\eta, \xi\rangle \\ &= \langle p, k_w\rangle\langle\eta, \varphi(w)^*\xi\rangle \\ &= \langle p\otimes\eta, k_w\otimes\varphi(w)^*\xi\rangle.\end{aligned}$$

Since the vectors $p\otimes\eta$ span a dense set, this shows that $M_\varphi^*(k_w\otimes\xi) = k_w\otimes\varphi(w)^*\xi$. ■

Lemma 1.10.5. *Let W be the operator introduced in Theorem 1.9.2, i.e.,*

$$Wh = \sum_{n=0}^{\infty} z^n\otimes D_{T^*}T^{*n}h.$$

The identity $WW^* + M_{\theta_T}M_{\theta_T}^* = I_{H^2_{\mathbb{D}}\otimes\mathcal{D}_{T^*}}$ holds for every contraction T.

Proof. Let us compute the action of W^* on an elementary tensor $k_w\otimes\xi$ for w in $\mathbb{D}$. Note that $\langle W^*(k_w\otimes\xi), h\rangle$ is $\langle k_w\otimes\xi, Wh\rangle$. Recall that k_w is the function in the Hardy space which has the power series expansion with the co-efficient of the nth. term being $\overline{w}^n$. On the other hand, Wh is the function in $H^2_{\mathbb{D}}\otimes\mathcal{D}_{T^*}$ that equals $\sum_{n=0}^{\infty} z^n\otimes D_{T^*}T^{*n}h$. So, the inner product $\langle W^*(k_w\otimes\xi), h\rangle$ is the inner product of the function $\sum_{n=0}^{\infty}\overline{w}^n z^n\otimes\xi$ with $\sum_{n=0}^{\infty} z^n\otimes D_{T^*}T^{*n}h$. Now,

$$\begin{aligned}\langle\sum_{n=0}^{\infty}\overline{w}^n z^n\otimes\xi, \sum_{n=0}^{\infty} z^n\otimes D_{T^*}T^{*n}h\rangle &= \sum_{n=0}^{\infty}\overline{w}^n\langle\xi, D_{T^*}T^{*n}h\rangle \\ &= \langle\sum_{n=0}^{\infty}\overline{w}^n T^n D_{T^*}\xi, h\rangle \\ &= \langle(I-\overline{w}T)^{-1}D_{T^*}\xi, h\rangle.\end{aligned}$$

Thus the operator W^* satisfies the identity

$$W^*(k_w\otimes\xi) = (I-\bar{w}T)^{-1}D_{T^*}\xi \quad \text{for } w\in\mathbb{D}, \xi\in\mathcal{D}_{T^*},$$

where $k_w(z) = (1 - \overline{w}z)^{-1}$. Therefore, for z, w in $\mathbb{D}$ and ξ, η in $\mathcal{D}_{T^*}$, we obtain that

$$\begin{aligned}
&\langle (WW^* + M_{\theta_T} M_{\theta_T}^*) k_z \otimes \xi, k_w \otimes \eta \rangle \\
&= \langle W^*(k_z \otimes \xi), W^*(k_w \otimes \eta) \rangle + \langle M_{\theta_T}^*(k_z \otimes \xi), M_{\theta_T}^*(k_w \otimes \eta) \rangle \\
&= \langle (I - \bar{z}T)^{-1} D_{T^*}\xi, (I - \bar{w}T)^{-1} D_{T^*}\eta \rangle + \langle k_z \otimes \theta_T(z)^*\xi, k_w \otimes \theta_T(w)^*\eta \rangle \\
&= \langle D_{T^*}(I - wT^*)^{-1}(I - \bar{z}T)^{-1} D_T^*\xi, \eta \rangle + \langle k_z, k_w \rangle \langle \theta_T(w)\theta_T(z)^*\xi, \eta \rangle \\
&= \langle k_z \otimes \xi, k_w \otimes \eta \rangle .
\end{aligned}$$

The last equality follows because we know from (1.10.2) that

$$1 - \theta_T(w)\theta_T(z)^* = (1 - w\bar{z}) D_{T^*}(1 - wT^*)^{-1}(1 - \bar{z}T)^{-1} D_{T^*}.$$

Using the fact that the vectors k_z form a total set in $H^2_{\mathbb{D}}$, we see that the set $\{k_z \otimes \xi : z \in \mathbb{D} \text{ and } \xi \in \mathcal{D}_{T^*}\}$ is a total set in $H^2_{\mathbb{D}} \otimes \mathcal{D}_{T^*}$ and hence the assertion follows. ■

Using the result above, a functional model for a pure contraction is immediate.

Theorem 1.10.6. *Every pure contraction T defined on a Hilbert space $\mathcal{H}$ is unitarily equivalent to the operator T_1 on the Hilbert space $\mathbb{H}_T = (H^2_{\mathbb{D}} \otimes \mathcal{D}_{T^*}) \ominus M_{\theta_T}(H^2_{\mathbb{D}} \otimes \mathcal{D}_T)$ defined as $T_1 = P_{\mathbb{H}_T}(M_z \otimes I)|_{\mathbb{H}_T}$.*

To prove the theorem, we give a name to the adjoint of the operator W considered above. We call the adjoint L. So, $L : H^2_{\mathbb{D}} \otimes \mathcal{D}_{T^*} \to \mathcal{H}$ is given by

$$L(f \otimes \xi) = f(T) D_{T^*}\xi, \quad \text{for all } f \in \mathbb{C}[z], \text{ and } \xi \in \mathcal{D}_{T^*}.$$

Proof. Since T is a pure contraction, it is evident from Lemma 1.10.5 that L^*L and $M_{\theta_T} M_{\theta_T}^*$ are orthogonal projections summing up to the identity operator. So, they are complementary projections. Hence,

$$L^*(\mathcal{H}) = W(\mathcal{H}) = \mathbb{H}_T = (H^2_{\mathbb{D}} \otimes \mathcal{D}_{T^*}) \ominus M_{\theta_T}(H^2_{\mathbb{D}} \otimes \mathcal{D}_T).$$

Let $V = M_z \otimes I$. For a vector of the form $z^n \otimes \xi$ of $H^2_{\mathbb{D}} \otimes \mathcal{D}_{T^*}$ and $h \in \mathcal{H}$ we have

$$\begin{aligned}
\langle L(z^n \otimes \xi), h \rangle &= \langle z^n \otimes \xi, \sum_{k=0}^{\infty} z^k \otimes D_{T^*} T^{*k} h \rangle \\
&= \langle \xi, D_{T^*} T^{*n} h \rangle = \langle T^n D_{T^*}\xi, h \rangle .
\end{aligned}$$

This implies that

$$L(z^n \otimes \xi) = T^n D_{T^*}\xi, \quad \text{for } n = 0, 1, 2, 3, \ldots$$

Therefore

$$\begin{aligned}\langle L(M_z \otimes I)(z^n \otimes \xi), h\rangle =& \langle z^{n+1} \otimes \xi, \sum_{k=0}^{\infty} z^k \otimes D_{T^*}T^{*k}h\rangle \\ =& \langle \xi, D_{T^*}T^{*n+1}h\rangle = \langle T^{n+1}D_{T^*}\xi, h\rangle.\end{aligned}$$

Consequently, $LV = TL$ on vectors of the form $z^n \otimes \xi$ which span $H^2_{\mathbb{D}} \otimes \mathcal{D}_{T^*}$ and hence

$$LV = TL.$$

Therefore V^* leaves the range of L^* (isometric copy of $\mathcal{H}$) invariant and $V^*|_{L^*\mathcal{H}} = L^*T^*L$ which is the isometric copy of the operator T^* on range of L^*. Thus $\mathbb{H}_T$ is co-invariant under $M_z \otimes I$. That completes the proof. ■

The space $\mathbb{H}_T$ obtained as above as a function space is sometimes called a *model space* and $P_{\mathbb{H}_T}(M_z \otimes I)|_{\mathbb{H}_T}$ is called the *model operator* for T.

1.11. The characteristic function as a complete unitary invariant

Definition 1.11.1. *Given two contractions T and R on Hilbert spaces $\mathcal{H}$ and $\mathcal{K}$ respectively, the characteristic functions of T and R are said to coincide if there exist unitary operators $\tau : \mathcal{D}_T \to \mathcal{D}_R$ and $\tau_* : \mathcal{D}_{T^*} \to \mathcal{D}_{R^*}$ such that the following diagram commutes for all z in $\mathbb{D}$:*

$$\begin{array}{ccc} \mathcal{D}_T & \xrightarrow{\theta_T(z)} & \mathcal{D}_{T^*} \\ \tau\downarrow & & \downarrow\tau_* \\ \mathcal{D}_R & \xrightarrow[\theta_R(z)]{} & \mathcal{D}_{R^*} \end{array}$$

The first result to note is the following.

Proposition 1.11.2. *The characteristic functions of two unitarily equivalent contractions coincide.*

Proof. Let T and R be two contractions on $\mathcal{H}$ and $\mathcal{K}$ respectively, such that there is a unitary operator $\sigma : \mathcal{H} \to \mathcal{K}$ satisfying $\sigma T \sigma^* = R$. Then it is easy to see that $\sigma D_T^2 \sigma^* = D_R^2$ and $\sigma D_{T^*}^2 \sigma^* = D_{R^*}^2$. Thus, $\sigma D_T \sigma^* = D_R$ and $\sigma D_{T^*} \sigma^* = D_R$. Hence, the restriction $\tau = \sigma|_{\mathcal{D}_T}$ is a unitary operator between $\mathcal{D}_T$ and $\mathcal{D}_R$. Similarly, the restriction $\tau_* = \sigma|_{\mathcal{D}_{T^*}}$ defines a unitary operator from $\mathcal{D}_{T^*}$ to $\mathcal{D}_{R^*}$. Finally, note that for an element of the form $D_T h$, we have

$$\begin{aligned}\theta_R(z)\underline{\tau} D_T h &= (-R + D_{R^*}(1 - zR^*)^{-1} z D_R)\sigma D_T h \\ &= (-\sigma T + D_{R^*}(1 - zR^*)^{-1} z\sigma D_T) D_T h \\ &= (-\sigma T + D_{R^*}(1 - zR^*)^{-1} \sigma z D_T) D_T h \\ &= (-\sigma T + \sigma D_{T^*}(1 - zT^*) z D_T) D_T h = \tau_* \theta_T(z) D_T h\end{aligned}$$

for all $z \in \mathbb{D}$. Since the operators $\theta_R(z)\underline{\tau}$ and $\tau_*\theta_T(z)$ agree on the range of D_T, they agree on the closure of it and hence the two characteristic functions θ_T and θ_R coincide.

Next we prove the converse of the above proposition for the case of pure contractions.

Proposition 1.11.3. *Let T and R be two pure contractions on $\mathcal{H}$ and $\mathcal{K}$ respectively. If their characteristic functions θ_T and θ_R coincide, then the operators T and R are unitarily equivalent.*

Proof. Let $\tau' : \mathcal{D}_T \to \mathcal{D}_R$ and $\tau'_* : \mathcal{D}_{T^*} \to \mathcal{D}_{R^*}$ be two unitary operators such that the diagram

$$\begin{array}{ccc} \mathcal{D}_T & \xrightarrow{\theta_T(z)} & \mathcal{D}_{T^*} \\ {\scriptstyle \tau'}\downarrow & & \downarrow{\scriptstyle \tau'_*} \\ \mathcal{D}_R & \xrightarrow[\theta_R(z)]{} & \mathcal{D}_{R^*} \end{array}$$

commutes for all z in $\mathbb{D}$. The operators τ' and τ'_* give rise to unitary operators $\tau = 1 \otimes \tau' : H^2_{\mathbb{D}} \otimes \mathcal{D}_T \to H^2_{\mathbb{D}} \otimes \mathcal{D}_R$ and $\tau_* = 1 \otimes \tau'_* : H^2_{\mathbb{D}} \otimes \mathcal{D}_{T^*} \to H^2_{\mathbb{D}} \otimes \mathcal{D}_{R^*}$ which satisfy the intertwining relation

$$M_{\theta_R}\tau = \tau_* M_{\theta_T}$$

We conclude that

$$\tau_*(\mathbb{H}_T) = \tau_*((\mathrm{Ran}\, M_{\theta_T})^{\perp}) = \tau_*((\mathrm{Ran}\, M_{\theta_T}))^{\perp} = (\mathrm{Ran}\, M_{\theta_R})^{\perp} = \mathbb{H}_R$$

where $\mathbb{H}_T$ and $\mathbb{H}_R$ are the model spaces for T and R as in the last section. Since the operator τ_* intertwines $(M_z \otimes 1_{\mathcal{D}_{T^*}})^*$ and $(M_z \otimes 1_{\mathcal{D}_{R^*}})^*$, the induced unitary operator $\tau_* : \mathbb{H}_T \to \mathbb{H}_R$ intertwines the adjoints of the restrictions, which are precisely the model operators $P_{\mathbb{H}_T}(M_z \otimes 1_{\mathcal{D}_{T^*}})|\mathbb{H}_T$ and $P_{\mathbb{H}_R}(M_z \otimes 1_{\mathcal{D}_{R^*}})|\mathbb{H}_R$. But then Theorem 1.10.6 shows that T and R are unitarily equivalent.

Summarizing the last two propositions we obtain the main result of this section.

Theorem 1.11.4. *Two pure contractions T and R on Hilbert spaces $\mathcal{H}$ and $\mathcal{K}$ are unitarily equivalent if and only if their characteristic functions coincide.*

We shall revisit and generalize this in Chapter 5. Sz.-Nagy and Foias proved a more general version of this, viz., for completely non-unitary contractions in [113].

1.12. Decomposition

A contraction T on a Hilbert space $\mathcal{H}$ is called *completely non-unitary* if there is no reducing subspace $\mathcal{M}$ such that $T|_{\mathcal{M}}$ is a unitary operator. We leave it to the reader to check that a unilateral shift (of any multiplicity) is completely non-unitary.

Theorem 1.12.1. *Given a contraction T on a Hilbert space $\mathcal{H}$, the space $\mathcal{H}$ can be decomposed into the direct sum of two subspaces $\mathcal{H}_0$ and $\mathcal{H}_1$ such that both of them reduce T, the restriction of T to $\mathcal{H}_0$ is a unitary operator and the restriction of T to $\mathcal{H}_1$ is a completely non-unitary operator.*

Proof. Since T is a contraction, we have

$$I \geq T^*T \geq T^{*2}T^2 \geq \cdots \geq 0.$$

So the sequence $\{T^{*n}T^n\}$ is a monotonically decreasing sequence of positive operators. Hence, there is a strong limit. We denote this strong limit by T_∞.

Note that T_∞ is a positive contraction. Let x belong to the null space $\mathcal{N}(I - T_\infty)$. Then $x = T_\infty x$. Thus,

$$\|x\|^2 = \langle T_\infty x, x\rangle \leq \langle T^{*n}T^n x, x\rangle = \|T^n x\|^2 \leq \|x\|^2$$

giving us that $\|x\|^2 = \|T^n x\|^2$ for all $n \geq 1$. Conversely, if $\|x\|^2 = \|T^n x\|^2$ for all $n \geq 1$, then

$$\|x\|^2 = \langle T^{*n}T^n x, x\rangle = \lim_{n\to\infty} \langle T^{*n}T^n x, x\rangle = \langle T_\infty x, x\rangle$$

implying that $\langle (I-T_\infty)x, x\rangle = 0$ which by virtue of the fact that $I-T_\infty \geq 0$ gives us that $(I - T_\infty)x = 0$. Hence,

$$\mathcal{N}(I - T_\infty) = \{x \in \mathcal{H} : \|x\| = \|T^n x\| \text{ for all } n \geq 1\}.$$

Similarly,

$$\mathcal{N}(I - (T^*)_\infty) = \{x \in \mathcal{H} : \|x\| = \|T^{*n} x\| \text{ for all } n \geq 1\}$$

where $(T^*)_\infty$ is the strong limit of T^nT^{*n}.

Now, let

$$\mathcal{H}_0 = \{x \in \mathcal{H} : \|x\| = \|T^n x\| = \|T^{*n}x\| \text{ for all } n \geq 1\}.$$

Note that $\mathcal{H}_0 = \mathcal{N}(I-T_\infty) \cap \mathcal{N}(I-(T^*)_\infty)$. Since, $\mathcal{N}(I-T_\infty)$ is invariant under T, $\mathcal{H}_0$ is invariant under both T and T^* and hence a reducing subspace for T. Let $\mathcal{H}_1 = \mathcal{H} \ominus \mathcal{H}_0$. This is a reducing subspace and if $\mathcal{M}$ is a reducing subspace for T such that $T_{\mathcal{M}}$ is a unitary operator, then for any x in $\mathcal{M}$, we have $\|x\| = \|T^n x\| = \|T^{*n} x\|$ for all $n \geq 1$. Hence x is in $\mathcal{H}_0$. Thus $T_{\mathcal{H}_1}$ is completely non-unitary. ∎

Theorem 1.12.2. *[Wold decomposition] Let V be an isometry on $\mathcal{H}$. Then the space $\mathcal{H}$ can be decomposed into the direct sum of two subspaces $\mathcal{H}_0$ and $\mathcal{H}_1$ such that both of them reduce V, the restriction of V to $\mathcal{H}_0$ is a unitary operator and the restriction of V to $\mathcal{H}_1$ is a unilateral shift.*

Proof. Since V is an isometry, VV^* is the orthogonal projection onto range of V. Let $\mathcal{E}$ = Range of $I - VV^*$. Since $\mathcal{E}$ is the orthocomplement of the range of V, it is the kernel of V^*. Hence, for x in $\mathcal{E}$, we have

$$\langle Vx, x\rangle = \langle x, V^*x\rangle = 0.$$

For the same reason, $V^k\mathcal{E}$ is orthogonal to $V^n\mathcal{E}$ for all $k \neq n$. Let

$$\mathcal{H}_1 = \mathcal{E} \oplus V\mathcal{E} \oplus \cdots .$$

Clearly, this is an invariant subspace for V and the restriction of V to this subspace is the unilateral shift of multiplicity equal to the dimension of the range of $I - VV^*$. Moreover, this is actually a reducing subspace because V^* maps $V^k\mathcal{E}$ onto $V^{k-1}\mathcal{E}$ for $k \geq 1$ and kills $\mathcal{E}$. Denote $\mathcal{H} \ominus \mathcal{H}_1$ by $\mathcal{H}_0$. On the reducing subspace $\mathcal{H}_0$, V is an isometry. Moreover, $\mathcal{H}_0$ is contained in the range of V. hence, if y is in $\mathcal{H}_0$, then $y = VV^*y$. Since, y is orthogonal to all $V^k\mathcal{E}$ for $k \geq 0$, it follows that V^*y is orthogonal to all $V^k\mathcal{E}$ for $k \geq 0$. Thus, V^*y is in $\mathcal{H}_0$. Hence, $V|_{\mathcal{H}_0}$ is a surjective isometry, i.e., a unitary operator. ∎

Note that if, to begin with, our isometry V was already a unilateral shift, then $\mathcal{H}_1$ would be the whole space so that the unitary part is zero.

1.13. The unitary dilation

Definition 1.13.1. *Let $\mathcal{H} \subset \mathcal{K}$ be two Hilbert spaces. Suppose V and U are bounded operators on $\mathcal{H}$ and $\mathcal{K}$ respectively such that*

$$U^n h = V^n h \textit{ for all } h \in \mathcal{H}.$$

Then U is called an extension of V. A unitary extension is an extension which is also a unitary operator.

Note that the block operator matrix of the operator U with respect to the decomposition $\mathcal{K} = \mathcal{H} \oplus (\mathcal{K} \ominus \mathcal{H})$ is

$$U = \begin{pmatrix} V & * \\ 0 & * \end{pmatrix}.$$

An extension U of a bounded operator V is also a dilation of V because $P_{\mathcal{H}} U^n h = P_{\mathcal{H}} V^n h = V^n h$ for any $h \in \mathcal{H}$ and $n \geq 1$. It moreover has the property that $\mathcal{H}$ is an invariant subspace for U. It is this second property that makes it clear that in general, contractions could not have isometric extensions.

Remark 1.13.2. *Let $\mathcal{H}$ be a Hilbert space and $T \in \mathcal{B}(\mathcal{H})$. Suppose V is a dilation of T on a Hilbert space $\mathcal{K}_1 \supset \mathcal{H}$ and U is an extension of V on a Hilbert space $\mathcal{K}_2 \supset \mathcal{K}_1$. Then U is a dilation of T. Indeed,*

$$\begin{aligned} P_{\mathcal{H}} U^n h &= P_{\mathcal{H}} V^n h \text{ because } U \text{ is an extension of } V \text{ and } h \in \mathcal{H} \subset \mathcal{K}_1 \\ &= T^n h \text{ because } V \text{ is a dilation of } T. \end{aligned}$$

Definition 1.13.3. *A unitary operator U on a Hilbert space $\mathcal{M}$ is called a bilateral shift if there is a subspace $\mathcal{L}$ of $\mathcal{M}$ satisfying*

(i) $U^n \mathcal{L} \perp \mathcal{L}$ for all integers $n \neq 0$ and

(ii) $\oplus_{n=-\infty}^{\infty} U^n \mathcal{L} = \mathcal{M}$.
The subspace $\mathcal{L}$ is called a generating subspace for U and $\dim \mathcal{L}$ is called the multiplicity of U.

Lemma 1.13.4. *A unilateral shift V on $\mathcal{M}$ always has an extension to a bilateral shift. Moreover, the extension preserves multiplicity.*

Proof: The generating subspace of V is $\mathcal{L} = \mathcal{M} \ominus V\mathcal{M}$. Define $\mathcal{K} = \oplus_{-\infty}^{\infty} \mathcal{L}_n$ where each $\mathcal{L}_n$ is the same as $\mathcal{L}$. For an element $(\ldots, l_{-2}, l_{-1}, l_0, l_1, l_2, \ldots)$ of $\mathcal{K}$ with $l_n \in \mathcal{L}_n$ for every $n \in \mathbb{Z}$, define

$$U(\ldots, l_{-2}, l_{-1}, l_0, l_1, l_2, \ldots) = (\ldots, l'_{-2}, l'_{-1}, l'_0, l'_1, l'_2, \ldots),$$

where now $l'_n \in \mathcal{L}_n$ and $l'_n = l_{n-1}$ for all $n \in \mathbb{Z}$. Clearly, U is unitary and $\{(\ldots, 0, 0, l_0, 0, 0, \ldots) : l_0 \in \mathcal{L}_0\}$ is a generating subspace for U. This subspace has the same dimension as that of $\mathcal{L}$. An element $\sum_0^{\infty} V^n l_n$ of $\mathcal{H}$ is identified with the element $(\ldots, 0, 0, l_0, l_1, l_2, \ldots)$ of $\mathcal{K}$. This is an isometric embedding.

Now for $h = \sum_0^\infty V^n l_n$, we have

$$\begin{aligned} Uh &= U(\ldots, 0, 0, l_0, l_1, l_2, \ldots) \\ &= (\ldots, 0, 0, 0, l_0, l_1, \ldots) = \sum_0^\infty V^n l_{n-1} = Vh. \end{aligned}$$

That completes the proof. ■

An immediate corollary is the following.

Corollary 1.13.5. *An isometry V on $\mathcal{H}$ always has an extension to a unitary.*

Proof: By Wold decomposition, we have $\mathcal{H} = \mathcal{H}_0 \oplus \mathcal{H}_1$ where $\mathcal{H}_0$ and $\mathcal{H}_1$ are reducing subspaces of V and $V = V_0 \oplus V_1$ where $V_0 = V|_{\mathcal{H}_0}$ is a unitary and $V_1 = V|_{\mathcal{H}_1}$ is a unilateral shift. By the lemma above, V_1 can be extended to a bilateral shift, say U_1. Now $V_0 \oplus U_1$ is a unitary extension of V. ■

Theorem 1.13.6. *For every contraction T on a Hilbert space $\mathcal{H}$, there is a unitary dilation U which is minimal in the sense that*

$$\overline{span}\{U^n h : h \in \mathcal{H}, n \in \mathbb{Z}\} = \hat{\mathcal{H}}.$$

This minimal unitary dilation is unique up to unitary equivalence.

Proof: Obtaining a unitary dilation is immediate from the Corollary above. We take an isometric dilation and then its unitary extension, say U_0. This is a unitary dilation, although may not be minimal. Let

$$\mathcal{K} = \overline{span}\{U_0^n h : h \in \mathcal{H} \text{ and } n = \ldots, -2, -1, 0, 1, 2, \ldots\}.$$

This is a reducing subspace for U_0 and the restriction U of U_0 to $\mathcal{K}$ is a minimal unitary dilation.

The uniqueness (up to unitary equivalence) proof is exactly on the same lines as the proof of uniqueness of isometric dilation. ■

Following Schäffer (See Section 1.6) a concrete representation of the minimal unitary dilation of a contraction T on a Hilbert space $\mathcal{H}$ can be given as follows. Define a map U on

$$\tilde{\mathcal{H}} := \cdots \oplus \mathcal{D}_{T^*} \oplus \mathcal{D}_{T^*} \oplus \mathcal{H} \oplus \mathcal{D}_T \oplus \mathcal{D}_T \oplus \cdots,$$

by $U(\ldots, h_{-2}, h_{-1}, [h_0], h_1, h_2, \ldots)$

$$= (\ldots, h_{-3}, h_{-2}, [D_{T^*}h_{-1} + Th_0], -T^*h_{-1} + D_T h_0, h_1, h_2, \ldots),$$

where $[\cdot]$ indicates the location of $\mathcal{H}$. Then U is a minimal unitary dilation of T. The block operator matrix of U is given by,

$$\begin{pmatrix} \ddots & & & & & & \\ & I & & & & & \\ & & I & & & & \\ & & & D_{T^*} & [T] & & \\ & & & -T^* & D_T & & \\ & & & & & I & \\ & & & & & & \ddots \end{pmatrix}.$$

Here zero entries are not shown and $[\cdot]$ is the location of operator from $\mathcal{H}$ to $\mathcal{H}$ and there we have T. The copies of identity, D_{T^*} and D_T are appearing just below the diagonal.

For a guided tour of dilation theory, see [110].

1.14. The Berger-Coburn-Lebow Theorem

This section contains an interesting application of the Wold decomposition, viz., the Berger-Coburn-Lebow Theorem which appeared in [23]. This theorem has found significant applications in recent research, see [34] and [47], for example.

We start with a preparatory lemma.

Lemma 1.14.1. *Let $\mathcal{F}$ be a Hilbert space and let E_1, E_2 be in $\mathcal{B}(\mathcal{F})$. Then E_1 and E_2 are partial isometries of the form $U^*P^\perp$ and PU respectively for some unitary U and projection P if and only if*

$$E_1E_2 = E_2E_1 = 0 \tag{1.14.1}$$

and

$$E_1E_1^* + E_2^*E_2 = I_{\mathcal{F}} = E_1^*E_1 + E_2E_2^*. \tag{1.14.2}$$

Proof: As usual, one way is easy, viz., if $E_1 = U^*P^\perp$ and $E_2 = PU$, then $E_1E_2 = E_2E_1 = 0$ and $E_1E_1^* + E_2^*E_2 = I_{\mathcal{F}} = E_1^*E_1 + E_2E_2^*$.

Conversely, let $E_1, E_2 \in \mathcal{B}(\mathcal{H})$ satisfy $E_1E_2 = E_2E_1 = 0$ and $E_1E_1^* + E_2^*E_2 = I_{\mathcal{F}} = E_1^*E_1 + E_2E_2^*$. Then $E_1E_1^*E_1 = E_1$ and $E_2E_2^*E_2 = E_2$ proving that $E_iE_i^*$ is the projection onto the range of E_i for $i = 1, 2$. Let $P = E_2E_2^*$. Set the unitary U to be $E_1^* + E_2$. The two given conditions show that this is a unitary. Also, $E_2 = PU$ and $E_1 = U^*P^\perp$. We may note here that $P^\perp = E_1^* E_1$ is the projection onto the the kernel of E_2^* which is also the same as range of E_1^*. We are done. ■

Theorem 1.14.2. *[Berger-Coburn-Lebow] Let V_1 and V_2 be two commuting isometries on a Hilbert space $\mathcal{H}$. Then there exists Hilbert spaces $\mathcal{F}$ and $\mathcal{L}$, a unitary operator $\omega : \mathcal{H} \to H^2(\mathcal{F}) \oplus \mathcal{L}$, a projection P in $\mathcal{B}(\mathcal{F})$, a unitary U in $\mathcal{B}(\mathcal{F})$ and commuting unitary operators W_1 and W_2 in $\mathcal{B}(\mathcal{L})$ such that*

$$\omega V_1 \omega^* = M_{U^*(P^\perp + zP)} \oplus W_1 \text{ and } \omega V_2 \omega^* = M_{(P+zP^\perp)U} \oplus W_2.$$

Proof: We shall need to dissect the Wold decomposition of the isometry $V = V_1 V_2$ in detail. Let D_{V^*} be the projection $I - VV^*$ and let $\mathcal{D}_{V^*}$ be the range of D_{V^*}. The Wold decomposition furnishes us the following: any $h \in \mathcal{H}$ decomposes as

$$h = (\oplus_{n=0}^\infty V^n D_{V^*} V^{*n} h) \oplus h_u$$

where $h_u = \lim_{n\to\infty} V^n V^{*n} h \in \mathcal{H}_u = \cap_{n=0}^\infty \mathrm{Ran} V^n$. Thus the space $\mathcal{H}$ becomes $(\oplus_{n=0}^\infty V^n \mathcal{D}_{V^*}) \oplus \mathcal{H}_u$. Define

$$\omega : \mathcal{H} \to H^2(\mathcal{D}_{V^*}) \oplus \mathcal{H}_u \text{ by}$$

$$h \to \sum_{n=0}^\infty (D_{V^*} V^{*n} h) z^n \oplus \lim_{n\to\infty} V^n V^{*n} h.$$

Then ω is a unitary operator and $\omega V = (M_z \oplus W)\omega$. Let $\tilde{V_i} = \omega V_i \omega^*$ for $i = 1, 2$. If $\tilde{V_i} = \begin{pmatrix} V_{i,11} & V_{i,12} \\ V_{i,21} & V_{i,22} \end{pmatrix}$ with respect to the decomposition $H^2(\mathcal{D}_{V^*}) \oplus \mathcal{H}_u$, then by the commutativity of $\tilde{V_i}$ with $M_z \oplus W$ (because each V_i commutes with V), we get $\tilde{V}_{i,12} W = M_z \tilde{V}_{i,12}$, for $i = 1, 2$.

Thus, $\tilde{V}_{i,12} W^n = M_z^n \tilde{V}_{i,12}$, for $i = 1, 2$ and for all $n = 1, 2, \ldots$. Taking adjoint and using the fact that $M_z^{*n} \to 0$ strongly as $n \to \infty$, we get $W^{*n} \tilde{V}_{i,12}^* \to 0$ strongly as $n \to \infty$. But W is a unitary. So $\tilde{V}_{i,12} = 0$. So,

$$\tilde{V_i} = \begin{pmatrix} V_{i,11} & 0 \\ V_{i,21} & V_{i,22} \end{pmatrix} \text{ for } i = 1, 2.$$

Since $\tilde{V_i}$ is an isometry, so is $V_{i,22}$. Moreover, $V_1 V_2 = V_2 V_1 = V$ implies that $V_{1,22} V_{2,22} = V_{2,22} V_{1,22} = W$, a unitary. So, each $V_{i,22}$ is a unitary. Again since $\tilde{V_i}$ is an isometry, we get

$$\begin{pmatrix} \tilde{V}_{i,11}^* & \tilde{V}_{i,21}^* \\ 0 & \tilde{V}_{i,22}^* \end{pmatrix} \begin{pmatrix} \tilde{V}_{i,11} & 0 \\ \tilde{V}_{i,21} & \tilde{V}_{i,22} \end{pmatrix} = \begin{pmatrix} I_{H^2(\mathcal{D}_{V^*})} & 0 \\ 0 & I_{\mathcal{H}_u} \end{pmatrix}.$$

In particular, $\tilde{V}_{i,21}^* \tilde{V}_{i,22} = 0$. Since $\tilde{V}_{i,22}$ is a unitary, this means that $\tilde{V}_{i,21} = 0$. Thus, each $\tilde{V_i}$ is block diagonal, i.e., $\tilde{V_i} = \begin{pmatrix} V_{i,11} & 0 \\ 0 & V_{i,22} \end{pmatrix}$.

Using commutativity of $\tilde{V}_i$ with $M_z \oplus W$, we get that $\tilde{V}_{i,11} = M_{\varphi_i}$ for $H^\infty(\mathcal{B}(\mathcal{D}_{V^*}))$ valued functions $\varphi_i, i = 1, 2$.

Call the unitary $\tilde{V}_{i,22}$ as W_i. So, we have shown that the V_i are unitarily equivalent to $M_{\varphi_i} \oplus W_i$. What remains is to decipher the structures of the φ_i. Here $V_1 = V_2^* V$ helps. This means that $M_{\varphi_1} = M_{\varphi_2}^* M_z$. Considering power eries expansions and equating terms, we get that

$$\varphi_1(z) = E_1 + zE_2^* \text{ and } \varphi_2(z) = E_2 + zE_1^*$$

for some $E_1, E_2 \in \mathcal{B}(\mathcal{D}_{V^*})$. Now, $V = V_1 V_2 = V_2 V_1$ implies that $M_z = M_{\varphi_1} M_{\varphi_2} = M_{\varphi_2} M_{\varphi_1}$ which readily implies (1.14.1). Moreover, the facts that M_{φ_1} and M_{φ_2} are isometries imply (1.14.2). Hence, by Lemma 1.14.1, we are done. ■

CHAPTER 2

C^*-Algebras and Completely Positive Maps

2.1. Banach algebras

In the first two sections of this chapter, we shall describe some basics of Banach and C^*-algebra theory which we are going to need in later chapters. Our presentation would be rather brief. For further details on the subject one may refer to standard books on C^*-algebra theory such as [83, 48, 70, 71]. In the last section, we study completely positive maps, a very useful class of maps on C^*-algebras.

Consider the space of all $n \times n$ matrices or the space $\mathcal{B}(\mathcal{H})$ of all bounded operators on a Hilbert space $\mathcal{H}$. It is not only a Banach space under the *operator norm*

$$\|A\| = \sup_{x \neq 0} \frac{\|Ax\|}{\|x\|}, \ A \in \mathcal{B}(\mathcal{H}),$$

but also has additional structures. It has a natural 'non-commutative' product coming from composition of linear maps. It has the notion of '*' or adjoint of an operator. Taking motivation from this, we make the following definitions.

Let $(\mathcal{A}, \|\cdot\|)$ be a normed linear space. It is said to be a *normed algebra* if we have a binary operation '.' on $\mathcal{A}$, with following properties:

(i) $(a.b).c = a.(b.c)$ for all $a, b, c \in \mathcal{A}$. (Associativity).

(ii) $\alpha(a.b) = (\alpha a).b = a.(\alpha b)$ for all $\alpha \in \mathbb{C}$, $a, b \in \mathcal{A}$. (Compatibility with scalar multiplication).

(iii) $a.(b + c) = a.b + a.c$ and $(a + b).c = a.c + b.c$ for all a, b, c in $\mathcal{A}$. (distributivity)

(iv) $\|a.b\| \leq \|a\|\|b\|$ for all a, b in $\mathcal{A}$.

The last condition ensures that the product map $(a, b) \mapsto a.b$ is jointly continuous. A normed algebra is said to be a *Banach algebra* if it is complete as a normed linear space, making it a Banach space. Every normed algebra can be completed to a Banach algebra. A Banach algebra is said to be *unital* if there exists $e \in \mathcal{A}$ such that $a.e = e.a = a$ for all $a \in \mathcal{A}$ and $\|e\| = 1$. It is clear that if a Banach algebra has a unit, then it is unique.

Moreover, every Banach algebra can be unitized in a natural way by adjoining a unit. Indeed, suppose $\mathcal{A}$ is a Banach algebra. Consider $\mathcal{A}^+ = \mathcal{A} \times \mathbb{C}$ with the natural vector space operations. Define multiplication by:

$$(a, \lambda).(b, \delta) = (ab + \lambda b + \delta a, \lambda\delta)\ , \quad \text{for all } (a, \lambda), (b, \delta) \in \mathcal{A}^+;$$

and norm by

$$\|(a, \lambda)\| = \|a\| + |\lambda|, \quad \text{for all } (a, \lambda) \in \mathcal{A}^+.$$

We leave it to the reader to verify that this yields a unital Banach algebra with $(0, 1)$ as the unit. Most results on non-unital Banach algebras can be deduced by first proving the result for the unitization and observing the consequences. In view of this, we restrict ourselves mostly to unital Banach algebras.

If there is no possibility of confusion we will identify the unit e of the Banach algebra with the scalar 1 of $\mathbb{C}$ and more generally we identify αe with scalar α for every complex number α. In the following, we would also denote $a.b$, simply as ab and this should cause no confusion as we are not going to consider more than one product in a Banach algebra.

Suppose $\mathcal{A}$ is a Banach algebra. Then $a \in \mathcal{A}$ is said to be *invertible* if there exists a^{-1} in $\mathcal{A}$, such that $aa^{-1} = a^{-1}a = 1$. Then a^{-1} is called an inverse of a. Once again it is clear that if an element has an inverse, the inverse is unique. The following elementary lemma has far reaching consequences. It is to be noted that it uses both the algebra structure as well as the norm on the algebra.

Lemma 2.1.1. *Let a be an element in a unital Banach algebra $\mathcal{A}$ with $\|a\| < 1$. Then $(1 - a)$ is invertible and $\|(1 - a)^{-1}\| \leq \frac{1}{1 - \|a\|}$.*

Proof. The geometric series $1 + a + a^2 + \cdots$, is Cauchy because for $m > n$, we have

$$\begin{aligned}
&\|(1 + a + a^2 + \cdots + a^m) - (1 + a + a^2 + \cdots + a^n)\| \\
&= \|a^{n+1} + a^{n+2} + \cdots + a^m\| \\
&\leq \|a^{n+1}\| + \|a^{n+2}\| + \cdots + \|a^m\| \\
&\leq \|a\|^{n+1} + \|a\|^{n+2} + \cdots + \|a\|^m
\end{aligned}$$

which goes to 0 as $n \to \infty$ since $\|a\| < 1$. Hence, the series converges, by completeness of the Banach algebra. We also have

$$(1-a)(1+a+a^2+\cdots+a^n) = 1-a^{n+1}.$$

Passing to the limit, $1-a$ is invertible and

$$(1-a)^{-1} = 1+a+a^2+\cdots.$$

The norm estimate follows because

$$\|(1-a)^{-1}\| = \|1+a+a^2+\cdots\| \le 1+\|a\|+\|a\|^2+\cdots = \frac{1}{1-\|a\|}.$$

■

The infinite geometric series $1+a+a^2+\cdots$ in a is sometimes called the *Neumann* series.

Theorem 2.1.2. *Let $\mathcal{A}$ be a unital Banach algebra. Then the set of invertible elements in $\mathcal{A}$ is an open set containing the identity and is a group under multiplication. Moreover the map $a \mapsto a^{-1}$ is continuous on the set of invertible elements.*

Proof. It is obvious that 1 is its own inverse and hence invertible. Similarly it is clear that if a and b are invertible so is ab with $(ab)^{-1} = b^{-1}a^{-1}$. Therefore the group property follows easily. Now if b is invertible in $\mathcal{A}$, consider any $c \in \mathcal{A}$ with $\|c\| < \frac{1}{\|b\|^{-1}}$. Then as $\|b^{-1}c\| < 1$, we have $(1-b^{-1}c)$ invertible and hence $b-c = b(1-b^{-1}c)$ is also invertible. Therefore elements in a certain neighbourhood of b are also invertible. In other words, invertible elements form an open set.

Continuity follows by a similar argument. Indeed, if a is invertible and $\|b-a\| < \|a^{-1}\|^{-1}$, then $\|a^{-1}(a-b)\| < 1$ and hence by Neumann series expansion, $1-a^{-1}(a-b)$ is invertible with

$$(1-a^{-1}(a-b))^{-1} = \sum_{j=0}^{\infty}(a^{-1}(a-b))^j.$$

Since $b = a(1-a^{-1}(a-b))$, we have

$$b^{-1}-a^{-1} = ((1-a^{-1}(a-b))^{-1}-1)a^{-1} = \sum_{j=1}^{\infty}(a^{-1}(a-b))^j a^{-1}.$$

Hence,

$$\|b^{-1}-a^{-1}\| \le \sum_{j=1}^{\infty}\|a^{-1}\|^{j+1}\|a-b\|^j = \frac{\|a^{-1}\|^2\|a-b\|}{1-\|a-b\|\|a^{-1}\|}.$$

Since this is true for any b in the neighbourhood of radius $\|a^{-1}\|^{-1}$ of a, if we take a sequence a_n that converges to a, there will be an N such that for $n > N$, we shall have

$$\|a_n^{-1} - a^{-1}\| \leq \frac{\|a^{-1}\|^2\|a - a_n\|}{1 - \|a - a_n\|\|a^{-1}\|} \to 0.$$

Thus, the sequence of inverses $\{a_n^{-1}\}$ converges to a^{-1}. ∎

Definition 2.1.3. *Suppose a is an element in a unital Banach algebra $\mathcal{A}$. Then the spectrum of a in $\mathcal{A}$ is defined as*

$$\sigma(a) = \{\lambda \in \mathbb{C} : (\lambda 1 - a) \text{ is not invertible}\}$$

The resolvent set of a is defined as $\mathbb{C} \setminus \sigma(a)$.

It is a deep fact that the spectrum of every element in a unital Banach algebra is non-empty. We will prove this later while proving the spectral radius formula. But with this in mind we make the following definition.

Definition 2.1.4. *Suppose a is an element in a unital Banach algebra with non-empty spectrum. Then the spectral radius of a is defined as*

$$r(a) = \sup\{|\lambda| : \lambda \in \sigma(a)\}.$$

At the moment, for argument's sake, we allow $r(a)$ to be $-\infty$, i.e., we allow the spectrum of a to be empty.

We note that for $|\lambda| > \|a\|$, we have $\lambda 1 - a = \lambda(1 - \frac{a}{\lambda})$ to be invertible as $\frac{\|a\|}{\lambda} < 1$. In other words $r(a) \leq \|a\|$. The openness of the set of invertible elements shows that $\sigma(a)$ is a closed set. So the spectrum of every element in a unital Banach algebra is a compact subset of the complex plane. The fact that it is non-empty is yet to be seen. Of course, for $n \times n$ matrices considered as elements of the Banach algebra $M_n(\mathbb{C})$ the spectrum is just the set of eigenvalues.

If a_1 and a_2 are two invertible elements in a unital C^*-algebra $\mathcal{A}$, then the product a_1a_2 is invertible with the inverse being $a_2^{-1}a_1^{-1}$. However, given any two elements a_1 and a_2 in a unital C^*-algebra $\mathcal{A}$, for the product a_1a_2 to be invertible, it is not necessary that a_1 and a_2 be invertible. We have seen in Chapter 1 that the unilateral forward shift is a non-invertible isometry, i.e., the product of the unilateral backward shift and the unilateral forward shift (in that order) is the identity operator which is invertible while the shifts are not. However, if we have an invertible product $a_1a_2\dots a_n$ of commuting elements, then we see that the element $a_1 \dots a_{j-1}a_{j+1} \dots a_n(a_1a_2 \dots a_n)^{-1}$ which is the same as $(a_1a_2\dots a_n)^{-1}a_1\dots a_{j-1}a_{j+1}\dots a_n$ is the inverse of a_j. This observation leads to the following theorem.

Theorem 2.1.5. *Polynomial spectral mapping theorem: Let a be an element in a unital Banach algebra with non-empty spectrum. Let p be a complex polynomial. Then*

$$\sigma(p(a)) = \{p(\lambda) : \lambda \in \sigma(a)\}.$$

Proof. If $\lambda \in \sigma(a)$, then

$$p(x) - p(\lambda) = (x - \lambda)q(x)$$

for some polynomial q. Thus,

$$p(a) - p(\lambda) = (a - \lambda)q(a)$$

where $q(a)$ commutes with $a-\lambda$. By the remark just before the theorem, $p(a)-p(\lambda)$ cannot be invertible. Conversely, for $\mu \in \mathbb{C}$, we factorize the polynomial to get

$$p(x) - \mu = \alpha(x - \beta_1)(x - \beta_2)\cdots(x - \beta_n),$$

for some complex numbers $\alpha \neq 0, \beta_1, \beta_2, \dots, \beta_n, n \geq 1$. Then

$$p(a) - \mu = \alpha(a - \beta_1)(a - \beta_2)\cdots(a - \beta_n)$$

and the factors $(a - \beta_1), (a - \beta_2), \dots, (a - \beta_n)$ mutually commute. So the non-invertibility of $(p(a) - \mu)$ implies that there is a β_i such that $a - \beta_i$ is not invertible. Now, $\mu = p(\beta_i)$. So, we are done. ∎

A matrix is called nilpotent if some power of it is the zero matrix. This happens if and only if 0 is its only eigenvalue. We know that there do exist non-zero nilpotent matrices. In other words, in general, $r(a)$ can be strictly smaller than $\|a\|$. The following theorem provides the exact relationship between the spectral radius and the norm. It is indeed a remarkable theorem, as it connects the spectrum, an algebraic notion with norm, which is an analytic notion. Most books on C^*-algebra provide a proof of this theorem using Liouville's theorem or some other tool from complex analysis. Here we present a proof through very elementary methods. This is a modified version of the proof in C. E. Rickart [100]. Note that the proof also shows that the spectrum is non-empty.

Theorem 2.1.6. *(Spectral radius formula): Let a be an element in a unital Banach algebra. Then $\sigma(a)$ is non-empty and*

$$r(a) = \lim_{n\to\infty} \|a^n\|^{\frac{1}{n}}.$$

Proof. As a first step we show that the required limit exists. Let $\nu = \inf_{n\in\mathbb{N}} \|a_n\|^{\frac{1}{n}}$. Now for $\epsilon > 0$, by the definition of the infimum, there exists $k \in \mathbb{N}$ such that

$$0 \leq \|a^k\|^{\frac{1}{k}} < \nu + \epsilon.$$

Given $m \geq k$, write $m = sk + r, s \in \mathbb{N}; 0 \leq r < k$. We have

$$\|a^m\| = \|a^{sk+r}\| \leq \|a^k\|^s.\|a\|^r < (\nu+\epsilon)^{sk}.\|a\|^r.$$

Therefore,

$$\|a^m\|^{\frac{1}{m}} < (\nu+\epsilon)^{\frac{sk}{sk+r}}.\|a\|^{\frac{r}{sk+r}}.$$

Now as $m \to \infty$, the choices of s, r are such that $\frac{sk}{sk+r} \to 1$ and $\frac{r}{sk+r} \to 0$. So

$$\limsup_{m\to\infty} \|a^m\|^{\frac{1}{m}} \leq (\nu+\epsilon).1 = \nu + \epsilon.$$

Consequently,

$$0 \leq \nu \leq \liminf_{n\to\infty} \|a^n\|^{\frac{1}{n}} \leq \limsup_{n\to\infty} \|a^n\|^{\frac{1}{n}} \leq \nu + \epsilon.$$

As this is true for every $\epsilon > 0$, we see that $\lim_{n\to\infty} \|a^n\|^{\frac{1}{n}}$ exists and equals the infimum ν. (This is a special case of a standard result. See Fekete's subadditivity lemma, Problem no 98 and its solution, Part I, Chapter 3 of Polya [93]).

If $\lambda \in \sigma(a)$, then by the polynomial spectral mapping theorem $\lambda^n \in \sigma(a^n)$ and therefore $|\lambda|^n \leq \|a\|^n$. Hence $|\lambda| \leq \|a^n\|^{\frac{1}{n}}$. This shows $r(a) \leq \nu$.

We wish to show that there exists $\lambda \in \sigma(a)$ such that $|\lambda| = \nu$ and that would prove the theorem. If $\nu = 0$, and $0 \notin \sigma(a)$, then for $n \geq 1$, $1 \leq \|a^n.a^{-1}.a^{-1}\cdots a^{-1}\| \leq \|a^n\|.(\|a^{-1}\|)^n$. Hence

$$\|a^n\|^{\frac{1}{n}} \geq \frac{1}{\|a^{-1}\|}$$

for all n and this contradicts $\nu = 0$. Hence, in this case $\sigma(a)$ contains 0 (and only 0).

Hereafter we assume $\nu > 0$. Firstly, we observe that we have

$$\frac{\|a^n\|}{\nu^n} \geq 1 \text{ for every } n.$$

Now, the fact that $\lim_{n\to\infty}(\|a^n\|)^{\frac{1}{n}}$ is ν immediately shows that

$$\frac{\|a^n\|}{|\lambda|^n} \to 0 \text{ as } n \to \infty \text{ for } |\lambda| > \nu.$$

We already know that if $|\lambda| > \nu$ then $\lambda \notin \sigma(a)$. Now also assume that there is no λ in $\sigma(a)$ with $|\lambda| = \nu$ and we would like to arrive at a contradiction.

For $n \geq 2$, let $\{1, \omega, \omega^2, \ldots, \omega^{n-1}\}$ be n-th roots of unity. For any complex number λ, consider the algebraic relation

$$\frac{1}{n}\sum_{j=0}^{n-1} \frac{1}{(x-\lambda\omega^j)} = \frac{x^{n-1}}{\prod_{j=0}^{n-1}(x-\lambda\omega^j)} = x^{n-1}(x^n-\lambda^n)^{-1}.$$

(The proof of this is by first observing that $x^n - \lambda^n = \prod_{j=0}^{n-1}(x-\lambda\omega^j)$ and then by differentiating both sides with respect to x.) Consequently,

$$\frac{1}{n}\sum_{j=0}^{n-1} x(x-\lambda\omega^j)^{-1} = x^n(x^n-\lambda^n)^{-1}.$$

Now replacing the variable x by a and λ with $|\lambda| \geq \nu$ (this is permitted as $\lambda w^j \notin \sigma(a)$ for all j and $\lambda^n \notin \sigma(a^n)$), we get

$$\frac{1}{n}\sum_{j=0}^{n} a(a-\lambda\omega^j)^{-1} = a^n(a^n-\lambda^n)^{-1}. \tag{2.1.1}$$

We note that the function $f(a) = a(a-\lambda)^{-1}$ is uniformly continuous in the annulus: $W = \{\lambda \in \mathbb{C} : \nu \leq |\lambda| \leq \nu + 1\}$. Therefore, given $\epsilon > 0$, there exists $0 < \delta < 1$, independent of n, such that

$$\|a(a-\nu\omega^j)^{-1} - a(a-\lambda\omega^j)^{-1}\| < \epsilon \text{ for all } j,$$

for all $\nu \leq |\lambda| \leq \nu + \delta$. Therefore, by (2.1.1), applied at ν and any λ satisfying $\nu < |\lambda| \leq \nu + \delta$,

$$\|a^n(a^n-\nu^n)^{-1} - a^n(a^n-\lambda^n)^{-1}\| \leq \epsilon.$$

Or equivalently,

$$\|\frac{a^n}{\nu^n}(\frac{a^n}{\nu^n}-1)^{-1} - \frac{a^n}{\lambda^n}(\frac{a^n}{\lambda^n}-1)^{-1}\| \leq \epsilon.$$

Now,

$$\lim_{n\to\infty} \|\frac{a^n}{\lambda^n}\| = 0,$$

which implies that

$$\lim_{n\to\infty} \frac{\|\frac{a^n}{\lambda^n}\|}{1-\|\frac{a^n}{\lambda^n}\|} = 0.$$

Hence,

$$\|\frac{a^n}{\nu^n}\left(\frac{a^n}{\nu^n}-1\right)^{-1}\| \leq 2\epsilon$$

for large n and for any $\epsilon > 0$. Recall that

$$\|\frac{a^n}{\nu^n}\| \geq 1.$$

So, if we set $b = a^n/\nu^n$, then $\|b\| \geq 1$ and $1 \notin \sigma(b)$. Then, we have

$$\|b\| \leq \|b(b-1)^{-1}\|\|(b-1)\| \leq \|b(b-1)^{-1}\|(\|b\|+1)$$

and hence

$$\|b(b-1)^{-1}\| \geq \frac{\|b\|}{\|b\|+1} \geq \frac{1}{2}.$$

We have reached a contradiction and hence there has to be a λ in $\sigma(a)$ such that $|\lambda| = \nu$. ■

It is to be noted that in general $(\|a^n\|)^{\frac{1}{n}}$ is not monotonic in n, and hence the existence of the limit in the previous theorem is not obvious. The following theorem states that in a unital Banach algebra, the spectrum of ab and that of ba are same, provided we exclude the point 0 from the spectrum. It is easy to get examples where $0 \in \sigma(ab)$, but $0 \notin \sigma(ba)$. Formally,

$$1 + ba + baba + \cdots = 1 + b(1 + ab + abab + \cdots)a,$$

and so we may expect $(1-ba)^{-1} = 1 + b(1-ab)^{-1}a$. This motivates the proof of this Theorem.

Theorem 2.1.7. *Let a, b elements in a unital Banach algebra $\mathcal{A}$. Then*

$$\sigma(ab) \cup \{0\} = \sigma(ba) \cup \{0\}.$$

Proof. It is enough to show one of the containments, the other follows by symmetry. Thus, suppose $\lambda \notin \sigma(ab)$ and $\lambda \neq 0$. We wish to show that $\lambda \notin \sigma(ba)$. By scaling a suitably, we may assume that $\lambda = 1$. Take $c = (1-ab)^{-1}$. Consider $d = 1 + bca$. We have

$$(1-ba)d = (1-ba)(1+bca) = 1+bca-ba-babca = 1+b[(1-ab)c]a-ba = 1.$$

Similarly, $d(1-ba) = 1$. Hence $d = (1-ba)^{-1}$ and $1 \notin \sigma(ba)$. ■

Definition 2.1.8. *Given two algebras $\mathcal{A}$ and $\mathcal{B}$, a map $\pi : \mathcal{A} \to \mathcal{B}$ is called an algebra homomorphism if it is linear and multiplicative, i.e., π satisfies*

$$\pi(c_1a_1 + c_2a_2) = c_1\pi(a_1) + c_2\pi(a_2)\ \forall c_1, c_2 \in \mathbb{C}, a_1, a_2 \in \mathcal{A};$$

$$\pi(a_1a_2) = \pi(a_1)\pi(a_2),\ \forall a_1, a_2 \in \mathcal{A}.$$

If, moreover, π is bijective, then it is called an isomorphism of the two algebras $\mathcal{A}$ and $\mathcal{B}$. An algebra homomorphism from a unital algebra $\mathcal{A}$ to a unital algebra $\mathcal{B}$ is said to be unital if it takes the identity of $\mathcal{A}$ to the identity of $\mathcal{B}$.

An immediate consequence of the non-emptiness of the spectrum for elements of unital Banach algebras is the following theorem.

Theorem 2.1.9. *(Gelfand-Mazur Theorem): Let $\mathcal{A}$ be a unital Banach algebra. Suppose $\mathcal{A}$ is a division algebra, that is, every non-zero element of $\mathcal{A}$ is invertible. Then $\mathcal{A}$ is isomorphic to $\mathbb{C}$ as a unital algebra. In particular $\mathcal{A}$ is commutative.*

Proof. Let a be any element of $\mathcal{A}$. As $\sigma(a)$ is non-empty, there exists $\lambda \in \mathbb{C}$ such that $\lambda \in \sigma(a)$. Now as $a - \lambda.1$ is not invertible, we must have $a - \lambda.1 = 0$ or $a = \lambda.1$ It follows that $\sigma(a)$ is singleton and $a \mapsto \lambda$ for $\lambda \in \sigma(a)$ is an algebra isomorphism. ∎

Definition 2.1.10. *Let $\mathcal{A}$ be a commutative unital Banach algebra. A subspace $\mathcal{I}$ of $\mathcal{A}$ is said to be an ideal of $\mathcal{A}$ if $ac = ca$ is in $\mathcal{I}$, for every $a \in \mathcal{A}$ and $c \in \mathcal{I}$. It is said to be closed if it is closed in the norm and it is said to be proper if it is not the whole of $\mathcal{A}$. Further, an ideal $\mathcal{I}$ of $\mathcal{A}$ is said to be maximal if it is not properly contained in some other proper ideal, that is, if $\mathcal{J}$ is another ideal of $\mathcal{A}$ such that $\mathcal{I} \subset \mathcal{J} \subseteq \mathcal{A}$ then $\mathcal{J} = \mathcal{A}$.*

If $\mathcal{I}$ is a closed ideal of $\mathcal{A}$, then the quotient space $\mathcal{A}/\mathcal{I}$ is a unital Banach algebra with canonical algebraic operations and the norm

$$\|\mathcal{I} + a\| = \inf\{\|c + a\| : c \in \mathcal{I}\}.$$

We leave the proof as an exercise to the reader.

We motivate the theory of commutative unital Banach algebras by looking at the following example. Consider the algebra $C(X)$ of continuous functions on a compact Hausdorff space X, with respect to the supremum norm and pointwise multiplication. It is a unital commutative Banach algebra. If S is any closed subset of X, we can see that

$$\mathcal{I}_S = \{f \in C(X) : f(x) = 0,\ x \in S\}$$

is a closed ideal of $\mathcal{A}$. It is a little exercise in topology that the converse is also true. Further, an ideal $\mathcal{I}$ of $C(X)$ is maximal if and only if $\mathcal{I} = \mathcal{I}_S$ for some singleton S. We also note that if $S = \{x\}$, then the map $\hat{x} : C(X) \to \mathbb{C}$ defined by $\hat{x}(f) = f(x)$ is a multiplicative linear map such that $\mathcal{I}_S = \ker(\hat{x})$.

Now consider a general unital commutative Banach algebra $\mathcal{A}$.

Definition 2.1.11. *For a unital Banach algebra $\mathcal{A}$, a non-zero linear functional $\phi : \mathcal{A} \to \mathbb{C}$ is said to be a complex homomorphism, or a character, or a multiplicative linear functional if ϕ is multiplicative, that is, $\phi(ab) = \phi(a)\phi(b)$ for all $a, b \in \mathcal{A}$.*

It is easily seen that a non-zero homomorphism from a unital Banach algebra into the algebra $\mathbb{C}$ of complex numbers is automatically unital.

Lemma 2.1.12. *Let ϕ be a complex homomorphism on a unital Banach algebra $\mathcal{A}$. Then $\|\phi\| = 1$. In particular, ϕ is a bounded linear functional. The kernel of ϕ is a maximal ideal $\mathcal{A}$.*

Proof. From the multiplicative property we observe that every unital homomorphism maps invertible elements to invertible elements. Now for $a \in \mathcal{A}$, by direct computation $(a - \phi(a).1)$ is in the kernel of ϕ. Hence it is not invertible. It follows that $\phi(a) \in \sigma(a)$.

Now from the very first inequality we proved for the spectral radius, namely, $r(a) \leq \|a\|$, we get $|\phi(a)| \leq \|a\|$. As $\phi(1) = 1$, we get $\|\phi\| = 1$.

By the boundedness of ϕ, its kernel, ker $(\phi) = \{a \in \mathcal{A} : \phi(a) = 0\}$ is a closed ideal of $\mathcal{A}$. We also note that by writing $a \in \mathcal{A}$ as $a = (a - \phi(a).1) + \phi(a).1$, and as already observed, $(a - \phi(a).1)$ is in the kernel of ϕ, ker (ϕ) has codimension 1 (this is true for any non-trivial linear functional, but for a general linear functional the kernel may not be an ideal). Hence it is maximal. ■

Suppose $\mathcal{I}$ is a proper ideal of $\mathcal{A}$. It is clear that if $1 \in \mathcal{I}$, then the ideal property would mean that every element $a = a.1$ of $\mathcal{A}$ is in $\mathcal{I}$. Hence $1 \notin \mathcal{I}$. This in turn implies that no invertible element can be in $\mathcal{I}$. We also recall that the set of invertible elements form an open set containing 1. Hence closure of any proper ideal $\mathcal{I}$ will also not have 1. So every maximal idea of $\mathcal{A}$ must be closed. In the converse direction, if $c \in \mathcal{A}$ is singular (not invertible), then the ideal generated by c, i.e., the set $\{ac : a \in \mathcal{A}\}$ (Here we are using the commutativity of $\mathcal{A}$) is a proper ideal in $\mathcal{A}$ containing a. It follows by Zorn's lemma that every singular element of $\mathcal{A}$ is contained in a maximal ideal of $\mathcal{A}$.

Theorem 2.1.13. *Let $\mathcal{A}$ be a unital Banach algebra and let $\mathcal{I}$ be a maximal ideal of $\mathcal{A}$. Then the quotient space $\mathcal{A}/\mathcal{I}$ is isomorphic to $\mathbb{C}$. Using this isomorphism, the quotient map $\phi : \mathcal{A} \to \mathcal{A}/\mathcal{I} \cong \mathbb{C}$ is a complex homomorphism.*

Proof. Suppose $\mathcal{I}+c$ is singular in the quotient space $\mathcal{A}/\mathcal{I}$, then the ideal generated by $\mathcal{I}$ and c:

$$\{b + ac : b \in \mathcal{I}, a \in \mathcal{A}\}$$

is a proper ideal of $\mathcal{A}$ containing $\mathcal{I}$. Then the maximality of $\mathcal{I}$ implies $c \in \mathcal{I}$. In other words, $\mathcal{A}/\mathcal{I}$ is a division algebra. Now the result follows by Gelfand-Mazur theorem (Theorem 2.1.9). ■

Finally, we are in a position to describe the famous Gelfand map or Gelfand transform. Let $\mathcal{A}$ be a unital commutative Banach algebra. Take

$$\Delta := \Delta\mathcal{A} := \{\phi : \phi \text{ is a complex homomorphism of } \mathcal{A}\}.$$

Since the complex homomorphisms are also known as characters, Δ is the space of characters on $\mathcal{A}$. One studies Banach spaces by looking at their dual spaces, which consists of morphisms to the field. One studies groups by looking at their characters, which are morphisms to the circle. In every case, one analyzes morphisms to a familiar object to understand the algebraic structures. A similar approach is seen here. We begin with the following Lemma.

Lemma 2.1.14. *If $\mathcal{A}$ is a commutative Banach algebra and a is an element in $\mathcal{A}$, then $\{\phi(a) : \phi \in \Delta\} = \sigma(a)$.*

Proof. We have seen that $\phi(a) \in \sigma(a)$ for every character ϕ. Now if $\lambda \in \sigma(a)$, then $a - \lambda.1$ is not invertible, and consequently there exists a maximal ideal $\mathcal{I}$ of $\mathcal{A}$, containing $a - \lambda.1$. Now the quotient map $\mathcal{A} \to \mathcal{A}/\mathcal{I}$, composed with the Gelfand-Mazur isomorphism, $\mathcal{A}/\mathcal{I} \cong \mathbb{C}$, shows that there exists a character ϕ such that $\phi(a) = \lambda$. ■

We have already noted that $\|\phi\| = 1$ for every character ϕ. In particular, Δ is a subset of the closed unit ball $B_1(\mathcal{A}^*)$ of the dual space $\mathcal{A}^*$ of continuous linear functionals on $\mathcal{A}$. We recall that by Banach-Alaoglu theorem (Theorem 1.23 in [50]), this space $B_1(\mathcal{A}^*)$ is compact under weak-$*$ topology. We consider the inherited/subspace topology on Δ. Since this topology is the topology of pointwise convergence, it is easy to see that Δ is a closed subset of $B_1(\mathcal{A}^*)$. Therefore Δ is compact. Consider the space of complex valued continuous functions $C(\Delta)$ with the usual supremum norm. The *Gelfand map* is the map $a \mapsto \hat{a}$ defined from $\mathcal{A}$ to $C(\Delta)$ by

$$\hat{a}(\phi) = \phi(a).$$

It is clear that $\hat{a} \in C(\Delta)$ and this map is an algebra homomorphism. The estimate $|\phi(a)| \leq \|a\|$ shows that the Gelfand map is contractive. We summarise these observations as follows.

Theorem 2.1.15. *(Gelfand's Theorem for commutative unital Banach algebra) Let $\mathcal{A}$ be a unital commutative Banach algebra. Then the map $x \mapsto \hat{x}$ from $\mathcal{A}$ to $C(\Delta)$ defined by*

$$\hat{x}(\phi) = \phi(x),$$

is a unital homomorphism of Banach algebras.

If the Gelfand map is injective, then we may realize the Banach algebra $\mathcal{A}$ as a subalgebra of the space of continuous functions on a compact metric space. Unfortunately this is not possible in general as the Gelfand map may not be injective. Here is an example.

Example 2.1.16. *Let $\mathcal{A}$ be the space of 3×3 complex matrices of the form*

$$\begin{bmatrix} a & b & c \\ 0 & a & b \\ 0 & 0 & a \end{bmatrix}.$$

Clearly $\mathcal{A}$ is a unital commutative Banach algebra with respect to usual operator norm. Since the matrix above has a as its only eigenvalue, this algebra has the unique complex homomorphism given by

$$\begin{bmatrix} a & b & c \\ 0 & a & b \\ 0 & 0 & a \end{bmatrix} \mapsto a.$$

So the space $C(\hat{\mathcal{A}})$ is isomorphic to $\mathbb{C}$ and the Gelfand map is not injective.

The Gelfand theory of commutative Banach algebras plays a very big role in harmonic analysis in the study of group algebras. It is, however, beyond the scope of this book.

2.2. C^*-algebras

The Gelfand map is injective, and is in fact is an isomorphism for a class of Banach algebras called C^*-algebras, which we are going to describe now. A Banach algebra $\mathcal{A}$ is said to be a Banach $*$-algebra if there is an involution, that is, a map $a \mapsto a^*$ on $\mathcal{A}$ with following properties:

(i) The map $a \mapsto a^*$ is anti-linear, that is, $(\alpha a + \beta b)^* = \bar{\alpha} a^* + \bar{\beta} b^*$ for all $\alpha, \beta \in \mathbb{C}$ and $a, b \in \mathcal{A}$.

(ii) $(a^*)^* = a$ for all $a \in \mathcal{A}$.

(iii) $(ab)^* = b^* a^*$ for all $a, b \in \mathcal{A}$.

(iv) $\|a^*\| = \|a\|$ for all $a \in \mathcal{A}$.

It is easy to see from the uniqueness of the unit, that in a unital Banach $*$-algebra $1^* = 1$. A Banach $*$-algebra $\mathcal{A}$ is called a C^**-algebra* if it satisfies the C^*-condition:

$$\|a^* a\| = \|a\|^2 \text{ for all } a \in \mathcal{A}.$$

For a compact Hausdorff space X, the algebra $C(X)$ of all continuous functions on X, is a C^*-algebra if we endow it with the supremum norm: $\|f\|_\infty = \sup\{|f(x)| : x \in X\}$; and adjoint $*$ as complex conjugate: $f^*(x) := \overline{f(x)}, \forall x \in X, f \in C(X)$. It is a unital commutative C^*-algebra. If $\mathcal{H}$ is a Hilbert space, then the algebra $\mathcal{B}(\mathcal{H})$ of all bounded operators on $\mathcal{H}$ is a non-commutative C^*-algebra.

Hereafter we consider only C^*-algebras. Once we have the notion of an adjoint we can make the following definition.

Definition 2.2.1. *Let $\mathcal{A}$ be a unital C^*-algebra and let a be an element of $\mathcal{A}$.*

(1) *a is said to be self-adjoint if $a = a^*$;*
(2) *a is said to be positive if $a = b^*b$ for some $b \in \mathcal{A}$;*
(3) *a is said to be a projection if $a = a^* = a^2$;*
(4) *a is said to be normal if $aa^* = a^*a$;*
(5) *a is said to be a unitary if $a^*a = aa^* = 1$.*

For a, b in a C^*-algebra, we write $0 \le b$ if b is positive and more generally we write $a \le b$ if $b - a$ is positive. Recall that in Section 1.2 of Chapter 1, we showed that for the C^*-algebra $\mathcal{B}(\mathcal{H})$, an element T is positive according to the above definition if and only if the quadratic form $\langle Th, h\rangle$ is non-negative for all h in the Hilbert space $\mathcal{H}$. Towards the end of this section, we shall have several characterizations of positive elements of a C^*-algebra. It will be seen that $\le$ defined above is a partial order on the C^*-algebra.

Theorem 2.2.2. *(Cartesian Decomposition) Let $\mathcal{A}$ be a C^*-algebra. Then every element $a \in \mathcal{A}$ decomposes uniquely as*

$$a = x + iy$$

where x, y are self-adjoint.

Proof. Take $x = \frac{a+a^*}{2}$ and $y = \frac{a-a^*}{2i}$. Then we see that x, y are self-adjoint and $a = x + iy$. Conversely, if $a = x + iy$ with $x = x^*$ and $y = y^*$, we get $a^* = x - iy$, and consequently $x = \frac{a+a^*}{2}$ and $y = \frac{a-a^*}{2i}$. ■

If $a = x + iy$ is the Cartesian decomposition of a, then x is known as real part of a and y is known as the imaginary part of a.

Theorem 2.2.3. *Let a be a self-adjoint element in a unital C^*-algebra. Then $\sigma(a) \subset \mathbb{R}$ and $r(a) = \|a\|$.*

Proof. First, by the C^*-condition and self-adjointness of a, we have $\|a^2\| = \|a^*a\| = \|a\|^2$. By induction, $\|a^{2^n}\| = (\|a\|)^{2^n}$. Now the spectral radius formula yields $r(a) = \|a\|$.

Suppose $\lambda \in \sigma(a)$. Then we know that $|\lambda| \leq \|a\|$ by Lemma 2.1.1. Note that for every real t, we have $\lambda + it \in \sigma(a + it)$. So we get

$$\begin{array}{rcl} |\lambda + it|^2 & \leq & \|(a + it)\|^2 \\ (\lambda + it)(\bar{\lambda} - it) & \leq & \|(a - it)(a + it)\| \\ |\lambda|^2 + it(\bar{\lambda} - \lambda) + t^2 & \leq & \|a^2\| + t^2(\text{ by triangle inequality}) \\ |\lambda|^2 + it(\bar{\lambda} - \lambda) & \leq & \|a\|^2 \end{array}$$

As this inequality is true for every real t, clearly $(\bar{\lambda} - \lambda) = 0$ or λ is real. ■

Definition 2.2.4. *Given two C^*-algebras $\mathcal{A}$ and $\mathcal{B}$, an algebra homomorphism π from $\mathcal{A}$ to $\mathcal{B}$ which also satisfies $\pi(a^*) = \pi(a)^*$ is called a C^*-algebra homomorphism or a $*$-homomorphism.*

If, moreover, π is bijective, then it is called a C^*-algebra isomorphism or a $*$-isomorphism between the C^*-algebras $\mathcal{A}$ and $\mathcal{B}$.

It is clear that $*$-homomorphisms preserve positivity, self-adjointness, normality and projection property. Further, unital $*$-homomorphisms preserve unitarity. Here we are proving the following result for unital $*$-homomorphisms, though the result is true for even non-unital $*$-homomorphisms.

Lemma 2.2.5. *Every unital $*$-homomorphism between unital C^*-algebras is contractive.*

Proof. Since every unital homomorphism maps invertible elements to invertible elements it is easy to see that if π is a unital homomorphism, $\sigma(\pi(a)) \subseteq \sigma(a)$, for every element a in the domain of π. In particular the spectral radius $r(\pi(a)) \leq r(a)$ for every a. Also recall that norm is same as the spectral radius for self-adjoint elements. Now for a in $\mathcal{A}$, we have

$$\|\pi(a)\|^2 = \|\pi(a)^*\pi(a)\| = \|\pi(a^*a)\| = r(\pi(a^*a)) \leq r(a^*a) = \|a\|^2.$$

■

Before we move on, here is a result that makes it much easier to talk about the spectrum of an element in a C^*-algebra. We mention at this point that such a result does not hold in general for Banach algebras is seen by considering the disk algebra as a subalgebra of all continuous functions on the circle (See [112]).

Theorem 2.2.6. *Let $\mathcal{A}$, $\mathcal{B}$ be unital C^*-algebras with $\mathcal{A} \subseteq \mathcal{B}$. Suppose $a \in \mathcal{A}$. Then a is invertible in $\mathcal{A}$ if and only if it is invertible in $\mathcal{B}$. Consequently the spectrum of a in $\mathcal{A}$ is same as the spectrum of a in $\mathcal{B}$.*

Proof. It is clear that the second half of this Theorem follows easily from the first half and that if a is invertible in $\mathcal{A}$ then it is invertible in $\mathcal{B}$. So the only non-trivial part is to show that if a is invertible in $\mathcal{B}$, then it is invertible in $\mathcal{A}$.

Case (i): $a = a^*$. Suppose b is the inverse of a in $\mathcal{B}$. As $a = a^*$ and the spectrum of a in any C^*-algebra is real, we see that $a_n = a + \frac{i}{n}$ is invertible in both $\mathcal{A}$ and $\mathcal{B}$, where i is $\sqrt{-1}$. Let the unique inverse of $a_n = a + \frac{i}{n}$ be b_n. All the b_n's have to be in the smaller C^*-algebra $\mathcal{A}$. As the mapping $x \mapsto x^{-1}$ is continuous in the set of invertible elements of $\mathcal{B}$, and $\lim_{n\to\infty} a_n = a$, we get $\lim_{n\to\infty} b_n = b$. But then, as $b_n \in \mathcal{A}$ for every n and $\mathcal{A}$ is closed, we have $b \in \mathcal{A}$.

Case (ii): General $a \in \mathcal{A}$. Suppose a is invertible in $\mathcal{B}$ and the inverse is b. Then a^* is also invertible in $\mathcal{B}$ with inverse b^*. As a^*a is a self-adjoint element of $\mathcal{A}$ and is invertible in $\mathcal{B}$, its inverse bb^* is in $\mathcal{A}$. Now we see that $b = bb^*a^*$ is in $\mathcal{A}$. ■

Hereafter, we can refer to the spectrum of an element a of a unital C^*-algebra as $\sigma(a)$ without referring to the C^*-algebra under consideration.

Now we arrive at the celebrated theorem of Gelfand showing that every commutative unital C^*-algebra is isometrically isometric to the C^*-algebra of continuous functions on a compact Hausdorff space.

Theorem 2.2.7. *(Gelfand's theorem) Let $\mathcal{A}$ be a unital commutative C^*-algebra. Then the Gelfand map $a \mapsto \hat{a}$ from $\mathcal{A}$ to $C(\hat{\mathcal{A}})$ is a $*$-isometric isomorphism of C^*-algebras.*

Proof. First we prove that $(\hat{a})^* = \widehat{a^*}$ for every $a \in \mathcal{A}$. We have seen that if a is self-adjoint, then $\sigma(a) \subset \mathbb{R}$. As $\phi(a) \in \sigma(a)$, for every character ϕ, we get $\hat{a}(\phi) = \phi(a) \in \mathbb{R}$ for every ϕ. In other words $\hat{a}$ is real valued and hence it is a self-adjoint element of $C(\Delta)$. Hence $(\hat{a})^* = \hat{a}$ for self-adjoint a.

Now the general case follows from the Cartesian decomposition, antilinearity of $*$ and linearity of the Gelfand map: For $a = x + iy$, with $x = x^*, y = y^*$,

$$(\hat{a})^* = (\widehat{(x+iy)})^* = (\hat{x} + i\hat{y})^* = (\hat{x})^* - i(\hat{y})^* = \hat{x} - i\hat{y} = \widehat{(x - iy)} = \widehat{(a^*)}.$$

Therefore the Gelfand map is a $*$-homomorphism.

Every element of the form a^*a, where $a \in \mathcal{A}$, is self-adjoint and therefore $r(a^*a) = \|a\|^2$. So we get

$$\begin{aligned}
\|\hat{a}\|_\infty^2 &= \|(\hat{a})^*\hat{a}\|_\infty \\
&= \sup\{|\widehat{a^*a}(\phi)| : \phi \in \Delta\} \\
&= \sup\{|\lambda| : \lambda \in \sigma(a^*a)\} \\
&= r(a^*a) \\
&= \|a^*a\| \\
&= (\|a\|)^2.
\end{aligned}$$

Therefore the Gelfand map is a $*$-homomorphism which is also isometric. Finally we show that the Gelfand map is surjective.

Let $\hat{\mathcal{A}} = \{\hat{a} : a \in \mathcal{A}\}$. We know that $\hat{\mathcal{A}}$ is contained in $C(\Delta)$. We want to show that it is whole of $C(\Delta)$. Clearly $\hat{\mathcal{A}}$ is a subalgebra of $C(\Delta)$. Due to isometry property of the Gelfand map we have proved, if a sequence $\{\widehat{a_n}\}$ is Cauchy then $\{a_n\}$ is Cauchy. It follows that $\hat{\mathcal{A}}$ is closed in $C(\Delta)$. Hence $\hat{\mathcal{A}}$ is a norm-closed subalgebra of $C(\Delta)$.

As $\hat{1}(\phi) = \phi(1) = 1$ for all ϕ, $\hat{\mathcal{A}}$ contains constants. If $\hat{a}(\phi_1) = \hat{a}(\phi_2)$ for all a, then $\phi_1(a) = \phi_2(a)$ for all a or $\phi_1 = \phi_2$. In other words, $\hat{\mathcal{A}}$ separates points of Δ.

Now recall the complex Stone-Weierstrass theorem (See W. Rudin: Functional Analysis), which says that, if a subalgebra of $\mathcal{A}$, (i) contains constants; (ii) closed under conjugation; (iii) norm-closed (iv) separates points of Δ, then it is whole of $C(\Delta)$. Therefore by a direct application of complex Stone-Weierstrass theorem, $\hat{\mathcal{A}} = C(\Delta)$ that is, the Gelfand map is surjective. This completes the proof of Gelfand's theorem. ∎

The following result is clear by direct verification.

Proposition 2.2.8. *Let X, Y be compact Hausdorff spaces. Consider unital C^*-algebras $C(X)$ and $C(Y)$. Suppose $\alpha : X \to Y$ is a continuous map. Then $\tilde{\alpha} : C(Y) \to C(X)$ defined by*

$$\tilde{\alpha}(g)(x) = g(\alpha(x)), \; g \in C(Y)$$

is a unital $*$-homomorphism. Further, $\tilde{\alpha}$ is a unital $*$-isomorphism if α is a homeomorphism.

Fix a unital C^*-algebra $\mathcal{A}$. Let S be a non-empty subset of $\mathcal{A}$. Consider the collection $\mathcal{E}$ of all unital C^*-subalgebras of $\mathcal{A}$ which contain

S. Clearly $\mathcal{A}$ is an element of this collection and hence $\mathcal{E}$ is non-empty. Let $\mathcal{B}$ be the intersection of all elements of $\mathcal{E}$. It is easily verified that $\mathcal{B}$ is a unital C^*-algebra. In particular, $\mathcal{B}$ is an element of $\mathcal{E}$. Clearly $\mathcal{B}$ is the smallest unital C^*-sub-algebra of $\mathcal{A}$, containing S.

Definition 2.2.9. *The C^*-algebra $\mathcal{B}$ constructed above is called the unital C^*-algebra generated by S in $\mathcal{A}$.*

It is to be noted that $\mathcal{B}$ is the norm-closure of the linear span of products of elements from $S \bigcup S^* \bigcup \{1\}$. In the special case, if S is a singleton say $S = \{a\}$ for some $a \in \mathcal{A}$, then the unital C^*-algebra generated by $\{a\}$ is denoted by $C^*\{a, 1\}$. Clearly $C^*\{a, 1\}$ is the norm-closure of linear combinations of all possible products of $1, a, a^*$.

Gelfand's theorem can be extended to general not necessarily unital commutative C^*-algebras to say that every commutative C^*-algebra is isometrically isometric to the algebra of continuous functions vanishing at infinity on a locally compact Hausdorff space. The algebra is unital if and only if the space is compact.

Now we move on to study non-commutative C^*-algebras. Note that in every unital C^*-algebra, a normal element (along with the unit) generates a commutative C^*-algebra and we can apply Gelfand's theorem to such subalgebras and we get the following theorem.

Theorem 2.2.10. *(Continuous functional calculus): Let a be a normal element in a unital C^*-algebra $\mathcal{A}$. Let $C^*\{a, 1\}$ be the C^*-subalgebra generated by a and 1 in $\mathcal{A}$. Then there exists a unique $*$-isomorphism π : $C(\sigma(a)) \to C^*\{a, 1\}$ such that $\pi(p) = p(a)$ for every polynomial p.*

Proof. In view of Gelfand's theorem, we know that the C^*-algbera $C^*\{a, 1\}$ is isomorphic to $C(\Delta)$ where Δ is the character space of $C^*\{a, 1\}$. Now we get a homeomorphism between the character space Δ and $\sigma(a)$. We have the map $\hat{a} : \Delta \to \sigma(a)$ defined by $\hat{a}(\phi) = \phi(a)$ (Note that a is the fixed given element). We have already seen that $\phi(a) \in \sigma(a)$ and $\{\phi(a) : \phi \in \Delta\} = \sigma(a)$. Hence $\hat{a}$ maps to $\sigma(a)$and is surjective.

If $\hat{a}(\phi_1) = \hat{a}(\phi_2)$, then $\phi_1(a) = \phi_2(a)$. Therefore $\phi_1(a^*) = \overline{\phi_1(a)} = \overline{\phi_2(a)} = \phi_2(a^*)$. Also $\phi_1(1) = 1 = \phi_2(1)$. Now by $*$-homomorphism properties of ϕ_1, ϕ_2 and continuity, $\phi_1(x) = \phi_2(x), \forall x \in C^*\{a, 1\}$. Hence $\phi_1 = \phi_2$. Therefore $\hat{a}$ is injective.

It has been noted earlier that the topology on Δ is the weak $*$-topology and it comes from pointwise convergence. Consequently the map $\hat{a}$ is continuous. Due to compactness of Δ, $\hat{a}$ is a homeomorphism (the inverse is automatically continuous). Therefore, the map $\psi := \widetilde{(\hat{a})}$ from $C(\sigma(a)) \to C(\Delta)$ defined by

$$\psi(g)(\phi) = g(\hat{a})(\phi) = g(\phi(a))$$

is a unital $*$-isomorphism. Let $\gamma : C^*\{a, 1\} \to C(\Delta)$ be the Gelfand map, defined by $\gamma(x) = \hat{x}$ for every $x \in C^*\{a, 1\}$. Now we have the diagram:

$$C^*\{a, 1\} \xrightarrow{\gamma} C(\Delta) \xleftarrow{\psi} C(\sigma(a)).$$

Take $\pi = \gamma^{-1} \circ \psi$. Being a composition of two unital $*$-isomorphisms, π is a unital $*$-isomorphism.

For any polynomial p, we have by the Definition of ψ,

$$\psi(p)(\phi) = p(\phi(a)), \ \forall \phi \in \Delta.$$

Also $\gamma(p(a))(\phi) = \widehat{p(a)}(\phi) = \phi(p(a)) = p(\phi(a))$, as ϕ is linear and multiplicative. Hence on applying γ^{-1}, $\pi(p) = p(a)$. Therefore π is a unital $*$-isomorphism satisfying $\pi(p) = p(a)$.

Now we go on to prove the uniqueness. Actually it follows from the Complex Stone-Weirstrass theorem applied to $C(\sigma(a))$ as the polynomials in $z, \bar{z}$, that is, functions of the form

$$\sum_{j,k=0}^{m,n} c_{j,k} z^j (\bar{z})^k$$

is dense in $C(\sigma(a))$. ■

Given a continuous function g on $\sigma(a)$, we use $g(a)$ for the element $\pi(g)$ in $C^*\{a, 1\}$. This is consistent with the notation $p(a)$ when p is a polynomial. This way we have extended the definition of $g(a)$ from polynomials to all continuous functions on $\sigma(a)$. This is known as continuous functional calculus. Note that this we are able to do only for normal elements a in $\mathcal{A}$. The continuous functional calculus is a very useful tool in C^*-algebra theory. We will be using it frequently. Here are a few applications.

There is measurable functional calculus when we study W^*-algebras or von Neumann algebras. We will not be exploring that.

Earlier we had proved the spectral mapping theorem for polynomials of arbitrary elements of Banach algebras.

Theorem 2.2.11. *(Continuous spectral mapping theorem): Let a be a normal element in a unital C^*-algebra $\mathcal{A}$. Then for any continuous function*

$f : \sigma(a) \to \mathbb{C}$,

$$\sigma(f(a)) = \{f(\lambda) : \lambda \in \sigma(a)\}.$$

Proof. This is clear from the definition of $f(a)$ as the spectrum of f in $C(\sigma(a))$ is the range of f. ■

We have observed that the spectrum of a self-adjoint element in a C^*-algebra is real. Here we have the results in the converse direction. It is to be noted that the assumption of normality of the element is crucial in this theorem.

Theorem 2.2.12. *Let a be a normal element in a unital C^*-algebra $\mathcal{A}$. Then*

(i) a is self-adjoint, that is, $a = a^*$ if and only if $\sigma(a) \subset \mathbb{R}$;

(ii) a is a projection, that is, $a = a^* = a^2$ if and only if $\sigma(a) \subseteq \{0, 1\}$.

(iii) a is a unitary, that is, $aa^* = a^*a = 1$ if and only if $\sigma(a) \subseteq \{z \in \mathbb{C} : |z| = 1\}$.

Proof. We know that the spectrum a is same as the spectrum of a in $C^*\{a, 1\}$, the unital C^*-algebra generated by a, which is commutative due to normality of a. Now by the continuous functional calculus, $C^*\{a, 1\}$ is isomorphic to the space of continuous functions on $\sigma(a)$ in canonical way. Since the spectrum of $f \in C(\sigma(a))$ is the range of f, the result follows easily. ■

Corollary 2.2.13. *Any element a of a unital C^*-algebra $\mathcal{A}$ can be written as a linear combination of four unitary elements.*

Proof. First note that by the Cartesian decomposition, we have

$$a = \frac{a + a^*}{2} + i\frac{a - a^*}{2i}.$$

Hence every element is a linear combination of two self-adjoint elements. So, to prove our original assertion, we need to show that any self-adjoint element is a linear combination of two unitary elements.

Let a be a self-adjoint element of $\mathcal{A}$. Without loss of generality, suppose $\|a\| \leq 1$. Then, $1-a^2$ has spectrum contained in $[0, 1]$. By the continuous functional calculus, $\sqrt{1 - a^2}$ exists. By straightforward computations, $u_1 = a+i\sqrt{1 - a^2}$ and $u_2 = a-i\sqrt{1 - a^2}$, have spectrum in the unit circle and hence are unitary elements. The proof is complete as $a = \frac{1}{2}u_1+\frac{1}{2}u_2$. ■

Corollary 2.2.14. *For any element a in a unital C^*-algebra $\mathcal{A}$, we have $a^*a \leq \|a\|^2 1$.*

Proof. We apply the continuous functional calculus with a^*a playing the role of the normal element. The unital C^*-algebra $C^*\{a^*a, 1\}$ is C^*-isomorphic to $C(\sigma(a^*a))$. Moreover, by this isomorphism, the identity element 1 is identified with the constant function 1 and a^*a is identified with the identity function f_0 defined by $f_0(t) = t$ on $\sigma(a^*a)$ (remember that $\sigma(a^*a)$ is a subset of the real line because of self-adjointness of a^*a, Theorem 2.2.3). Since,

$$f_0 \leq \sup\{|s| : s \in \sigma(a^*a)\}1 = r(a^*a)1 = \|a^*a\|1 = \|a\|^2 1$$

and since a C^* isomorphism preserves the order relation among self-adjoint elements, we have $a^*a \leq \|a\|^2 1$. ■

Now we wish to show that the spectrum of a positive element in a C^*-algebra is contained in the positive real line. This is rather tricky and requires some preparation.

Lemma 2.2.15. *Let a, b self-adjoint elements in a unital C^*-algebra $\mathcal{A}$. Suppose there exist positive real numbers s, t such that $\sigma(a) \subseteq [0, s]$ and $\sigma(b) \subseteq [0, t]$ then $\sigma(a + b) \subseteq [0, s + t]$.*

Proof. Since a is a self-adjoint element, making use of Gelfand isomorphism we can treat it as the identity function in $C(\sigma(a))$. As $\sigma(a) \subseteq [0, s]$, we see $\|s - a\| \leq s$. Similarly $\|t - b\| \leq t$.

Consequently, $\|s + t - (a + b)\| \leq \|s - a\| + |t - b\| \leq s + t$ and therefore $\sigma(a + b) \subseteq [0, 2(s + t)]$. We also have $\|a + b\| \leq \|a\| + \|b\| \leq s + t$ and hence $\sigma(a + b) \subseteq [-(s + t), (s + t)]$. So we get $\sigma(a + b) \subseteq [0, 2(s + t)] \cap [-(s + t), (s + t)] = [0, s + t]$. ■

Since we do not yet know that spectrum of positive elements contain only positive real numbers, therefore in the following proposition the spectral information about a_+ and a_- are part of the conclusion. The decomposition here is also unique, but we do not discuss it now.

Proposition 2.2.16. *Let a be a self-adjoint element in a unital C^*-algebra $\mathcal{A}$. Then a decomposes as $a = a_+ - a_-$, where a_+, a_- are self-adjoint elements, $\sigma(a_+) \subseteq \mathbb{R}_+$, $\sigma(a_-) \subseteq \mathbb{R}_+$ and $a_+a_- = a_-a_+ = 0$.*

Proof. Define functions f_+, f_- on $\sigma(a) \subset \mathbb{R}$ by $f_+(t) = \max\{t, 0\}$ and $f_-(t) = \max\{-t, 0\}$. Then f_+, f_- are non-negative continuous functions satisfying $f_+(t) - f_-(t) = t$ for all t, and $f_+.f_- = f_-f_+ = 0$. Using continuous functional calculus take $a_+ = f_+(a)$ and $a_- = f_-(a)$ and the result follows. ■

Theorem 2.2.17. *Let a be a positive element in a unital C^*-algebra $\mathcal{A}$. Then $\sigma(a) \subseteq \mathbb{R}_+$.*

Proof. Let $a = a_+ - a_-$ be the decomposition of a as in the Proposition 2.2.16. We wish to show that $a_- = 0$ and that would conclude the proof. Since a is positive, $a = b^*b$ for some $b \in \mathcal{A}$. Now take $d = ba_-$. We have $d^*d = a_-b^*ba_- = a_-(a_+ - a_-)a_- = -(a_-)^3$. In particular $\sigma(-d^*d) \subset \mathbb{R}_+$. From this, we wish to infer that $d = 0$.

We know from Theorem 2.1.7 that $\sigma(d^*d) \cup \{0\} = \sigma(dd^*) \cup \{0\}$. Hence $\sigma(-dd^*) \subset \mathbb{R}_+$. Then by Lemma 2.2.15, $\sigma(-dd^* - d^*d) \subset \mathbb{R}_+$. Now suppose $d = x + iy$ is the Cartesian decomposition of d (See Theorem 2.2.2). By direct computation $dd^* + d^*d = 2(x^2 + y^2)$. As x, y are self-adjoint, we first get that $\sigma(x^2 + y^2) \subset \mathbb{R}_+$ by Theorem 2.2.11. Thus, once again by Lemma 2.2.15, $\sigma(d^*d) = \sigma(2(x^2 + y^2) + (-dd^*)) \subset \mathbb{R}_+$. Consequently $\sigma(d^*d) \subset \mathbb{R}_+ \cap \mathbb{R}_- = \{0\}$. As d^*d is self-adjoint it follows that $d^*d = 0$. Then by C^*-condition $d = 0$. Now $d = ba_- = 0$, implies $b^*ba_- = (a_+ - a_-)a_- = 0$ or $-(a_-)^2 = 0$, which implies $a_- = 0$, as a_- is self-adjoint. ■

We studied commutative C^*-algebras by analyzing complex homomorphisms on them. Non-commutative C^*-algebras may not have non-trivial complex homomorphisms. So it becomes important to study a bigger class of linear functionals called positive functionals.

Definition 2.2.18. *Let $\mathcal{A}$ be a unital C^* algebra. Then a linear functional $\phi : \mathcal{A} \to \mathbb{C}$ is said to be a positive linear functional if $\phi(a) \geq 0$ for $a \geq 0$. A positive linear functional is said to be a state if $\|\phi\| = 1$.*

The following fact is quite interesting and also will be crucial to show the existence of sufficiently many states in a C^*-algebra to recover full information about the C^*-algebra.

Lemma 2.2.19. *Let $\mathcal{A}$ be a unital C^*-algebra and let $\phi : \mathcal{A} \to \mathbb{C}$ be a linear functional. Then ϕ is positive if and only if the norm of ϕ is attained at the identity, that is, $\|\phi\| = \phi(1)$. In particular, it is a state if and only if $\|\phi\| = \phi(1) = 1$.*

Proof. There is nothing to show if $\phi = 0$. Suppose ϕ is a non-zero positive linear functional. Define a sesquilinear form $[\cdot, \cdot]$ on $\mathcal{A} \times \mathcal{A}$ by

$$[b, a] = \phi(a^*b).$$

By positivity of ϕ, this is a semi-inner product. Thus, Cauchy-Schwarz inequality can be applied. Hence, we get

$$|\phi(a^*b)|^2 \leq \phi(b^*b)\phi(a^*a) \text{ for all } a, b \in \mathcal{A}.$$

So, for a in $\mathcal{A}$, we have $|\phi(a)|^2 \leq \phi(1)\phi(a^*a) \leq \|a\|^2\phi(1)^2$ since $a^*a \leq \|a\|^2 1$. This implies that $\|\phi\| \leq \phi(1)$. Hence, $\|\phi\| = \phi(1)$.

Conversely, let ϕ be a linear functional on $\mathcal{A}$ and let $\|\phi\| = \phi(1)$. Without loss of generality, assume $\|\phi\| = \phi(1) = 1$. Take $a \in \mathcal{A}$ such that $a \geq 0$. Assume, without loss of generality that, $\|a\| \leq 1$. So, $0 \leq a \leq 1$. Suppose $\phi(a) = x + iy$.

Claim: $y = 0$.

Proof of claim: For any real t, we have $\phi(a + it) = x + i(y + t)$. So,

$$\begin{aligned} |x + i(y+t)|^2 &\leq \|a + it\|^2 \\ \text{or , } x^2 + (y+t)^2 &\leq \|(a - it)(a + it)\| = \|a\|^2 + |t|^2 \\ \text{or , } x^2 + y^2 + 2ty &\leq \|a\|^2. \end{aligned}$$

This inequality can hold for all real t only if $y = 0$.

Since $\phi(a)$ is real, using $\|a\| \leq 1$, we have $-1 \leq \phi(a) \leq 1$. Also, since a is a self-adjoint contraction, $\sigma(a) \subseteq [-1, 1]$. In fact by Theorem 2.2.17, $\sigma(a) \subseteq [0, 1]$. Consequently, we have $0 \leq 1 - a \leq 1$ and hence replacing a by $(1 - a)$ in the result proved just now, $|1 - \phi(a)| \leq 1$, which is the same as $0 \leq \phi(a) \leq 2$. So, $0 \leq \phi(a) \leq 1$. Thus, $\phi(a)$ is non-negative. ■

Lemma 2.2.20. *Let $\mathcal{A}$ be a unital C^*-algebra and let $a \in \mathcal{A}$. Then there exists a state ϕ on $\mathcal{A}$ such that $\phi(a^*a) = \|a\|^2$.*

Proof. Take $b = a^*a$. Let $\mathcal{A}_0 := C^*\{b, 1\}$ be the unital C^*-subalgebra of $\mathcal{A}$ generated by b. Since b is positive (in particular $b^* = b$) and the generated C^*-algebra is just the closure of the polynomials in b in the norm of $\mathcal{A}$, we see that $\mathcal{A}_0$ is a commutative C^*-algebra. By the continuous functional calculus (Theorem 2.2.10), we have $\mathcal{A}_0$ to be isomorphic to $C(\sigma(b))$. This isomorphism maps b to the identity function $z \mapsto z$ on $\sigma(b)$. Note further that self-adjointness of b implies that the spectral radius of b is same as its norm. Hence there exists $\lambda \in \sigma(b)$ such that $|\lambda| = \|b\|$. The evaluation functional $f \mapsto f(\lambda)$ is a state on $C(\sigma(b))$ and maps the identity function $z \mapsto z$ to λ. Now the Gelfand isomorphism means that we have a state ϕ_0 on $\mathcal{A}_0$ with $\phi_0(b) = \lambda$. In particular $\|\phi_0(b)\| = |\lambda| = \|b\|$.

By Hahn-Banach theorem, ϕ_0 can be extended to a bounded linear functional ϕ on $\mathcal{A}$, retaining the norm, that is, $\|\phi\| = \|\phi_0\|$. However, as ϕ_0

is a state, $\phi_0(1) = 1$. Also, as ϕ is an extension of ϕ_0, we have $\phi(1) = 1$. Now by Lemma 2.2.19, ϕ is positive. It follows that ϕ is a state on $\mathcal{A}$ with $\phi(b) = \|b\|$. ■

For any element $a \in \mathcal{A}$, as $0 \leq a^*a \leq \|a\|^2$ and consequently for any state ϕ on $\mathcal{A}$,

$$0 \leq \phi(a^*a) \leq \|a\|^2. \tag{2.2.1}$$

We will be frequently using this fact.

Definition 2.2.21. *A representation of a C^*-algebra $\mathcal{A}$ is a pair $(\mathcal{H}, \pi)$, where $\mathcal{H}$ is a Hilbert space and $\pi : \mathcal{A} \to \mathcal{B}(\mathcal{H})$ is a $*$-homomorphism.*

A representation $(\mathcal{H}, \pi)$ is said to be faithful if π is injective.

A cyclic representation is a representation $(\mathcal{H}, \pi)$ with a vector v in $\mathcal{H}$ such that $\pi(\mathcal{A})v$ is dense in $\mathcal{H}$. The vector v is then called cyclic.

Two representations $(\mathcal{H}_1, \pi_1)$ and $(\mathcal{H}_2, \pi_2)$ are (unitarily) equivalent if there is a unitary operator (i.e., a Hilbert space isomorphism) $U : \mathcal{H}_1 \to \mathcal{H}_2$ such that $U\pi_1(a) = \pi_2(a)U$ for all a in $\mathcal{A}$.

We have arrived at the all important Gelfand-Naimark-Segal (GNS) Construction for states on C^*-algebras.

Theorem 2.2.22. *(GNS Construction): Let $\mathcal{A}$ be a unital C^*-algebra and let ϕ be a state on $\mathcal{A}$. Then there exists a triple $(\mathcal{H}, \pi, u)$ where $\mathcal{H}$ is a Hilbert space, $\pi : \mathcal{A} \to \mathcal{B}(\mathcal{H})$ is a unital $*$-homomorphism and $u \in \mathcal{H}$ is a unit vector such that*

$$\phi(a) = \langle \pi(a)u, u \rangle \text{ for all } a \in \mathcal{A}. \tag{2.2.2}$$

The triple $(\mathcal{H}, \pi, u)$ can be chosen such that

$$\mathcal{H} = \overline{\text{span}}\, \{\pi(a)u : a \in \mathcal{A}\}. \tag{2.2.3}$$

Moreover such a triple is unique up to unitary equivalence, that is, given another such triple $(\mathcal{H}', \pi', u')$ satisfying (2.2.2) and (2.2.3), there exists a unitary $U : \mathcal{H} \to \mathcal{H}'$ such that $Uu = u'$ and

$$\pi'(a) = U\pi(a)U^* \ \forall a \in \mathcal{A}.$$

Proof. We have seen in the proof of Lemma 2.2.19 that sesqui-linear form $[\cdot, \cdot]$ defined on $\mathcal{A} \times \mathcal{A}$ by

$$[b, a] = \phi(a^*b)$$

defines a semi-inner product.

It follows that K defined on $\mathcal{A} \times \mathcal{A}$ by

$$K(a,b) = \phi(a^*b)$$

is a positive kernel on $\mathcal{A}$. By the Gelfand-Naimark-Segal construction for a positive kernel done in Chapter 1, there exists a Hilbert space $\mathcal{H}$ and a map $\lambda : \mathcal{A} \to \mathcal{H}$ such that

$$K(a,b) = \langle \lambda(b), \lambda(a) \rangle = \phi(a^*b),$$

and

$$\mathcal{H} = \overline{\text{span}}\lambda(\mathcal{A}) = \overline{\text{span}}\{\lambda(b) : b \in \mathcal{A}\}.$$

Alternatively, we can take $\mathcal{H}$ as the completion of the inner product space got by quotienting the semi-inner product space $(\mathcal{A}, \langle \cdot, \cdot \rangle)$ by null vectors

$$\mathcal{N} := \{b \in \mathcal{A} : [b,b] = 0\} = \{b \in \mathcal{A} : [b,a] = 0 \text{ for all } a \in \mathcal{A}\}.$$

Here the equality follows from the Cauchy-Schwarz inequality for semi-inner products, and it is needed to see that $\mathcal{N}$ is a vector space. And then define the map λ as the quotient map $a \mapsto \mathcal{N} + a$.

For $a \in \mathcal{A}$, define $\pi(a) : \lambda(\mathcal{A}) \to \mathcal{H}$ by $\pi(a)\lambda(c) = \lambda(ac)$. Then by (2.2.1), $\pi(a)$ is bounded on $\lambda(\mathcal{A})$ and extends uniquely to a bounded linear operator on $\mathcal{H}$. Moreover,

$$\langle \pi(a)\lambda(c), \pi(b)\lambda(d) \rangle = \langle \lambda(ac), \lambda(bd) \rangle = \phi((bd)^*ac) \;\forall a,b,c,d \in \mathcal{A}$$

It is easily seen that $\pi(ab) = \pi(a)\pi(b)$ on $\lambda(\mathcal{A})$ and hence on whole of $\mathcal{H}$. Similarly $\pi(a)^* = \pi(a^*)$ by direct computation. Now take $u = \lambda(1)$ and we can easily verify all the properties required of the triple $(\mathcal{H}, \pi, u)$. For the second part, define U by $U(\pi(a)u) = \pi'(a)u'$ and extend continuously to whole of $\mathcal{H}$. The verification of the properties required of U is straight-forward. ■

Given a state ϕ, any triple $(\mathcal{H}, \pi, u)$ satisfying (2.2.2) is called a GNS triple for ϕ. It is said to be minimal if (2.2.3) is also satisfied. The theorem says that every state admits a minimal GNS triple which is unique up to unitary equivalence. We illustrate the GNS construction for states through some simple examples.

Example 2.2.23. *Let $M_n(\mathbb{C})$ be the C^*-algebra of $n \times n$ matrices. Define $\psi : M_n(\mathbb{C}) \to \mathbb{C}$ by $\psi(X) = x_{11}$, where x_{11} is the first diagonal entry of X. Then ψ is a state. To do the GNS construction, we set the semi-inner product*

$$\langle Y, X \rangle = (X^*Y)_{11} = \sum_{k=1}^{n} \overline{x_{k1}} y_{k1},$$

on $n \times n$ matrices. We see that only the first columns of X, Y matter. Here is a concrete realization of GNS construction:

$$\mathcal{H} = \mathbb{C}^n, \pi(X) = X, \; \forall X \in M_n(\mathbb{C}).$$

$$[X] = \begin{bmatrix} x_{11} \\ x_{21} \\ \vdots \\ x_{n1} \end{bmatrix}, \; u = [I] = \begin{bmatrix} 1 \\ 0 \\ \vdots \\ 0 \end{bmatrix},$$

where $[X]$ denotes the equivalence class of X. Clearly,

$$\psi(X) = \langle Xu, u \rangle, \; \forall X \in M_n(\mathbb{C}).$$

Example 2.2.24. *Let $\phi : C[0,1] \to \mathbb{C}$ be the state $\phi(f) = f(\frac{1}{2})$. Here a possible identification of GNS construction is: $\mathcal{H} = \mathbb{C}$, $\pi(f)(z) = f(\frac{1}{2})z$, $[f] = f(\frac{1}{2}), u = 1 \in \mathbb{C}$.*

Example 2.2.25. *Fix $n \geq 1$. Consider $\mathbb{C}^n$ as a C^*-algebra, the usual way. Let $(p_1, p_2, \ldots, p_n)$ be a probability vector, that is, $p_j \geq 0$ for $1 \leq j \leq n$ and $\sum_{j=1}^n p_j = 1$. Define $\phi : \mathbb{C}^n \to \mathbb{C}$ by*

$$\phi((x_1, x_2, \ldots, x_n)) = \sum_{j=1}^n p_j x_j.$$

Then ϕ is a state. Assume $p_j > 0$ for every j. A possible identification of the GNS construction is given by: $\mathcal{H} = \mathbb{C}^n$, $\pi(x) = D(x)$, where $D(x)$ is the diagonal matrix with entries coming from the co-ordinates of x,

$$[x] = \begin{bmatrix} x_1\sqrt{p_1} \\ x_2\sqrt{p_2} \\ \vdots \\ x_n\sqrt{p_n} \end{bmatrix}, u = \begin{bmatrix} \sqrt{p_1} \\ \sqrt{p_2} \\ \vdots \\ \sqrt{p_n} \end{bmatrix}.$$

Moreover this is a minimal representation if $p_j > 0$ for every j.

Example 2.2.26. *Let $\alpha : C[0,1] \to \mathbb{C}$ be the state*

$$\alpha(f) = \int_0^1 f(t)dt, \; f \in C[0,1].$$

Then a possible identification of the GNS construction is given by: $\mathcal{H} = L^2[0,1]$, with $\langle g, f \rangle = \int_0^1 \overline{f(t)} g(t)dt$, $\pi(f)g = fg$ (the operator 'multiplication by f'), and $[f] = f, u = 1$ (the constant function 1). This is an example where, there is no need to do quotienting but we need to consider completion to get a Hilbert space.

Example 2.2.27. *Define $\tau : M_n(\mathbb{C}) \to \mathbb{C}$ by*

$$\tau(X) = \frac{1}{n}\text{trace}(X), \text{ for all } X \in M_n(\mathbb{C}).$$

Then τ is a state (known as normalized trace). For this state one way of realizing the GNS construction is as follows: $\mathcal{H} = M_n(\mathbb{C})$ with inner product:

$$\langle Y, X\rangle = \frac{1}{n}\, trace(X^*Y),\ for\ all\ X, Y \in M_n(\mathbb{C}).$$

$\pi(X)(Y) = XY$, $X, Y \in M_n(\mathbb{C})$ (So $\pi(X)$ is left-multiplication by X. $[X] = X$, and $u = I$. It is easy to see that the state is recovered as $\langle \pi(X)u, u\rangle$.

The next result shows that any representation of a C^*-algebra is a direct sum of cyclic representations. Now we need to consider an uncountable direct sum of Hilbert spaces. Suppose $\{\mathcal{H}_i : i \in I\}$ is a family of Hilbert spaces over some index set I. Then the direct sum $\oplus_{i\in I}\mathcal{H}_i$ consists of elements of the form $\oplus_{i\in I}h_i$ with $h_i \in \mathcal{H}_i$, where $h_i = 0$ except for a countable set and $\sum_{i\in I}\|h_i\|^2 < \infty$. Note that if $\{A_i : i \in I\}$ is a family of bounded operators on $\mathcal{H}_i$ with $\sup_{i\in I}\|A_i\| < \infty$, then $\oplus_{i\in I}A_i$ defined by

$$(\oplus_{i\in I}A_i) \oplus_{i\in I} h_i = \oplus_{i\in I}A_ih_i$$

defines a bounded operator on $\oplus_{i\in I}\mathcal{H}_i$ with

$$\|(\oplus_{i\in I}A_i)\| = \sup_{i\in I}\|A_i\|.$$

This allows us to define *arbitrary direct sums of representations*, needed in the following theorem, as follows. Consider a family of representations $\{(\mathcal{H}_i, \pi_i)\}$. The *direct sum representation* is $(\mathcal{H}, \pi)$ where the space $\mathcal{H}$ is $\oplus_i\mathcal{H}_i$ and for a in $\mathcal{A}$, the operator $\pi(a)$ is $\oplus_i\pi_i(a)$ in the sense defined above.

Theorem 2.2.28. *If $(\mathcal{H}, \pi)$ is a representation of a C^*-algebra $\mathcal{A}$, then there is a family of cyclic representations $\{(\mathcal{H}_i, \pi_i)\}$ of $\mathcal{A}$ such that π is the same as $\oplus_i\pi_i$.*

Proof. If π is not cyclic already, then consider the collection $\mathcal{E}$ of all subsets E of nonzero vectors in $\mathcal{H}$ such that $\pi(\mathcal{A})h \perp \pi(\mathcal{A})h'$ where h and h' are in E and $h \neq h'$. We order $\mathcal{E}$ and it is easy to check that every chain has an upper bound. So, by Zorn's Lemma, $\mathcal{E}$ has a maximal element E_0. We claim that

$$\mathcal{H} = \oplus\{\overline{\pi(\mathcal{A})e} : e \in E_0\}.$$

If not, then there is a non-zero vector h in $\mathcal{H}$ such that $\langle \pi(a)e, h\rangle = 0$ for every $a \in \mathcal{A}$ and every $e \in E_0$. Let a and b be elements of $\mathcal{A}$. Then for e in E_0, we have

$$0 = \langle \pi(b^*a)e, h\rangle = \langle \pi(b^*)\pi(a)e, h\rangle = \langle \pi(a)e, \pi(b)h\rangle.$$

So, $\pi(\mathcal{A})\mathcal{E} \perp \pi(\mathcal{A})h$ for all e in E_0 implying that $E_0 \cup \{h\} \in \mathcal{E}$. Since h is non-zero, this contradicts the maximality of E_0.

Now, for e in E_0, let $\mathcal{H}_e = \overline{\pi(\mathcal{A})e}$. It is easy to see that $\mathcal{H}_e$ reduces $\pi(a)$ for every $a \in \mathcal{A}$. So, define $\pi_e : \mathcal{A} \to \mathcal{B}(\mathcal{H}_e)$ by

$$\pi_e(a) = \pi(a)|_{\mathcal{H}_e}.$$

This is a representation of $\mathcal{A}$ and π is the direct sum over all π_e as e runs over E_0. ■

The following theorem is also due to Gelfand, Naimark and Segal. It shows that every unital C^*-algebra can be embedded as a unital subalgebra of the algebra of all bounded operators on a Hilbert space. It is to be noted that the Hilbert space considered here could be non-separable.

Theorem 2.2.29. *(Gelfand, Naimark and Segal): Let $\mathcal{A}$ be a unital C^*-algebra. Then there exists a pair $(\mathcal{H}, \pi)$ where $\mathcal{H}$ is a complex Hilbert space and $\pi : \mathcal{A} \to \mathcal{B}(\mathcal{H})$ is an isometric $*$-homomorphism.*

Proof. Let $a \in \mathcal{A}$. We have seen that there exists a state ϕ_a on $\mathcal{A}$ such that $\phi_a(a^*a) = \|a\|^2$. Now by GNS construction, there exists a triple $(\mathcal{H}_a, \pi_a, u_a)$ where $\mathcal{H}_a$ is a Hilbert space, $\pi_a : \mathcal{A} \to \mathcal{B}(\mathcal{H}_a)$ is a unital $*$-homomorphism and u_a is a unit vector in $\mathcal{H}_a$ such that

$$\phi_a(x) = \langle \pi_a(x)u_a, u_a \rangle \text{ for all } x \in \mathcal{A}.$$

Observe that

$$\|\pi_a(a)\|^2 = \|\pi_a(a)^*\pi_a(a)\| = \|\pi_a(a^*a)\|$$

Consequently

$$\|\pi_a(a)\|^2 \geq \langle \pi_a(a^*a)u_a, u_a \rangle = \phi_a(a^*a) = \|a\|^2.$$

Now let $\mathcal{H}$ be the direct sum of all $\mathcal{H}_a$ with a varying in $\mathcal{A}$, that is,

$$\mathcal{H} = \oplus_{a\in\mathcal{A}}\mathcal{H}_a.$$

Taking $\pi = \oplus_{a\in\mathcal{A}}\pi_a$ we get a unital $*$-homomorphism $\pi : \mathcal{A} \to \mathcal{B}(\mathcal{H})$ such that for every $a \in \mathcal{A}$

$$\|\pi(a)\|^2 \geq \|a\|^2 \text{ for all } a \in \mathcal{A}.$$

Since, a $*$-homomorphism is always contractive, we have that π is an isometric $*$-homomorphism. ■

If the C^*-algebra $\mathcal{A}$ is separable, then each of the GNS Hilbert spaces $\mathcal{H}_a$ are separable. We can consider the direct sum of the previous proof, varying 'a' on a countable dense set in $\mathcal{A}$. This way we can ensure that $\mathcal{H}$ is separable. The isometric $*$-homomorphism π is also constructed by taking the direct sum only over the chosen countable dense set in $\mathcal{A}$. However, note that even for a separable infinite dimensional Hilbert space $\mathcal{H}$, the C^*-algebra $\mathcal{B}(\mathcal{H})$ is non-separable. So, in general, separability of a C^*-algebra is a rather stringent condition.

Recall that in Theorem 2.2.12, we did not handle positive elements. This was because more care was required. We are now ready to discuss positive elements at greater length. We start with a demonstration of the techniques developed in this chapter to reprove a result that we proved in Section 1.2 of Chapter 1. In the constructive proof provided there, we alluded to a more general situation, viz., continuous functional calculus for an element of a commutative C^*-algebra and that is what we use now.

Lemma 2.2.30. *Let T be a bounded operator in the C^*-algebra $\mathcal{B}(\mathcal{H})$ of all bounded operators on a Hilbert space $\mathcal{H}$. Then $T = S^*S$ for some $S \in \mathcal{B}(\mathcal{H})$ if and only if $\langle Th, h\rangle \geq 0$ for all $h \in \mathcal{H}$. Moreover, for such T, $\sigma(T) \subset \mathbb{R}_+$.*

Proof. Suppose $T = S^*S$ for some $S \in \mathcal{B}(\mathcal{H})$, clearly we have $\langle Th, h\rangle = \|Sh\|^2 \geq 0$. In the converse direction suppose $\langle Th, h\rangle \geq 0$ for all h, clearly T is self-adjoint. First we wish to show that $\sigma(T) \subset \mathbb{R}_+$. We already know that $\sigma(T) \subseteq \mathbb{R}$. Now suppose $t > 0$. We want to show that $T+t$ is invertible and that would show that $(-t)$ is not in the spectrum. Equivalently we need to show $(T+t)$ is one to one and onto. For $h \in \mathcal{H}$, by expanding the inner product,

$$\|(T+t)h\|^2 = \langle (T+t)h, (T+t)h\rangle \geq t^2\|h\|^2. \tag{2.2.4}$$

So $(T+t)$ is bounded below. This not only proves that $(T+t)$ is one to one, but also shows that the range is closed. Indeed, if $\{(T+t)h_n\}$ is a convergent (hence Cauchy) sequence from the range, then by (2.2.4), we have the sequence $\{h_n\}$ to be Cauchy, and hence convergent to h, say. So, $\{(T+t)h_n\}$ converges to $\{(T+t)h\}$ which is in the range. Also, $[(T+t)(\mathcal{H})]^\perp = \text{Ker } (T+t)^* = \text{Ker } (T+t) = \{0\}$ showing that the range is dense. Hence, $T+t$ is one to one and onto. Thus, $(T+t)^{-1}$ is bounded.

Now the C^*-algebra $C^*\{T, I\}$ is commutative, and the spectrum of T is contained in $\mathbb{R}_+$. Here $g(t) = t^{\frac{1}{2}}$ is a continuous function on the spectrum of T, and so by functional calculus, taking $S = g(T) := T^{\frac{1}{2}}$, we write $T = S^*S$. ∎

Theorem 2.2.31. *Let a be an element in a unital C^*-algebra $\mathcal{A}$. Then the following are equivalent.*

(i) $a = c^2$ for some $c \in \mathcal{A}$ with $c = c^*$.

(ii) a is positive, that is, $a = b^*b$ for some $b \in \mathcal{A}$;

(iii) For any Hilbert space $\mathcal{H}$ and any $*$-homomorphism $\pi : \mathcal{A} \to \mathcal{B}(\mathcal{H})$, $\langle \pi(a)h, h\rangle \geq 0$, for all $h \in \mathcal{H}$.

(iv) There exists a Hilbert space $\mathcal{H}$ and an isometric unital $*$-homomorphism $\pi : \mathcal{A} \to \mathcal{B}(\mathcal{H})$ such that $\langle \pi(a)h, h\rangle \geq 0$ for all $h \in \mathcal{H}$.

(v) $a = a^*$ and $\sigma(a) \subset \mathbb{R}_+$;

(vi) $a = a^*$ and for every positive linear functional ϕ on $\mathcal{A}$, $\phi(a) \geq 0$;

Proof. Here (i) implies (ii) is a triviality by taking $b = c$. Similarly (ii) implies (iii) follows easily as $\pi(a) = \pi(b)^*\pi(b)$. From Theorem 2.2.29, we get an isometric$*$-homomorphism from the C^*-algebra to some $\mathcal{B}(\mathcal{H})$. So (iii) implies (iv). Then (iv) implies (v) from the previous theorem because first we get that $\sigma(\pi(a)) \subset \mathbb{R}_+$. Then we note that $\sigma(\pi(a)) \subset \sigma(a)$. However, if $\pi(a) - \lambda$ has an inverse in $\mathcal{B}(\mathcal{H})$, then the inverse must be in $\pi(\mathcal{A})$ (that π is isometry is being used here). Hence, $a - \lambda$ would be invertible showing that $\sigma(\pi(a)) = \sigma(a)$. Other implications follow easily from the functional calculus. ∎

As simple consequences we get the following corollaries.

Corollary 2.2.32. *Let $\mathcal{A}, B$ be unital C^*-algebras with $\mathcal{A} \subseteq \mathcal{B}$. Suppose $a \in A$. Then a is positive as an element of $\mathcal{A}$ if and only if it is positive as an element of $\mathcal{B}$.*

Corollary 2.2.33. *Sums and convex combinations of positive elements in a C^*-algebra are positive.*

Now it is easy to see the uniqueness in the decomposition of a self-adjoint element as positive and negative part (Theorem 2.2.16). Recall the definition of a cone from Chapter 1, Section 1.2. The results above shows that the positive elements in a C^*-algebra form a cone.

Corollary 2.2.34. *If x is a positive element in $\mathcal{A}$ and s is a positive number, then $\|x + s\| = \|x\| + s$.*

Here we have polar decomposition for invertible elements in a C^*-algebra. For non-invertible elements one may require partial isometries instead of unitaries and it may not be possible within the C^*-algebra (One requires von Neumann algebra).

Theorem 2.2.35. *(Polar decomposition): Let a be an invertible element in a unital C^*-algebra $\mathcal{A}$. Then there exists a unique unitary u such that*

$$a = u|a|$$

where $|a| = (a^*a)^{\frac{1}{2}}$.

Proof. As a is invertible, a^* is also invertible with $(a^*)^{-1} = (a^{-1})^*$. It follows that a^*a and hence $|a|$ is also invertible. Now take $u = a(|a|)^{-1}$. We need to show that u is a unitary. This is clear from direct computation as:

$$uu^* = a(|a|)^{-1}(|a|)^{-1}a^* = a(a^*a)^{-1}a^* = a(a^*a)^{-1}a^*aa^{-1} = aa^{-1} = 1.$$

Similarly $u^*u = 1$. The uniqueness is obvious, due to invertibility of $|a|$. ■

Here is a theorem due to Kadison and Pedersen, see [69].

Theorem 2.2.36. *Let $\mathcal{A}$ be a unital C^*-algebra. Let $a \in \mathcal{A}$ be in the unit ball of $\mathcal{A}$ such that*

$$\|a\| < 1 - \frac{2}{n},$$

for some $n \in \mathbb{N}$ with $n \geq 3$. Then there exist n unitaries in $\mathcal{A}$ such that

$$a = \frac{u_1 + u_2 + \cdots + u_n}{n}.$$

Proof. Suppose v, t are in $\mathcal{A}$ where v is a unitary and $\|t\| < 1$. Take $b = \frac{v+t}{2}$. Then clearly $\|b\| < 1$. Further $b = v(\frac{1+v^*t}{2})$. As $\|v^*t\| = \|t\| < 1$, $1+v^*t$ is invertible and hence b is also invertible. Therefore, $b = u|b|$ for some unitary u by the polar decomposition. Note that by C^*-condition, $\|(|b|)\| = \|b\| < 1$. So $0 \leq |b| \leq 1$.

As in the proof of the 'four unitary result' Corollary 2.2.13, we can now write $|b|$ as mid-point of two unitaries. Consequently, $b = \frac{1}{2}(w_1 + w_2)$ for some unitaries w_1, w_2. We re-write this as:

$$v + t = w_1 + w_2. \tag{2.2.5}$$

Then $v + 2t = (v + t) + t$ is a sum of 3 unitaries. By induction, we get $v + (n-1)t = u_1 + u_2 + \cdots + u_n$ for some unitaries, $u_1, u_2, \ldots, u_n$. The proof of the theorem is complete, by taking $v = 1$ and $t = \frac{na-1}{n-1}$, so that $na = v + (n-1)t$ and verifying that

$$\|t\| = \|\frac{na-1}{n-1}\| < \frac{1}{n-1}(n(1-\frac{2}{n})+1) = \frac{1}{n-1}(n-2+1) = 1.$$

■

We note that every point in the open unit disc in the complex plane is a convex combination of two points on the unit circle, where one of the points can be chosen arbitrarily. Here is a result of similar nature.

Corollary 2.2.37. *Let c be an element of a unital C^*-algebra $\mathcal{A}$. If $\|c-1\| < 1$ then c is a sum of two unitaries. If d is an element of $\mathcal{A}$ with $\|d\| < 1$ then d is a sum of three unitaries, where one of the unitaries can be taken to be any given unitary. In particular, every element in a unital C^*-algebra is a positive multiple of a sum of three unitaries.*

Proof. The first part follows from the proof of the previous theorem by taking $v = 1$ and $t = c - 1$. The second part follows from the first: Suppose a unitary $u \in \mathcal{A}$ is given. Take $c = 1 - du^*$. Then $\|c-1\| < 1$ and hence $1 - du^* = v_1 + v_2$ for some unitaries v_1, v_2. So $d = u + (-v_1u) + (-v_2u)$. ■

A direct application of this observation to frame theory was found by Casazza [39]. The next result is from [105].

Theorem 2.2.38. *(Russo Dye Theorem) Let $\mathcal{A}$ be a unital C^*-algebra. Then the closed unit ball of $\mathcal{A}$ is the closed convex hull of its unitaries.*

Proof. Clearly convex combination of unitaries are in the closed unit ball of $\mathcal{A}$. For every element in the open unit ball, that is, a with $\|a\| < 1$, the theorem of Kadison and Pedersen (Theorem 2.2.36) is applicable for suitable n. Therefore the closed convex hull is whole of the closed unit ball. ■

We remark that the unit ball of a unital C^*-algebra can have extreme points different from unitaries. In fact it is a result due to Kadison ([106] that the set of extreme points of the closed unit ball of a unital C^*-algebra $\mathcal{A}$ consists of partial isometries $v \in \mathcal{A}$ satisfying $(1-vv^*)\mathcal{A}(1-v^*v) = \{0\}$.

Definition 2.2.39. *Let $\mathcal{A}, \mathcal{B}$ be two C^*-algebras. A linear map $\tau : \mathcal{A} \to \mathcal{B}$ is to said to be positive if $\tau(a) \geq 0$ in $\mathcal{B}$ for all $a \geq 0$ in $\mathcal{A}$.*

Theorem 2.2.40. *Let $\mathcal{A}$ and $\mathcal{B}$ be unital C^*-algebras and let $\tau : \mathcal{A} \to \mathcal{B}$ be a positive map. Then τ is bounded.*

Proof. We will show that $\|\tau\| \leq 2\|\tau(1)\|$. Suppose a is a self-adjoint element in $\mathcal{A}$. Using Corrollary 2.2.16, write $\tau(a) = \tau(a_+) - \tau(a_-)$. As $\tau(a_+)$ and $\tau(a_-)$ are positive it follows that $\tau(a)$ is self-adjoint. In particular there exists $\lambda \in \sigma(\tau(a))$ such that $|\lambda| = \|\tau(a)\|$. It follows that there exists a state ϕ on $\mathcal{B}$ such that $\phi(\tau(a)) = \lambda$, in particular, $|\phi(\tau(a))| = \|\tau(a)\|$. Now as τ is positive and ϕ is a state, $\phi \circ \tau$ is a positive linear functional on $\mathcal{A}$ and hence it attains its norm at the identity. So we get, $\|\tau(a)\| = |\phi(\tau(a))| \leq \|a\||\phi(\tau(1))| \leq \|a\|\|\phi\|\|\tau(1)\| = \|a\|\|\tau(1)\|$.

Now if b is an arbitrary element of $\mathcal{A}$, $\|\tau(b)\| \leq \|\tau(\frac{b+b^*}{2})\| + \|\tau(\frac{b-b^*}{2i})\| \leq 2\|b\|\|\tau(1)\|$, by triangle inequality. ■

Later we will see that we actually have $\|\tau\| = \|\tau(1)\|$ for positive linear maps, similar to positive linear functionals. The proof we have requires the notion of completely positive maps.

2.3. Completely positive maps

Consider a unital C^*-algebra $\mathcal{A}$. Let $M_n(\mathcal{A})$ be the algebra of $n \times n$ matrices with entries in $\mathcal{A}$. It is clearly a vector space with respect to natural operations and is also a $*-$algebra, where the multiplication is given by the 'matrix multiplication':

$$(AB)_{ij} = \sum_{k=1}^{n} a_{ik} b_{kj}, \ \forall 1 \leq i, j \leq n;$$

for $A = [a_{ij}], B = [b_{ij}]$ in $M_n(\mathcal{A})$ and involution $*$, by 'matrix involution':

$$(A^*)_{ij} = a_{ji}^*, \ \forall 1 \leq i, j \leq n.$$

By the Gelfand-Naimark-Segal theory, we may consider $\mathcal{A}$ as a C^*-subalgebra of $\mathcal{B}(\mathcal{H})$ for some Hilbert space $\mathcal{H}$. Then any element $A = [a_{ij}]_{1 \leq i,j \leq n}$ of $M_n(\mathcal{A})$ can be thought of as the operator:

$$\begin{pmatrix} h_1 \\ h_2 \\ \vdots \\ h_n \end{pmatrix} \mapsto \begin{bmatrix} a_{11} & a_{12} & \dots & a_{1n} \\ a_{21} & a_{22} & \dots & a_{2n} \\ \vdots & \vdots & \ddots & \vdots \\ a_{n1} & a_{n2} & \dots & a_{nn} \end{bmatrix} \begin{pmatrix} h_1 \\ h_2 \\ \vdots \\ h_n \end{pmatrix} = \begin{pmatrix} \sum_j a_{1j} h_j \\ \sum_j a_{2j} h_j \\ \vdots \\ \sum_j a_{nj} h_j \end{pmatrix},$$

on the direct sum of n-copies of $\mathcal{H}$, and this provides a norm on $M_n(\mathcal{A})$ by considering the operator norm. Thus $M_n(\mathcal{A})$ is a C^*-algebra. Since any unital Banach $*$-algebra can admit at most one C^*-norm,[1] the norm on $M_n(\mathcal{A})$ is uniquely determined. In particular, this norm does not depend upon the choice of embedding we made of $\mathcal{A}$ in some $\mathcal{B}(\mathcal{H})$.

Recall that a positive element in a C^*-algebra $\mathcal{A}$ is an element of the form b^*b for some $b \in \mathcal{A}$. The following characterization of positive elements in $M_n(\mathcal{A})$ is very useful.

Proposition 2.3.1. *Let $\mathcal{A}$ be a C^*-algebra. Consider a matrix $A = [a_{ij}]_{1\leq i,j\leq n}$ in $M_n(\mathcal{A})$. Then the following are equivalent:*

(i) A is positive;

(ii) A is a finite sum of matrices of the form $[c_i^* c_j]_{1\leq i,j\leq n}$;

(iii) $\sum_{i,j=1}^n b_i^* a_{ij} b_j \geq 0$ for all $b_1, b_2, \ldots, b_n$ in $\mathcal{A}$.

Proof. (i)$\Rightarrow$ (ii). Suppose $A \in M_n(\mathcal{A})$ is positive. So $A = C^*C$ for some $C = [c_{ij}] \in M_n(\mathcal{A})$. We get $a_{ij} = \sum_{k=1}^n (C^*)_{ik}(C)_{kj} = \sum_{k=1}^n c_{ki}^* c_{kj}$.

(ii) $\Rightarrow$ (iii). If $a_{ij} = c_i^* c_j$, then

$$\sum_{i,j=1}^n b_i^* a_{ij} b_j = \sum_{i,j}^n b_i^* c_i^* c_j b_j = (\sum_{i=1}^n c_i b_i)^* (\sum_{j=1}^n c_j b_j) \geq 0.$$

The map

$$[a_{ij}]_{1\leq i,j\leq n} \mapsto \sum_{i,j=1}^n b_i^* a_{ij} b_j$$

respects linearity. Thus, if A is a finite sum of matrices of the form $[c_i^* c_j]_{1\leq i,j\leq n}$, then $\sum_{i,j=1}^n b_i^* a_{ij} b_j$ is a sum of positive elements and hence is positive.

(iii)$\Rightarrow$ (i). Let $\pi : \mathcal{A} \to \mathcal{B}(\mathcal{H})$ be a cyclic representation of $\mathcal{A}$ with a cyclic vector $u \in \mathcal{H}$ as obtained from Theorem 2.2.22. From (iii), we get that

$$\pi(\sum_{i,j=1}^n b_i^* a_{ij} b_j) = \sum_{i,j=1}^n \pi(b_i)^* \pi(a_{ij}) \pi(b_j)$$

is a positive operator on $\mathcal{H}$. This in turn implies that

$$\left\langle [\pi(a_{ij})] \begin{pmatrix} \pi(b_1)u \\ \pi(b_2)u \\ \vdots \\ \pi(b_n)u \end{pmatrix}, \begin{pmatrix} \pi(b_1)u \\ \pi(b_2)u \\ \vdots \\ \pi(b_n)u \end{pmatrix} \right\rangle \geq 0$$

[1]This is because for any x, $\|x\|$ is the square root of the spectral radius of x^*x, and the spectral radius is algebraically determined, without referring to the norm.

for any $b_1, b_2, \ldots, b_n \in \mathcal{A}$. By cyclicity, the vectors

$$\left\{ \begin{pmatrix} \pi(b_1)u \\ \pi(b_2)u \\ \vdots \\ \pi(b_n)u \end{pmatrix} : b_1, b_2, \ldots, b_n \in \mathcal{A} \right\}$$

form a total set. So $[\pi(a_{ij})]$ is a positive operator on the direct sum of n copies of $\mathcal{H}$. Now let $(\mathcal{K}, \rho)$ be any representation of $\mathcal{A}$. Theorem 2.2.28 shows that ρ is a direct sum of cyclic representations. Thus, $[\rho(a_{ij})]$ is a positive operator on the direct sum of n copies of $\mathcal{K}$. That completes the proof by part (iii) of Theorem 2.2.31. ■

With these preliminaries, we can define completely positive maps between C^*-algebras.

Definition 2.3.2. *Let $\mathcal{A}, \mathcal{B}$ be two C^*-algebras. For $n \geq 1$, a linear map $\tau : \mathcal{A} \to \mathcal{B}$ is to said to be n-positive if $\tau^{(n)} : M_n(\mathcal{A}) \to M_n(\mathcal{B})$ defined by*

$$\tau^{(n)}([a_{ij}]_{1 \leq i,j \leq n}) = [\tau(a_{ij})]_{1 \leq i,j \leq n}$$

is positive. A linear map τ which is n-positive for every $n \geq 1$ is said to be completely positive.

It is clear that 1-positivity is the same as positivity. For $n \geq 1$, in defining $\tau^{(n)}$ we are acting τ entry-wise in matrices having entries from $\mathcal{A}$. We are mostly interested in unital C^*-algebras and unital linear maps, i.e., those maps which send the unit (or identity) of the first algebra to the unit of the second algebra. Let us look at some examples.

For $n \geq 1$, $\mathbb{C}^n$ with coordinate-wise addition, product and point-wise complex conjugation and supremum (l^∞) norm is a unital commutative C^*-algebra. We may also think of $\mathbb{C}^n$ as the space of continuous functions on discrete metric space $\{1, 2, \ldots, n\}$. Note that an element in $\mathbb{C}^n$ is positive if and only if all its coordinates are non-negative. With this in mind, we have the following example.

Example 2.3.3. *(i) Take $\mathcal{A} = \mathcal{B} = \mathbb{C}^n$. Then any linear map $\tau : \mathcal{A} \to \mathcal{B}$ is given by $\tau(x) = Px$ for some $n \times n$ matrix $P = [p_{ij}]$. Such a map τ is positive (and completely positive) if and only if entries p_{ij} are non-negative. The completely positive map τ is unital if and only if P is a stochastic matrix. That is, $p_{ij} \geq 0\ \forall i, j$, and $\sum_j p_{ij} = 1$ for every i.*

The following is a standard example to see that there are linear maps which are positive but not 2-positive.

Example 2.3.4. *Take $\mathcal{A} = \mathcal{B} = M_n, n \geq 2$ and $\tau(X) = X^{tr}$ (Transpose of X). Then τ is positive. If $(E_{ij})_{i\leq 1, j\leq n}$ are matrix units of M_n, i.e., E_{ij} is the matrix with 1 at the (i,j)th. place and 0 elsewhere, then*

$$\begin{bmatrix} E_{11} & E_{12} \\ E_{21} & E_{22} \end{bmatrix} \geq 0 \text{ but } \begin{bmatrix} E_{11} & E_{21} \\ E_{12} & E_{22} \end{bmatrix} \text{ is not positive.}$$

Hence τ is not even 2-positive.

Recall that an element B in the C^*-algebra $M_n(\mathbb{C})$ is positive if and only if $B = C^*C$ for some C in $M_n(\mathbb{C})$. In fact, even if $B = C^*C$ for some $k \times n$ matrix for some k, B is positive.

Example 2.3.5. *Take $\mathcal{A} = M_n, \mathcal{B} = M_m$ for some m, n. Let L be an $n \times m$ matrix. Define $\tau : \mathcal{A} \to \mathcal{B}$ by $\tau(X) = L^*XL$. Then τ is a completely positive map.*

Proof. If X is an $n \times n$ positive matrix, then $X = Y^*Y$ for some $n \times n$ matrix Y and hence $\tau(X) = L^*Y^*YL = (YL)^*YL$ is positive. Hence τ is positive. Suppose $X = [X_{ij}]$ is a $k \times k$ matrix with entries from $M_n(\mathbb{C})$ and X is positive. Thus, each X_{ij} is an $n \times n$ matrix and $\tau^{(k)}(X) = [\tau(X_{ij})] = [L^*X_{ij}L] = \tau(X) = L^*XL$ which can be written as

$$\begin{pmatrix} L^* & 0 & \cdots & 0 \\ 0 & L^* & \cdots & 0 \\ \vdots & \ddots & \cdots & \vdots \\ 0 & 0 & \cdots & L^* \end{pmatrix} [X_{ij}] \begin{pmatrix} L & 0 & \cdots & 0 \\ 0 & L & \cdots & 0 \\ \vdots & \ddots & \cdots & \vdots \\ 0 & 0 & \cdots & L \end{pmatrix}.$$

Now we see that τ is k-positive for every $k \geq 1$ for the same reason for which it is positive. ■

Here we wish to present a characterization of n-positivity.

Theorem 2.3.6. *Let $\mathcal{A}, \mathcal{B}$ be two unital C^*-algebras and let $\tau : \mathcal{A} \to \mathcal{B}$ be a linear map. Then τ is n-positive if and only if for all $a_1, a_2, \ldots, a_n \in \mathcal{A}$ and $b_1, b_2, \ldots, b_n \in B$,*

$$\sum_{i,j} b_i^* \tau(a_i^* a_j) b_j \geq 0. \tag{2.3.1}$$

Consequently τ is completely positive if and only if (2.3.1) holds for all $n \geq 1$ and all $a_1, a_2, \ldots, a_n$ in $\mathcal{A}$ and $b_1, b_2, \ldots, b_n$ in $\mathcal{B}$.

Proof. This is an immediate consequence of Proposition 2.3.1. ■

Example 2.3.7. *[Schur Maps] Take $\mathcal{A} = \mathcal{B} = M_n$. Define $\tau_A : \mathcal{A} \to \mathcal{B}$ by 'Schur product' (entrywise product)*

$$\tau_A([x_{ij}]) = [a_{ij}x_{ij}].$$

Then τ_A is positive if and only if A is positive and in such a case it is also completely positive.

Proof. It is easily seen that a square matrix with all entries equal to 1 is positive. By the action of Schur map on this matrix it follows that if τ_A is positive then A is positive.

In the converse direction, first suppose $A = [a_{ij}] = [c_i\bar{c}_j]$ for some $c_1, c_2, \ldots, c_n$. Then for any $X = [x_{ij}]$,

$$\tau_A(X) = [c_i\bar{c}_j x_{ij}] = CXC^*,$$

where C is the diagonal matrix with i-th diagonal entry equal to c_i. Hence $\tau_A(X)$ is positive if X is positive. Since by Spectral theorem every positive matrix is a sum of matrices of the form $[c_i\bar{c}_j]$, we have that $\tau_A(X)$ is positive if X is any positive matrix. We leave it to the reader to show that the same technique in fact implies that τ is completely positive (see previous Example). ∎

Example 2.3.8. *Take $\mathcal{A} = \mathcal{B} = M_n$. Define α, β on M_n by $\alpha([x_{ij}]) = x_{11}I, \beta(X) =$ trace $(X) \cdot I$, then α, β are completely positive.*

Proof. First observe that $X \mapsto x_{11}$ and $X \mapsto \text{trace}(X)$ are positive linear functionals. Now the result follows easily. ∎

As we have already noted, positive elements in a C^*-algebra form a cone. So it is easily verified that sums, convex combinations and compositions of completely positive maps are completely positive.

Proposition 2.3.9. *Suppose $\tau : \mathcal{A} \to \mathcal{B}$ is a $*$-homomorphism. Then τ is completely positive.*

Proof. For any $n \geq 1$, suppose $a_1, a_2, \ldots, a_n \in \mathcal{A}$ and $b_1, b_2, \ldots, b_n \in B$. Then

$$\sum_{i,j} b_i^*\tau(a_i^*a_j)b_j = \sum_{i,j} b_i^*\tau(a_i^*)\tau(a_j)b_j = (\sum_i \tau(a_i)b_i)^*(\sum_j \tau(a_j)b_j) \geq 0.$$

∎

The following result is due to Stinespring and Arveson.

Proposition 2.3.10. *Suppose $\tau : \mathcal{A} \to \mathcal{B}$ is a positive linear map. If $\mathcal{A}$ or $\mathcal{B}$ is commutative then τ is completely positive.*

Proof. First we consider the case, when $\mathcal{B}$ is commutative. So we can take $B = C(X)$ for some compact Hausdorff space X. Now for any integer $n \geq 1$, elements $a_1, a_2, \ldots, a_n$ in $\mathcal{A}$, functions $f_1, f_2, \ldots, f_n$ in $C(X)$ and for any $x \in X$, we get

$$\sum_{k,j} [\overline{f_k}\tau(a_k^* a_j) f_j](x) = \sum_{k,j} \overline{f_k(x)}\tau(a_k^* a_j)(x) f_j(x) = \tau(b^* b) \geq 0,$$

where $b = \sum_j f_j(x) a_j$. This shows complete positivity of τ by Theorem 2.3.6.

Now suppose $\mathcal{A}$ is commutative. As before, we can take $\mathcal{A} = C(X)$ for some compact Hausdorff space. Without loss of generality, $\mathcal{B} \subseteq \mathcal{B}(\mathcal{H})$ for some Hilbert space $\mathcal{H}$ (i.e., we are applying GNS construction to represent $\mathcal{B}$ on some Hilbert space $\mathcal{H}$). Now, consider an integer $n \geq 1$, functions $f_1, f_2, \ldots, f_n$ in $\mathcal{A}$ and elements $b_1, b_2, \ldots, b_n$ in $\mathcal{B}$. By Theorem 2.3.6 again, we need to show that $\sum_{k,j} b_k^* \tau(\overline{f_k} f_j) b_j$ is a positive operator on $\mathcal{H}$. Suppose $h_1, h_2, \ldots, h_n$ are any n-vectors in $\mathcal{H}$. We need to show that

$$\sum_{k,j} \langle b_k^* \tau(\overline{f_k} f_j) b_j h_j, h_k \rangle \geq 0.$$

Considering the vectors $b_j h_j$ and continuing to call them h_j, we see that it suffices to show that

$$\sum_{k,j} \langle \tau(\overline{f_k} f_j) h_j, h_k \rangle \geq 0$$

for every $f_1, f_2, \ldots, f_n$ in $\mathcal{A}$ and any $h_1, h_2, \ldots, h_n$ in $\mathcal{H}$.

This proof depends on the celebrated Riesz Representation Theorem or the Riesz-Markov theorem (See [104]). It states that any positive linear functional on $C(X)$ is implemented by a unique positive regular Borel measure on X. In the same vein, there is a Riesz representation theorem for bounded linear functionals on $C(X)$. It states that any bounded linear functional on $C(X)$ is implemented by a unique complex regular Borel measure on X.

We will use the polarization identity to express $\langle \tau(f) h_j, h_k \rangle$ as a linear combination of $\langle \tau(f)(h_j + i^r h_k), (h_j + i^r h_k) \rangle, 0 \leq r \leq 3$. For every j, k, r

$$b \mapsto \langle b(h_j + i^r h_k), (h_j + i^r h_k) \rangle$$

is a positive linear functional. Consequently, it follows that

$$f \mapsto \langle \tau(f)(h_j + i^r h_k), (h_j + i^r h_k) \rangle$$

is a positive linear functional on $\mathcal{A}$. Thus, there exists a positive regular Borel measure $\mu_{j,k,r}$ on X such that

$$\langle \tau(f)(h_j + i^r h_k), (h_j + i^r h_k)\rangle = \int_X f(t) d\mu_{j,k,r}(t).$$

Similarly, the sum of these linear functionals

$$f \mapsto \sum_{j,k=1}^{n} \sum_{r=0}^{3} \langle \tau(f)(h_j + i^r h_k), (h_j + i^r h_k)\rangle$$

is also a positive linear functional on $\mathcal{A}$. Thus, there exists a positive regular Borel measure μ on X such that

$$\sum_{j,k=1}^{n} \sum_{r=0}^{3} \langle \tau(f)(h_j + i^r h_k), (h_j + i^r h_k)\rangle = \int_X f(t) d\mu(t).$$

In fact by the uniqueness of the associated measure in Riesz-Markov theorem, $\mu = \sum_{j,k,r} \mu_{j,k,r}$. In particular every $\mu_{j,k,r}$ is absolutely continuous with respect to μ. It follows, by Radon-Nikodym theorem that there exist positive measurable functions $\{g_{j,k,r} : 1 \leq j, k \leq n; 0 \leq r \leq 3\}$ on X such that

$$\langle \tau(f)(h_j + i^r h_k), (h_j + i^r h_k)\rangle = \int_X f(t) g_{j,k,r}(t) d\mu(t).$$

By polarization identity there exist complex measurable functions $\{g_{k,j} : 1 \leq j, k \leq n\}$ such that

$$\langle \tau(f) h_j, h_k\rangle = \int_X f(t) g_{k,j}(t) d\mu(t).$$

Denoting by $[g_{k,j}(t)]$ the $n \times n$ matrix with (k, j)th entry equal to $g_{k,j}(t)$, for any vector $v \in \mathbb{C}^n$, we have

$$\begin{aligned}
\int_X f(t)\langle [g_{k,j}(t)]v, v\rangle d\mu(t) &= \int_X f(t) \sum_{j,k=1}^{n} g_{k,j}(t) v_j \overline{v_k} d\mu(t) \\
&= \sum_{j,k=1}^{n} v_j \overline{v_k} \int_X f(t) g_{k,j}(t) d\mu(t) \\
&= \sum_{j,k=1}^{n} v_j \overline{v_k} \langle \tau(f) h_j, h_k\rangle \\
&= \langle \tau(f) \sum_j v_j h_j, \sum_k v_k h_k\rangle \geq 0
\end{aligned}$$

Since this is true for any continuous $f \geq 0$, there exists a measurable subset N_v of X such that $\mu(N_v) = 0$ and $\langle [g_{k,j}(t)]v, v\rangle \geq 0$ for $t \notin N_v$.

Since a countable union of measure zero sets has measure zero, given any countable subset S of $\mathbb{C}^n$, we can get a set $N \subset X$ such that $\mu(N) = 0$ and for $v \in S$, $\langle [g_{k,j}(t)]v, v\rangle \geq 0$ for $t \notin N$. As $\mathbb{C}^n$ is separable, it has a countable dense set and the matrix valued function, $t \mapsto [g_{k,j}(t)]_{1\leq k,j\leq n}$ is positive almost everywhere μ. In other words, there exists a μ-null subset N of X such that for $t \notin N$,

$$[g_{k,j}(t)]_{1\leq k,j\leq n} \geq 0.$$

This means that the quadratic form $\sum_{k,j} \overline{f_k(t)} f_j(t) g_{k,j}(t)$ is non-negative almost everywhere. Consequently,

$$\sum_{k,j} \langle \tau(\overline{f_k} f_j) h_j, h_k\rangle = \int_X \sum_{k,j} (\overline{f_k(t)} f_j(t) g_{k,j}(t)) d\mu(t) \geq 0$$

■

Theorem 2.3.11. *(Stinespring) Let $\mathcal{A}$, B be two unital C^*- algebras, and let $\mathcal{B} \subseteq \mathcal{B}(\mathcal{H})$ for some Hilbert space $\mathcal{H}$. Suppose $\tau : \mathcal{A} \to \mathcal{B}$ is a linear map. Then τ is completely positive if and only if there exists a triple $(\mathcal{K}, \pi, V)$, where $\mathcal{K}$ is a Hilbert space, $\pi : \mathcal{A} \to \mathcal{B}(\mathcal{K})$ is a unital $*$-homomorphism, $V : \mathcal{H} \to \mathcal{K}$ is a bounded linear map such that*

$$\tau(a) = V^*\pi(a)V, \text{ for all } a \in \mathcal{A}. \tag{2.3.2}$$

Further, V is an isometry if and only if τ is unital.

Proof. Suppose τ is completely positive. Let $\mathcal{M}$ be the set $\mathcal{A} \times \mathcal{H}$. Define $K : \mathcal{M} \times \mathcal{M} \to \mathbb{C}$ by

$$K((a, u), (b, v)) = \langle \tau(a^*b)v, u\rangle.$$

For arbitrary $(a_1, u_1), \ldots, (a_n, u_n) \in \mathcal{M}$ and $c_1, \ldots, c_n \in \mathbb{C}$,

$$\begin{aligned}\sum \bar{c}_i c_j K((a_i, u_i), (a_j, u_j)) &= \sum \bar{c}_i c_j \langle \tau(a_i^* a_j) u_j, u_i\rangle \\ &= \langle [\tau(a_i^* a_j)] \left(\bigoplus_{j=1}^n c_j u_j\right), \left(\bigoplus_{i=1}^n c_i u_i\right)\rangle,\end{aligned}$$

where that last inner product is in direct sum of n copies of $\mathcal{H}$. The operator $[\tau(a_i^* a_j)]$ is positive in view of complete positivity of τ and positivity of $[a_i^* a_j] = [a_1, \ldots, a_n]^*[a_1, \ldots, a_n]$. Hence K is a positive definite kernel. By GNS construction we have a Hilbert space $\mathcal{K}$ with a map $\lambda : \mathcal{M} \to \mathcal{K}$ such that

$$\langle \lambda(b, v), \lambda(a, u)\rangle = \langle \tau(a^*b)v, u\rangle, \text{ for } (a, u), (b, v) \in \mathcal{M},$$

and $\{\lambda(a, u) : (a, u) \in \mathcal{M}\}$ is total in $\mathcal{K}$. Observe that λ is linear in both the variables. For unitary $w \in \mathcal{A}$, define $\pi(w)$ by $\pi(w)\lambda(a, u) = \lambda(wa, u)$. Then it is clear that $\pi(w)$ extends to a unitary operator of $\mathcal{K}$.

Recall that as a consequence of the continuous functional calculus in Corollary 2.2.13 we obtained that every element in the C^*-algebra $\mathcal{A}$ is a linear combination of at most four unitary elements. Now, extending π linearly we obtain a unital $*$-homomorphism $\pi : \mathcal{A} \to \mathcal{B}(K)$, satisfying $\pi(a)\lambda(b,u) = \lambda(ab,u)$, for all $a,b \in A$ and $u \in \mathcal{H}$. Define a linear map $V : \mathcal{H} \to \mathcal{K}$ by $Vu = \lambda(1,u), u \in \mathcal{H}$. ($V$ is bounded and in fact $\|V\| = \|\tau(1)\|^{1/2}$ as $\|Vu\|^2 = \|\lambda(1,u)\|^2 = \langle \tau(1)u,u\rangle = \|\tau(1)^{1/2}u\|^2$ for $u \in \mathcal{H}$.) Now,

$$\begin{aligned} \langle V^*\pi(a)Vv,u\rangle &= \langle \pi(a)Vv,Vu\rangle \\ &= \langle \pi(a)\lambda(1,v),\lambda(1,u),\rangle \\ &= \langle \lambda(a,v),\lambda(1,u)\rangle \\ &= \langle \tau(a)v,u\rangle,\ a \in \mathcal{A}, u,v \in \mathcal{H}. \end{aligned}$$

Hence $\tau(a) = V^*\pi(a)V$. The converse is easy to see as $*$-homomorphisms are completely positive. The second statmenet is obvious as $\tau(1) = V^*V$. ■

Definition 2.3.12. *Given a completely positive map $\tau : \mathcal{A} \to \mathcal{B}(\mathcal{H})$, a triple $(\mathcal{K},\pi,V)$ satisfying (2.3.2) is called a Stinespring triple associated with τ. A Stinespring triple $(\mathcal{K},\pi,V)$ is said to be minimal if*

$$\overline{span}\,\{\pi(a)Vu : a \in \mathcal{A}, u \in \mathcal{H}\} = \mathcal{K}.$$

Theorem 2.3.13. *The triple $(\mathcal{K},\pi,u)$ constructed above in Theorem 2.3.11 is minimal. If $(\tilde{\mathcal{K}},\tilde{\pi},\tilde{V})$ is another minimal triple then there exists a unitary $U : \mathcal{K} \to \tilde{\mathcal{K}}$ such that $\tilde{\pi}(a) = U\pi(a)U^*\ \forall a \in \mathcal{A}$, and $\tilde{V} = UV$.*

The minimality follows from the Gelfand Naimark Segal theory because if $\pi(a)Vu = \lambda(a,u)$. If $(\tilde{\mathcal{K}},\tilde{\pi},\tilde{V})$ is another minimal Stinespring representation, define $U : \mathcal{K} \to \tilde{\mathcal{K}}$ by defining

$$U\pi(a)Vu = \tilde{\pi}(a)\tilde{V}u, a \in \mathcal{A}, u \in \mathcal{H}$$

and extending linearly. It is clear that U is a unitary. It is easily verified that U has requisite properties. □

Let τ be a completely positive map with Stinespring representation $(\mathcal{K},\pi,V)$. Then as every $*$-homomorphism is contractive (Lemma 2.2.5), we have

$$\|\tau(a)\| = \|V^*\pi(a)V\| \leq \|V^*\|\|V\|\|\pi(a)\| \leq \|V\|^2\|a\|.$$

Moreover $\|\tau(1)\| = \|V^*V\| = \|V\|^2$. Hence every completely positive map (on a unital C^*-algebra) is bounded and the norm is attained at the identity.

It is be noted that the Stinespring's construction is a generalization of GNS construction for states. So examples given earlier for GNS construction serve as examples. Here are some additional examples.

Example 2.3.14. *Let $\tau : M_2(\mathbb{C}) \to M_2(\mathbb{C})$ be the completely positive map defined by*

$$\tau(X) = x_{11}I, \ \forall X \in M_2(\mathbb{C}),$$

where x_{11} is the first diagonal entry of X in standard basis. Then a realization of the minimal Stinespring tuple for τ is given by $(\mathbb{C}^4, \pi, V)$, where

$$\pi(X) = \begin{bmatrix} X & 0 \\ 0 & X \end{bmatrix}, \ \forall X \in M_2(\mathbb{C}),$$

and

$$V\begin{pmatrix} u_1 \\ u_2 \end{pmatrix} = \begin{pmatrix} u_1 \\ 0 \\ u_2 \\ 0 \end{pmatrix}, \ \begin{pmatrix} u_1 \\ u_2 \end{pmatrix} \in \mathbb{C}^2.$$

Example 2.3.15. *Suppose $\mathcal{G}$ and $\mathcal{H}$ are Hilbert spaces and $\tau : B(\mathcal{G}) \to B(\mathcal{H})$ is a map of the form*

$$\tau(X) = \sum_{j=1}^{n} L_j^* X L_j, \ \forall X \in B(\mathcal{G}),$$

for some bounded operators $L_j \in B(\mathcal{H}, \mathcal{G}), n \in \mathbb{N}$. Then τ is completely positive. Moreover, τ admits a Stinespring tuple $(\mathcal{K}, \pi, V)$, where $\mathcal{K} = \mathcal{G} \otimes \mathbb{C}^n$,

$$\pi(X) = X \otimes I_{\mathbb{C}^n}, \ X \in B(\mathcal{G})$$

and $V \in B(\mathcal{H}, \mathcal{K})$ is given by $Vh = \oplus_{j=1}^n (L_j h \otimes e_j), \ h \in \mathcal{H}$ where $\{e_j : 1 \leq j \leq n\}$ is an ortho-normal basis of $\mathbb{C}^n$. Some additional arguments show that this Stinespring representation is minimal if and only if $\{L_j : 1 \leq j \leq n\}$ is a linearly independent family.

Actually all completely positive maps from $B(\mathcal{G})$ to $B(\mathcal{H})$ are of the form described in the previous example when $\mathcal{G}$ and $\mathcal{H}$ are finite dimensional. The coefficients L_j are known as Choi-Kraus coefficients of the completely positive map. Even in the infinite dimensional case, the result is true, under some topological conditions (We need to assume normality of the completely positive map and also instead of finite sum we may have to deal with a series convergent in SOT).

Example 2.3.16. *Consider the Schur product map τ_A of Example 2.3.7. Using the spectral theorem for positive matrices, we write A as a sum of positive rank one operators. Say $A = [a_{ij}]$ with $a_{ij} = \sum_{k=1}^r \overline{b_{ki}} b_{kj}$, where r is* *the rank of A. Then*

$$\tau_A(X) = \sum_{k=1}^{r} L_k^* X L_k, \ X \in M_n(\mathbb{C}),$$

where L_k is the diagonal matrix with diagonal entries $(b_{k1}, b_{k2}, \dots, b_{kn})$. Hence the method of previous example can be applied to get the Stinespring representation.

Corollary 2.3.17. *(Kadison - Schwarz inequality) Let $\tau : \mathcal{A} \to \mathcal{B}$ be a compeltely positive map, where $\mathcal{A}, \mathcal{B}$ are unital C^*-algebras. Then for every $a \in \mathcal{A}$,*

$$\tau(a^*)\tau(a) \leq \tau(a^*a)\|\tau(1)\|.$$

Proof. Let $(\mathcal{K}, \pi, V)$ be a Stinespring representation of τ. Then as $VV^* \leq \|V\|^2 1$, we have

$$\begin{aligned} \tau(a^*)\tau(a) &= V^*\pi(a^*)VV^*\pi(a)V \\ &\leq V^*\pi(a^*)\pi(a)V(\|V\|^2) \\ &= \tau(a^*a)\|\tau(1)\|. \end{aligned}$$

■

Actually Kadison-Schwarz inequality is also true for 2-positive maps, and this can be proved using simple 2×2 matrix tricks. The following result is due to Russo and Dye.

Corollary 2.3.18. *Let $\tau : \mathcal{A} \to \mathcal{B}$ be a positive map, where $\mathcal{A}$,$\mathcal{B}$ are unital C^*-algebras. Then $\|\tau\| = \|\tau(1)\|$.*

Proof. Consider any unitary $u \in \mathcal{A}$. Let $\mathcal{A}_0 = C^*\{u, 1\}$ be the unital C^*-algebra generated by u. Then $\mathcal{A}_0$ is commutative. Then by Proposition 2.3.10, τ restricted $\mathcal{A}_0$ is completely positive and hence Kadison-Schwarz inequality is applicable. We get

$$\tau(u^*)\tau(u) \leq \tau(u^*u)\|\tau(1)\| \leq \|\tau(1)\|^2.$$

Hence $\|\tau(u)\| \leq \|\tau(1)\|$ for every unitary u. Consequently, $\|\tau(a)\| \leq \|\tau(1)\|$, whenever a is a convex combination of unitaries. The proof is complete by the Russo-Dye Theorem 2.2.38.

■

As before, let $(\mathcal{K}, \pi, V)$ be a minimal Stinespring representation of a completely positive map $\tau : \mathcal{A} \to \mathcal{B}(\mathcal{H})$. Assume that τ is unital, hence

V is an isometry. Identify $u \in \mathcal{H}$ with $Vu \in \mathcal{K}$. This way $\mathcal{H}$ becomes a subspace of $\mathcal{K}$, and Stinespring representation becomes

$$\tau(a) = P_{\mathcal{H}}\pi(a)_{\mathcal{H}}, \quad \text{for all } a \in \mathcal{A},$$

where $P_{\mathcal{H}}$ is the projection of $\mathcal{K}$ onto $\mathcal{H}$. In other words Stinespring representation theorem tells us that every unital completely positive map is a compression of a unital $*$-homomorphism. Of course, compressions of $*$-homomorphisms are always completely positive. More generally it is true that all contractive completely positive maps are compressions of (not necessarily unital) $*$-homomorphisms.

We observe that if $\tau : \mathcal{A} \to \mathcal{B} \subseteq \mathcal{B}(\mathcal{H})$ is a non-degenerate representation, that is, if τ is a $*$-homomorphism and $\overline{\text{span}}\{\tau(X)u : X \in \mathcal{A}, u \in \mathcal{H}\} = \mathcal{H}$, then $(\mathcal{H}, \tau, 1)$ is the minimal Stinespring representation of τ.

Here is an interesting result of Kirchberg. We don't really need this in the sequel and so we do not bother to explain the notion of multiplier algebra which can be found in standard texts on C^*-algebras.

Proposition 2.3.19. *Let $\mathcal{B}$ be a C^*-algebra such that the multiplier algebra $M(\mathcal{B})$ of $\mathcal{B}$ contains n isometries $S_1, S_2, \ldots, S_n$ such that $S_i^* S_j = 0$ for $i \neq j$, for some n. Suppose $\tau : M_n \to \mathcal{B}$ is a completely positive map. Then there exist $T_1, \ldots, T_n$ in $\mathcal{B}$ such that $\tau([x_{ij}]) = \sum_{i,j} x_{ij} T_i^* T_j$, for $[x_{ij}] \in M_n$.*

Proof. Let $E_{ij}, 1 \leq i, j \leq n$ be matrix units of M_n. We have $F = [\tau(E_{ij})] \geq 0$, by complete positivity of τ. Now as $S_i^* S_j = \delta_{ij}, 1 \leq i, j \leq n$, taking $[T_1, \ldots, T_n] = [S_1, \ldots, S_n]F^{1/2}$, we have $[T_1, \ldots, T_n]^*[T_1, \ldots, T_n]$ $= F^{1/2}[S_1, \ldots, S_n]^*[S_1, \ldots, S_n]F^{1/2} = F$. Hence $\tau(E_{ij}) = T_i^* T_j$ and by linearity of τ we have the result. ∎

CHAPTER 3

Dilation Theory in Two Variables - The Bidisc

We start this chapter from where we ended the first one. Since the dilation theory for a single contraction is so well-developed, it is natural to ask whether the same is true in multi-variable generalizations. The rest of this book will be concerned with that. As we shall see, this question will have a great deal of relationship with geometry. We shall progress in small steps. In this chapter, we deal with two commuting contractions. Ando gave a beautiful generalization of Sz.-Nagy and Foias's theorem for two commuting contractions.

3.1. Tuples of commuting contractions

The concept of dilation for a tuple of operators is a natural generalization of dilation of a single operator.

Definition 3.1.1. *Let $\mathcal{H} \subset \mathcal{K}$ be two Hilbert spaces. Suppose*

$$\underline{T} = (T_1, T_2, \ldots, T_d) \text{ and } \underline{V} = (V_1, V_2, \ldots, V_d)$$

are tuples of bounded operators acting on $\mathcal{H}$ and $\mathcal{K}$ respectively, i.e., $T_i \in \mathcal{B}(\mathcal{H})$ and $V_i \in \mathcal{B}(\mathcal{K})$. The operator tuple $\underline{V}$ is called a dilation of the operator tuple $\underline{T}$ if

$$T_{i_1} T_{i_2} \ldots T_{i_k} h = P_{\mathcal{H}} V_{i_1} V_{i_2} \ldots V_{i_k} h$$

for all $h \in \mathcal{H}$, $k \geq 1$ and all $1 \leq i_1, i_2, \ldots, i_k \leq d$. If V_i are isometries, then $\underline{V}$ is called an isometric dilation.

If the components of the operator tuple $\underline{T}$ are commuting operators, then one usually looks for a commuting dilation. We shall see in this section to what extent that becomes possible. First, we prove Ando's theorem which fulfils this aim in the case of two commuting contractions.

Theorem 3.1.2. *For a pair $\underline{T} = (T_1, T_2)$ of commuting contractions on a Hilbert space $\mathcal{H}$, there is a commuting isometric dilation $\underline{V} = (V_1, V_2)$.*

Proof: Let $\mathcal{H}_+ = \mathcal{H} \oplus \mathcal{H} \oplus \cdots$ be the direct sum of infinitely many copies of $\mathcal{H}$. Define two isometries W_1 and W_2 on $\mathcal{H}_+$ as follows. For $h = (h_0, h_1, h_2, \ldots) \in \mathcal{H}_+$, set

$$W_i h = (T_i h_0, D_{T_i} h_0, 0, h_1, h_2, \ldots), \text{ for } i = 1, 2.$$

Here D_{T_i} are the defect operators. Recall that for a contraction A, the defect operator D_A is the unique positive square root of $I - A^*A$. Clearly, W_1 and W_2 are isometries. However, they need not commute. We shall modify them to get commuting isometries.

Let $\mathcal{H}_4 = \mathcal{H} \oplus \mathcal{H} \oplus \mathcal{H} \oplus \mathcal{H}$ and let U be a unitary operator on $\mathcal{H}_4$. We shall specify U later. Identify $\mathcal{H}_+$ and $\mathcal{H} \oplus \mathcal{H}_4 \oplus \mathcal{H}_4 \oplus \cdots$ by the following identification:

$$h = (h_0, h_1, h_2, \ldots) \to (h_0, (h_1, h_2, h_3, h_4), (h_5, h_6, h_7, h_8), \ldots).$$

Now define a unitary operator $W : \mathcal{H}_+ \to \mathcal{H}_+$ by

$$Wh = (h_0, U(h_1, h_2, h_3, h_4), U(h_5, h_6, h_7, h_8), \ldots),$$

where $h = (h_0, h_1, h_2, \ldots)$. The unitarity of W is clear since U is a unitary and

$$W^*h = W^{-1}h = (h_0, U^{-1}(h_1, h_2, h_3, h_4), U^{-1}(h_5, h_6, h_7, h_8), \ldots).$$

We define $V_1 = WW_1$ and $V_2 = W_2W^{-1}$. These are isometries because they are products of isometries. These act on $\mathcal{H}_+$ and

$$V_i^*(h_0, 0, \ldots) = (T_i^* h_0, 0, \ldots), \text{ for } i = 1, 2.$$

Now we shall see that U can be chosen so that V_1 and V_2 commute.

To choose such an U, we first compute V_1V_2 and V_2V_1.

$$\begin{aligned}
V_1V_2(h_0, h_1, \ldots) &= WW_1W_2W^{-1}(h_0, h_1, \ldots) \\
&= WW_1W_2((h_0, U^{-1}(h_1, h_2, h_3, h_4), U^{-1}(h_5, h_6, h_7, h_8), \ldots) \\
&= WW_1(T_2h_0, D_{T_2}h_0, 0, U^{-1}(h_1, h_2, h_3, h_4), U^{-1}(h_5, h_6, h_7, h_8), \ldots) \\
&= W(T_1T_2h_0, D_{T_1}T_2h_0, 0, D_{T_2}h_0, 0, U^{-1}(h_1, h_2, h_3, h_4), \\
&\qquad U^{-1}(h_5, h_6, h_7, h_8), \ldots) \\
&= (T_1T_2h_0, U(D_{T_1}T_2h_0, 0, D_{T_2}h_0, 0), (h_1, h_2, h_3, h_4), \\
&\qquad (h_5, h_6, h_7, h_8), \ldots)
\end{aligned}$$

and

$$\begin{aligned} V_2V_1(h_0, h_1, \ldots) &= W_2W_1(h_0, h_1, \ldots) \\ &= W_2(T_1h_0, D_{T_1}h_0, 0, h_1, h_2, \ldots) \\ &= (T_2T_1h_0, D_{T_2}T_1h_0, 0, D_{T_1}h_0, 0, h_1, h_2, \ldots). \end{aligned}$$

Since T_1 and T_2 commute, V_1V_2 will be equal to V_2V_1 if

$$U(D_{T_1}T_2h, 0, D_{T_2}h, 0) = (D_{T_2}T_1h, 0, D_{T_1}h, 0) \tag{3.1.1}$$

for all $h \in \mathcal{H}$. Now a simple calculation shows that

$$\|(D_{T_1}T_2h, 0, D_{T_2}h, 0)\| = \|(D_{T_2}T_1h, 0, D_{T_1}h, 0)\|, \text{ for all } h \in \mathcal{H}.$$

Hence one can consider the two closed subspaces $\mathcal{L}_1$ and $\mathcal{L}_2$ defined as the closure of the subspaces

$$\{(D_{T_1}T_2h, 0, D_{T_2}h, 0) : h \in \mathcal{H}\} \text{ and } \{(D_{T_2}T_1h, 0, D_{T_1}h, 0) : h \in \mathcal{H}\}$$

respectively and define an isometry $U : \mathcal{L}_1 \to \mathcal{L}_2$ by (3.1.1). To extend U to the whole of $\mathcal{H}_4$ as a unitary operator, i.e., an isometry of $\mathcal{H}_4$ onto itself, one just needs to check that $\mathcal{H}_4 \ominus \mathcal{L}_1$ and $\mathcal{H}_4 \ominus \mathcal{L}_2$ have the same dimension. If $\mathcal{H}$ is finite dimensional, this is obvious because $\mathcal{L}_2$ is an isometric image of $\mathcal{L}_1$. When $\mathcal{H}$ is infinite dimensional, note that $\mathcal{L}_1^\perp$ and $\mathcal{L}_2^\perp$ have dimension at least as large as dim$\mathcal{H}$ because each of $\mathcal{L}_1^\perp$ and $\mathcal{L}_2^\perp$ contain a subspace isomorphic to $\mathcal{H}$, for example the subspace $\{(0, h, 0, 0) : h \in \mathcal{H}\}$. Thus

$$\dim \mathcal{H}_4 \geq \dim(\mathcal{H}_4 \ominus \mathcal{L}_i) \geq \dim \mathcal{H} = \dim \mathcal{H}_4 \text{ for } i = 1, 2.$$

So they have the same dimension. This completes the proof that U can be so defined that V_1 and V_2 commute. ■

Later in this chapter, we shall give a function theoretic construction of the Ando dilation for a special class of commuting contractions.

Theorem 3.1.3. *Let V_1 and V_2 be two commuting isometries on a Hilbert space $\mathcal{H}$. Then there is a Hilbert space $\mathcal{K}$ and two commuting unitaries U_1 and U_2 on $\mathcal{K}$ such that*

$$U_1h = V_1h \text{ and } U_2h = V_2h \text{ for all } h \in \mathcal{H}.$$

In other words, two commuting isometries can be extended to two commuting unitaries.

We remark here that simple modifications of the proof of this theorem yield that the same is true for any number (finite or infinite) of commuting isometries.

Proof: We have seen in Chapter 1 that any isometry can be extended to a unitary. So, we first find a unitary extension U_1 on a Hilbert space $\tilde{\mathcal{H}}$ of the isometry V_1. Without loss of generality we may assume that this extension is *minimal*, i.e.,

$$\tilde{\mathcal{H}} = \overline{\text{span}}\{U_1^n h : n \in \mathbb{Z} \text{ and } h \in \mathcal{H}\}.$$

We want to define an isometric extension $\tilde{V}_2$ on $\tilde{\mathcal{H}}$ of V_2 which

(1) would commute with U_1,
(2) would be a unitary on $\tilde{\mathcal{H}}$ if V_2 already is a unitary.

Assume for a moment that this has been accomplished, i.e., we have found a $\tilde{V}_2$ satisfying the above conditions. Then if V_2 happens to be unitary, we are done. If not, then just repeat the construction by extending $\tilde{V}_2$ to a unitary. Since U_1 is already a unitary, the resulting extensions that we shall get will be commuting unitaries.

Now we get down to finding a $\tilde{V}_2$ satisfying (1) and (2). To that end, note that

$$\begin{aligned}
&\|\sum_{n=-\infty}^{\infty} U_1^n V_2 h_n\|^2 \\
&= \sum_{n,m} \langle U_1^n V_2 h_n, U_1^m V_2 h_m\rangle \\
&= \sum_{n\geq m} \langle U_1^{n-m} V_2 h_n, V_2 h_m\rangle + \sum_{n<m} \langle V_2 h_n, U_1^{m-n} V_2 h_m\rangle \\
&= \sum_{n\geq m} \langle V_1^{n-m} V_2 h_n, V_2 h_m\rangle + \sum_{n<m} \langle V_2 h_n, V_1^{m-n} V_2 h_m\rangle \text{ as } U_1 \text{ extends } V_1 \\
&= \sum_{n\geq m} \langle V_2 V_1^{n-m} h_n, V_2 h_m\rangle + \sum_{n<m} \langle V_2 h_n, V_2 V_1^{m-n} h_m\rangle \text{ as } V_2,\ V_1 \text{ commute} \\
&= \sum_{n\geq m} \langle V_1^{n-m} h_n, h_m\rangle + \sum_{n<m} \langle h_n, V_1^{m-n} h_m\rangle \text{ as } V_2 \text{ is an isometry} \\
&= \|\sum_{n=-\infty}^{\infty} U_1^n h_n\|^2 \text{ tracing the steps back with } V_2 = I_{\mathcal{H}}.
\end{aligned}$$

Thus on $\tilde{\mathcal{H}}$, one can unambiguously define an isometry $\tilde{V}_2$ by

$$\tilde{V}_2\left(\sum_{n=-\infty}^{\infty} U_1^n h_n\right) = \sum_{n=-\infty}^{\infty} U_1^n V_2 h_n.$$

Of course, $\tilde{V}_2$ extends V_2 and commutes with U_1. Finally if V_2 is a unitary, i.e., a surjective isometry, then it follows from the definition of $\tilde{V}_2$ that it has a dense range. An isometry with a dense range is surjective and hence unitary. We have finished the proof. ■

This theorem of course immediately produces the unitary dilation theorem for two commuting contractions:

Theorem 3.1.4. *[Ando's dilation theorem] Given two commuting contractions T_1 and T_2 on a Hilbert space $\mathcal{H}$, there is a Hilbert space $\mathcal{K} \supset \mathcal{H}$ and two commuting unitaries U_1 and U_2 on $\mathcal{K}$ such that*

$$T_1^m T_2^n h = P_{\mathcal{H}} U_1^m U_2^n h, \quad \text{for all } h \in \mathcal{H} \text{ and } m, n = 0, 1, 2, \ldots. \tag{3.1.2}$$

Proof First get a commuting isometric dilation $\underline{V} = (V_1, V_2)$ on a space $\mathcal{H}_+$ by Theorem 3.1.2. Then get a commuting unitary extension $\underline{U} = (U_1, U_2)$ of $\underline{V}$ by Theorem 3.1.3. Now $\underline{U}$ is the required dilation because U_1 and U_2 commute and for any $h \in \mathcal{H}$ and $m, n = 0, 1, 2, \ldots$, we have

$$\begin{aligned} P_{\mathcal{H}} U_1^m U_2^n h &= P_{\mathcal{H}} V_1^m V_2^n h \text{ because } \underline{U} \text{ extends } \underline{V} \text{ and } h \in \mathcal{H} \subset \mathcal{H}_+ \\ &= T_1^m T_2^n h \text{ because } \underline{V} \text{ dilates } \underline{T}. \end{aligned}$$

■

The simultaneous unitary dilation theorem immediately produces an analogue of von Neumann's inequality.

Corollary 3.1.5. *[Ando's inequality] Let T_1 and T_2 be two commuting contractions acting on a Hilbert space $\mathcal{H}$. Suppose $p(z_1, z_2)$ is any polynomial in two variables. Then*

$$\|p(T_1, T_2)\| \leq \sup\{|p(z_1, z_2)| : |z_1| \leq 1, |z_2| \leq 1\}.$$

Proof Let $\mathcal{K} \supset \mathcal{H}$ be a unitary dilation space for $\underline{T}$ and U_1 and U_2 be two commuting unitaries on $\mathcal{K}$ as obtained from Theorem 3.1.4. Since U_1 and U_2 are commuting unitaries, all four of U_1, U_2, U_1^* and U_2^* commmute. Thus the C^*-algebra $\mathcal{C}$ generated by U_1 and U_2 is commutative. So by Gelfand theory for commutative C^*-algebras developed in Chapter 2, $\mathcal{C}$ is isometrically $*$-isomorphic to $C(\mathcal{M}_{\mathcal{C}})$ where $\mathcal{M}_{\mathcal{C}}$ is the set of all multiplicative linear functionals (characters) χ on $\mathcal{C}$. Such functionals satisfy $\|\chi\| = \chi(1_{\mathcal{K}}) = 1$.

Thus $|\chi(U_i)|^2 = \chi(U_i)\chi(U_i^*) = \chi(U_iU_i^*) = \chi(1_\mathcal{K}) = 1$. So for any polynomial $p(z_1, z_2)$, we have

$$\begin{aligned}\|p(U_1, U_2)\| &= \sup_{\chi\in\mathcal{M}_\mathcal{C}} |\chi(p(U_1, U_2))| \\ &= \sup_{\chi\in\mathcal{M}_\mathcal{C}} |p(\chi(U_1), \chi(U_2))| \text{ as } \chi \text{ is multiplicative \& linear} \\ &\le \sup_{|z_1|=1,|z_2|=1} |p(z_1, z_2)|.\end{aligned}$$

Hence

$$\|p(T_1, T_2)\| = \|P_\mathcal{H} p(U_1, U_2)|_\mathcal{H}\| \le \|p(U_1, U_2)\| \le \sup_{|z_1|=1,|z_2|=1} |p(z_1, z_2)|.$$

■

The simultaneous unitary dilation of a pair of contractions is due to Ando [9] and the proofs given here are essentially the same as his original ones. We shall end this section with the rather striking fact that Ando's theorem does not generalize to more than two commuting contractions. The unitary extension theorem of isometries holds good, as remarked above, for any number of commuting isometries. It is the isometric dilation of contractions which fails for more than a pair of contractions.

Perhaps the easiest way to see it is to construct a triple of commuting contractions which do not have a commuting unitary dilation. This is what was done in [86] and we reproduce it here. To that end, let $\mathcal{L}$ be a Hilbert space and let A_1, A_2, A_3 be three unitary operators on $\mathcal{L}$ such that

$$A_1A_2^{-1}A_3 \neq A_3A_2^{-1}A_1.$$

(For example, $A_2 = I$ and A_1, A_3 two non-commuting unitaries will do.) Let $\mathcal{H} = \mathcal{L} \oplus \mathcal{L}$ and let $T_i \in \mathcal{B}(\mathcal{H})$ for $i = 1, 2, 3$ be defined as

$$T_i(h_1, h_2) = (0, A_ih_1) \text{ where } h_1, h_2 \in \mathcal{L}.$$

Clearly $\|T_i\| = \|A_i\| = 1$ for $i = 1, 2, 3$ and $T_iT_j = T_jT_i = 0$ for $i, j = 1, 2, 3$. So (T_1, T_2, T_3) is a commuting triple of contractions on $\mathcal{H}$. Suppose there exist commuting unitary operators U_1, U_2, U_3 on some Hilbert space $\mathcal{K} \supset \mathcal{H}$ such that $T_i = P_\mathcal{H} U_i|_\mathcal{H}$ for $i = 1, 2, 3$. Then

$$P_\mathcal{H} U_i(h, 0) = T_i(h, 0) = (0, A_ih), h \in \mathcal{H} \quad i = 1, 2, 3.$$

Note that $\|U_i(h, 0)\| = \|h\|$ and $\|(0, A_ih)\| = \|A_ih\| = \|h\|$. So $U_i(h, 0) = (0, A_ih)$. Hence

$$\begin{aligned}U_kU_j^{-1}U_i(h, 0) &= U_kU_j^{-1}(0, A_ih) = U_kU_j^{-1}(0, A_j(A_j^{-1}A_ih)) \\ &= U_kU_j^{-1}U_j(0, A_j^{-1}A_ih) = U_k(0, A_j^{-1}A_ih) = (0, A_kA_j^{-1}A_ih).\end{aligned}$$

Since the U_i commute, $U_k U_j^{-1} U_i = U_i U_j^{-1} U_k$ for all $i, j, k = 1, 2, 3$. So $A_k A_j^{-1} A_i = A_i A_j^{-1} A_k$ for all $i, j, k = 1, 2, 3$. That is a contradiction. So there is no commuting dilation.

Note that a von Neumann type inequality is an immediate corollary of unitary dilation. So one way to show non-existence of dilation is to show that von Neumann's inequality is violated. This is what Crabb and Davie did with a triple of operators acting on an eight-dimensional space [44]. The literature over the years is full of a lot of discussions and considerations of many aspects of the issue originating from this spectacular failure of von Neumann's inequality. The survey article of Drury [60] is very insightful, so is the monograph by Pisier [92]. The reader is referred to Varopoulos [117] for his probabilistic arguments and establishing connection with Grothedieck's inequality, and Parrott [86] who gave an example which satisfies von Neumann's inequality but does not have a unitary dilation.

3.2. Two applications

A beautiful application of Ando's dilation theorem was discovered by Sarason in [107] in relation to interpolation problems. It was later generalised by Sz.-Nagy and Foias in [113]. The following is called the *commutant lifting theorem* because if a bounded operator R commutes with a bounded operator T, then R is called a commutant of T and the following $lifts$ a commutant to a bigger space.

Theorem 3.2.1. *Let T be a contraction acting on a Hilbert space $\mathcal{H}$ with minimal unitary dilation $(\mathcal{K}, U)$. If $R \in \mathcal{B}(\mathcal{H})$ commutes with T, then there is an $S \in \mathcal{B}(\mathcal{K})$ commuting with U such that $\|R\| = \|S\|$ and $RT^n = P_{\mathcal{H}} SU^n|_{\mathcal{H}}$ for $n = 0, 1, 2, \ldots$.*

Proof. Assume, without loss of generality that $\|R\| = 1$. So, we can apply Ando's theorem to get commuting unitary operators U_1 and U_2 on a Hilbert space $\mathcal{K}_1$ containing $\mathcal{H}$ such that

$$T^n R^m = P_{\mathcal{H}} U_1^n U_2^m|_{\mathcal{H}} \text{ for } n, m = 0, 1, 2, \ldots.$$

Let $\mathcal{K}_1{}'$ be the closed linear span of $\{U_1^n h : n \in \mathbb{Z}, h \in \mathcal{H}\}$. Then it is easy to see that $\mathcal{K}_1{}'$ is a reducing subspace for U_1 and moreover $U_1' = P_{\mathcal{K}_1'} U_1|_{\mathcal{K}_1{}'}$ is a minimal unitary dilation for T. Since the minimal unitary dilation is unique up to unitary equivalence, $(\mathcal{K}, U)$ is unitarily equivalent to $(\mathcal{K}_1{}', U_1')$ by a unitary that preserves $\mathcal{H}$. We shall make this unitary identification and will henceforth assume that $(\mathcal{K}, U)$

= $(\mathcal{K}_1{}', U_1')$. Now, decompose $\mathcal{K}_1 = \mathcal{K} \oplus (\mathcal{K}_1 \ominus \mathcal{K})$. In this decomposition, the operator matrices of U_1 and U_2 are

$$U_1 = \begin{pmatrix} U & 0 \\ 0 & V \end{pmatrix} \text{ and } U_2 = \begin{pmatrix} S & B \\ C & D \end{pmatrix}$$

where we have denoted by S the operator $P_{\mathcal{K}} U_2|_{\mathcal{K}}$ on $\mathcal{K}$. The operators U and S commute by commutativity of U_1 and U_2. Moreover,

$$P_{\mathcal{H}} S U^n = P_{\mathcal{H}} P_{\mathcal{K}} U_2 U_1^n = P_{\mathcal{H}} U_2 U_1^n = R T^n$$

for $n = 0, 1, 2, \dots$. The operator S is a contraction too by being a part of a unitary and $\|S\|$ is at least as big as $\|R\|$ because $R = P_{\mathcal{H}} S|_{\mathcal{H}}$. Hence $\|S\| = 1$. ■

A two cross two matrix trick can immediately be used to generalize the theorem above to intertwiners. Given two bounded operators T_1 and T_2, a bounded operator A is said to *intertwine* T_1 and T_2 if $AT_1 = T_2 A$. We have an *intertwining lifting theorem* as follows.

Theorem 3.2.2. *Let $T_1 \in \mathcal{B}(\mathcal{H}_1)$ and $T_2 \in \mathcal{B}(\mathcal{H}_2)$ be contractions. Let $(\mathcal{K}_1, U_1)$ and $(\mathcal{K}_2, U_2)$ be their minimal unitary dilations respectively. If A intertwines T_1 and T_2, then there is an R intertwining U_1 and U_2 such that $\|A\| = \|R\|$ and*

$$AT_1^n = T_2^n A = P_{\mathcal{H}_2} R U_1^n|_{\mathcal{H}_1} = P_{\mathcal{H}_2} U_2^n R|_{\mathcal{H}_1}$$

for all $n = 0, 1, 2, \dots$.

Proof. Consider $\hat{T} = \begin{pmatrix} T_1 & 0 \\ 0 & T_2 \end{pmatrix}$ on $\mathcal{H}_1 \oplus \mathcal{H}_2$ with its minimal unitary dilation $\hat{U} = \begin{pmatrix} U_1 & 0 \\ 0 & U_2 \end{pmatrix}$ acting on $\mathcal{K}_1 \oplus \mathcal{K}_2$. Note that $\hat{A} = \begin{pmatrix} 0 & 0 \\ A & 0 \end{pmatrix}$ on $\mathcal{H}_1 \oplus \mathcal{H}_2$ commutes with $\hat{T}$. So, applying the commutant lifting theorem, we get a norm preserving lifting $\hat{R} = \begin{pmatrix} A & B \\ R & D \end{pmatrix}$ on $\mathcal{K}_1 \oplus \mathcal{K}_2$ of $\hat{A}$ that satisfies

$$\hat{A}\hat{T}^n = P_{\mathcal{H}_1 \oplus \mathcal{H}_2} \hat{R} \hat{U}^n|_{\mathcal{H}_1 \oplus \mathcal{H}_2}.$$

We have $AT_1^n = P_{\mathcal{H}_2} R U_1^n|_{\mathcal{H}_1}$ by the $(2, 1)$ entry of the operator matrix above. The norm preserving condition is straightforward. ■

Ando's general construction does not lend itself to any function theoretic interpretation. In some special cases, there are beautiful function theoretic constructions of the dilation. This endeavour was started by Agler and McCarthy in [3]. Later, a simpler approach was found by Das and Sarkar [46] in a special case. This is what we shall pursue now. We begin with a remarkably simple folklore result that has had a far reaching impact.

3.3. The realization formula in the unit disc

The following result is known as the Realization formula and this forms the basis of much of operator theory and complex geometry in two or more dimensions. It is difficult to trace its origin and it has gained the status of folklore.

Theorem 3.3.1. *A function f is in $H^\infty(\mathbb{D})$ and satisfies $\|f\|_\infty \le 1$ if and only if there is a Hilbert space H and a unitary operator*

$$U = \begin{pmatrix} A & B \\ C & D \end{pmatrix} : \mathbb{C} \oplus H \to \mathbb{C} \oplus H$$

such that

$$f(z) = A + zB(I - zD)^{-1}C.$$

Proof. The easy part of the proof is to show that an f which has such a realization formula is indeed in the norm unit ball of $H^\infty(\mathbb{D})$. So, suppose there is a unitary $U = \left(\begin{smallmatrix} A & B \\ C & D \end{smallmatrix}\right) : \mathbb{C} \oplus H \to \mathbb{C} \oplus H$ such that $f(z) = A + zB(I - zD)^{-1}C$. Since U is a unitary, we have

$$\begin{pmatrix} 1 & 0 \\ 0 & I \end{pmatrix} = U^*U = \begin{pmatrix} \overline{A} & C^* \\ B^* & D^* \end{pmatrix}\begin{pmatrix} A & B \\ C & D \end{pmatrix} = \begin{pmatrix} |A|^2 + C^*C & \overline{A}B + C^*D \\ B^*A + D^*C & B^*B + D^*D \end{pmatrix}$$

and hence

$$|A|^2 + C^*C = 1, B^*B + D^*D = I \text{ and } \overline{A}B + C^*D = 0. \tag{3.3.3}$$

Now,

$$\begin{aligned}
&\overline{f(z)}f(z)\\
=&(\overline{A} + \overline{z}C^*(I - \overline{z}D^*)^{-1}B^*)(A + zB(I - zD)^{-1}C)\\
=&|A|^2 + \overline{A}zB(I - zD)^{-1}C + \overline{z}C^*(I - \overline{z}D^*)^{-1}B^*A\\
+&|z|^2C^*(I - \overline{z}D^*)^{-1}B^*B(I - zD)^{-1}C\\
=&|A|^2 - C^*(zD)(I - zD)^{-1}C - C^*(I - \overline{z}D^*)^{-1}(\overline{z}D^*)C\\
+&|z|^2C^*(I - \overline{z}D^*)^{-1}(I - zD)^{-1}C\\
-&|z|^2C^*(I - \overline{z}D^*)^{-1}D^*D(I - zD)^{-1}C \text{ by virtue of equation (3.3.3)}
\end{aligned}$$

Hence,

$$\begin{aligned}
&1-\overline{f(z)}f(z)\\
=&1-|A|^2+C^*(zD)(I-zD)^{-1}C+C^*(I-\overline{z}D^*)^{-1}(\overline{z}D^*)C\\
-&|z|^2C^*(I-\overline{z}D^*)^{-1}(I-zD)^{-1}C\\
+&|z|^2C^*(I-\overline{z}D^*)^{-1}D^*D(I-zD)^{-1}C\\
=&C^*C-C^*C+C^*(I-zD)^{-1}C-C^*C+C^*(I-\overline{z}D^*)^{-1}C\\
-&|z|^2C^*(I-\overline{z}D^*)^{-1}(I-zD)^{-1}C\\
+&C^*(I-\overline{z}D^*)^{-1}(zD)^*(zD)(I-zD)^{-1}C\\
=&C^*(I-zD)^{-1}C-C^*C+C^*(I-\overline{z}D^*)^{-1}C\\
-&|z|^2C^*(I-\overline{z}D^*)^{-1}(I-zD)^{-1}C\\
+&C^*(I-\overline{z}D^*)^{-1}(zD)^*(zD)(I-zD)^{-1}C\\
=&C^*((I-zD)^{-1}-I+(I-\overline{z}D^*)^{-1}-|z|^2(I-\overline{z}D^*)^{-1}(I-zD)^{-1}\\
+&(-I+(I-\overline{z}D^*)^{-1})(-I+(I-zD)^{-1}))C\\
=&(1-|z|^2)C^*(I-\overline{z}D^*)^{-1}(I-zD)^{-1}C
\end{aligned}$$

which is clearly a positive operator (on $\mathbb{C}$, which means it is a scalar). Hence we have proved that $1 \geq |f(z)|$ for all z in $\mathbb{D}$.

It is the other implication that requires the fact that $H^\infty(\mathbb{D})$ is the multiplier algebra of $H^2(\mathbb{D})$. This means that $f \in H^\infty(\mathbb{D})$ if and only if the multiplication operator

$$M_f(\xi)(z) = f(z)\xi(z)$$

is a bounded operator on $H^2(\mathbb{D})$ with $\|M_f\| = \|f\|_\infty$. This was shown in Chapter 1.

For any $n \in \mathbb{N}$, any n scalars $c_1, c_2, \ldots, c_n$ and any n points $w_1, w_2, \ldots, w_n$ in $\mathbb{D}$, consider the vector $\sum_{i=1}^n c_i k(\cdot, w_i)$ where $k(\cdot, w)$ is the function $(1-z\overline{w})^{-1}$. From the definition of the inner product on the Hardy space, we have $\langle f, k(\cdot, w)\rangle = f(w)$ for every f in the Hardy space and every w in $\mathbb{D}$. If f is in the norm unit ball of $H^\infty(\mathbb{D})$, then M_f is a contraction. Hence

$$\|M_f^* \sum_{i=1}^n \overline{c_i} k(\cdot, w_i)\|^2 \leq \|\sum_{i=1}^n \overline{c_i} k(\cdot, w_i)\|^2. \tag{3.3.4}$$

A quick calculation reveals that $M_f^* k(\cdot, w) = \overline{f(w)} k(\cdot, w)$. Indeed,

$$\begin{aligned}&\langle M_f^* k(\cdot, w), \xi\rangle \\ =&\langle k(\cdot, w), M_f \xi\rangle \\ =&\langle k(\cdot, w), f\xi\rangle \\ =&\overline{f(w)\xi(w)} \\ =&\overline{f(w)}\langle k(\cdot, w), \xi\rangle.\end{aligned}$$

Hence (3.3.4) implies that

$$\sum_{i=1}^{n} c_i \bar{c}_j f(w_i)\overline{f(w_j)}(1 - w_i \bar{w}_j)^{-1} \leq \sum_{i=1}^{n} c_i \bar{c}_j (1 - w_i \bar{w}_j)^{-1}.$$

We have used the fact that $\langle k(\cdot, w), k(\cdot, z)\rangle = (1 - z\bar{w})^{-1}$. In other words, $(1 - f(z)\overline{f(w)})(1 - z\bar{w})^{-1}$ is a positive definite kernel on the open unit disc.

By what we learnt in Chapter 1, we can apply GNS construction to this kernel to get a Hilbert space $\mathcal{H}$ and a function $g : \mathbb{D} \to \mathcal{H}$ such that

$$(1 - f(z)\overline{f(w)})(1 - z\bar{w})^{-1} = \langle g(z), g(w)\rangle$$

where the inner product is in $\mathcal{H}$. Thus,

$$1 - f(z)\overline{f(w)} = (1 - z\overline{w})\langle g(z), g(w)\rangle$$

which on rearrangement of terms, gives us

$$1 + \langle zg(z), wg(w)\rangle = f(z)\overline{f(w)} + \langle g(z), g(w)\rangle.$$

Now, define $U : \mathbb{C} \oplus H \to \mathbb{C} \oplus H$ by

$$U\begin{pmatrix} 1 \\ zg(z) \end{pmatrix} = \begin{pmatrix} f(z) \\ g(z) \end{pmatrix}$$

and then extending linearly to the closure of span of vectors of the form $\begin{pmatrix} 1 \\ zg(z) \end{pmatrix}$ as z varies over $\mathbb{D}$. The inner product equality above says that U is an isometry. Now, there is a standard technique of extending U to the whole of $\mathbb{C} \oplus \mathcal{H}$. If the dimension of the orthocomplement of the span mentioned above is not more than the dimension of the orthocomplement of the range of U, then we can extend U isometrically to the whole of $\mathbb{C} \oplus \mathcal{H}$, else we add an infinite dimensional Hilbert space to $\mathcal{H}$ to make the orthocomplements of the same dimension (infinity) and then extend U. Now, writing U as $\left(\begin{smallmatrix} A & B \\ C & D \end{smallmatrix}\right)$, we get

$$A + Bzg(z) = f(z) \text{ and } C + Dzg(z) = g(z).$$

Eliminating g, we get that $f(z) = A + zB(I - zD)^{-1}C$. ■

Such an U which occurs in the realization formula is sometimes called a *unitary colligation*.

3.4. Contractions with finite dimensional defect spaces

This section is based on the work contained in [46].

Recall that a contraction T is called *pure* if $\lim_{m\to 0}\|T^{*m}f\| = 0$ for all $f \in \mathcal{H}$. We have seen that if T is a pure contraction on a Hilbert space $\mathcal{H}$, then $W : \mathcal{H} \to H^2 \otimes \mathcal{D}_{T^*}$ defined by $Wh = \sum_{n=0}^{\infty} z^n \otimes D_{T^*}T^{*n}h$ is an isometry. Moreover, this isometry serves the purpose of embedding $\mathcal{H}$ in $H^2 \otimes \mathcal{D}_{T^*}$ in such a way that the corresponding isometric copy of the operator T has $M_z \otimes I$ as its minimal isometric dilation. In the picture where $H^2 \otimes \mathcal{D}_{T^*}$ is identified with $H^2_{\mathbb{D}}(\mathcal{D}^*_T)$, the isometry W is

$$(Wh)(z) = D_{T^*}(I_{\mathcal{H}} - zT^*)^{-1}h \qquad (z \in \mathbb{D}, h \in \mathcal{H}). \tag{3.4.5}$$

This is straightforward from the fact that the identification between $H^2 \otimes \mathcal{D}_{T^*}$ and $H^2_{\mathbb{D}}(\mathcal{D}^*_T)$ is that $z^n \otimes h$ is mapped to the function $z^n h$ for h in $\mathcal{D}_{T^*}$. The isometry W also has the property that $WT^* = M_z^*W$. In particular, $\mathcal{Q} := \text{ran}W$ is an M_z^*-invariant subspace of $H^2_{\mathbb{D}}(\mathcal{D}^*_T)$ and T is unitarily equivalent to $P_{\mathcal{Q}}M_z|_{\mathcal{Q}}$.

This section is devoted to the study of contractions with finite dimensional defect spaces. Let (T_1, T_2) be a pair of commuting contractions and $\dim \mathcal{D}_{T_j^*} < \infty$, $j = 1, 2$. Since

$$(I_{\mathcal{H}} - T_1T_1^*) + T_1(I_{\mathcal{H}} - T_2T_2^*)T_1^* = T_2(I_{\mathcal{H}} - T_1T_1^*)T_2^* + (I_{\mathcal{H}} - T_2T_2^*),$$

it follows that

$$\|D_{T_1^*}h\|^2 + \|D_{T_2^*}T_1^*h\|^2 = \|D_{T_1^*}T_2^*h\|^2 + \|D_{T_2^*}h\|^2 \quad (h \in \mathcal{H}).$$

Thus

$$U : \{D_{T_1^*}h \oplus D_{T_2^*}T_1^*h : h \in \mathcal{H}\} \to \{D_{T_1^*}T_2^*h \oplus D_{T_2^*}h : h \in \mathcal{H}\}$$

defined by

$$U\left(D_{T_1^*}h, D_{T_2^*}T_1^*h\right) = \left(D_{T_1^*}T_2^*h, D_{T_2^*}h\right) \qquad (h \in \mathcal{H}), \tag{3.4.6}$$

is an isometry. Moreover, since $\dim \mathcal{D}_{T_j^*} < \infty$, $j = 1, 2$, it follows that U extends to a unitary, denoted again by U, on $\mathcal{D}_{T_1^*} \oplus \mathcal{D}_{T_2^*}$. We write this unitary operator in the block matrix form.

$$U := \begin{bmatrix} A & B \\ C & D \end{bmatrix} : \mathcal{D}_{T_1^*} \oplus \mathcal{D}_{T_2^*} \to \mathcal{D}_{T_1^*} \oplus \mathcal{D}_{T_2^*}, \tag{3.4.7}$$

such that (3.4.6) holds.

Let us make an elementary observation.

Lemma 3.4.1. *Let (T_1, T_2) be a pair of commuting contractions on a Hilbert space $\mathcal{H}$. Let T_1 be pure and dim $\mathcal{D}_{T_j^*} < \infty$, $j = 1, 2$. Then with $U = \begin{bmatrix} A & B \\ C & D \end{bmatrix}$ as above we have*

$$D_{T_1^*}T_2^* = AD_{T_1^*} + \sum_{n=0}^{\infty} BD^n CD_{T_1^*}T_1^{*n+1},$$

where the series converges in the strong operator topology.

Proof. For each $h \in \mathcal{H}$ we have

$$\begin{bmatrix} A & B \\ C & D \end{bmatrix} \begin{bmatrix} D_{T_1^*}h \\ D_{T_2^*}T_1^*h \end{bmatrix} = \begin{bmatrix} D_{T_1^*}T_2^*h \\ D_{T_2^*}h \end{bmatrix},$$

that is,

$$D_{T_1^*}T_2^*h = AD_{T_1^*}h + BD_{T_2^*}T_1^*h, \tag{3.4.8}$$

and

$$D_{T_2^*}h = CD_{T_1^*}h + DD_{T_2^*}T_1^*h. \tag{3.4.9}$$

By replacing h by T_1^*h in (3.4.9), we have

$$D_{T_2^*}T_1^*h = CD_{T_1^*}T_1^*h + DD_{T_2^*}T_1^{*2}h.$$

Then by (3.4.8), we have

$$\begin{aligned} D_{T_1^*}T_2^*h &= AD_{T_1^*}h + BD_{T_2^*}T_1^*h \\ &= AD_{T_1^*}h + B(CD_{T_1^*}T_1^*h + DD_{T_2^*}T_1^{*2}h), \end{aligned}$$

that is,

$$D_{T_1^*}T_2^*h = AD_{T_1^*}h + BCD_{T_1^*}T_1^*h + BDD_{T_2^*}T_1^{*2}h. \tag{3.4.10}$$

In (3.4.10), again replacing $D_{T_2^*}T_1^{*2}h$ by $CD_{T_1^*}T_1^{*2}h + DD_{T_2^*}T_1^{*3}h$, we have

$$D_{T_1^*}T_2^*h = AD_{T_1^*}h + BCD_{T_1^*}T_1^*h + BDCD_{T_1^*}T_1^{*2}h + BD^2D_{T_2^*}T_1^{*3}h.$$

Going on in this way, we obtain

$$D_{T_1^*}T_2^*h = AD_{T_1^*}h + \sum_{n=0}^{m} BD^n CD_{T_1^*}T_1^{*n+1}h + (BD^{m+1}D_{T_2^*}T_1^{*m+2}h)$$

for all m in $\mathbb{N}$. By $\|D\| \leq 1$ and

$$\lim_{m\to\infty} T_1^{*m}h = 0,$$

it follows that

$$\lim_{m\to\infty} (BD^m D_{T_2^*}T_1^{*m+1}h) = 0.$$

Finally, the convergence of the desired series follows from the fact that

$$\begin{aligned}\|D_{T_1^*}T_2^*h - AD_{T_1^*}h - \sum_{n=0}^{m} BD^nCD_{T_1^*}T_1^{*n+1}h\| &= \|BD^{m+1}CD_{T_2^*}T_1^{*m+2}h\| \\ &\leq \|T_1^{*m+2}h\|,\end{aligned}$$

for all $m \in \mathbb{N}$ and again that $\lim_{m\to\infty} T_1^{*m}h = 0$. This completes the proof of the lemma. ∎

Remark 3.4.2. *With the above assumptions, the conclusion of Lemma 3.4.1 remains valid if we add the possibility*

$$\dim \mathcal{D}_{T_1^*} = \infty, \quad or \quad \dim \mathcal{D}_{T_2^*} = \infty.$$

To this end, let $\dim \mathcal{D}_{T_1^*} = \infty$*, or* $\dim \mathcal{D}_{T_2^*} = \infty$*. Then there exists an infinite dimensional Hilbert space* $\mathcal{D}$ *such that the isometry*

$$U : \{D_{T_1^*}h \oplus D_{T_2^*}T_1^*h : h \in \mathcal{H}\} \oplus \{0_{\mathcal{D}}\} \to \{D_{T_1^*}T_2^*h \oplus D_{T_2^*}h : h \in \mathcal{H}\} \oplus \{0_{\mathcal{D}}\}$$

defined by

$$U\left(D_{T_1^*}h, D_{T_2^*}T_1^*h, 0_{\mathcal{D}}\right) = \left(D_{T_1^*}T_2^*h, D_{T_2^*}h, 0_{\mathcal{D}}\right) \qquad (h \in \mathcal{H}),$$

extends to a unitary, denoted again by U*, on* $\mathcal{D}_{T_1^*} \oplus (\mathcal{D}_{T_2^*} \oplus \mathcal{D})$*. Now we proceed similarly with the unitary matrix*

$$U = \begin{bmatrix} A & B \\ C & D \end{bmatrix} \in \mathcal{B}(\mathcal{D}_{T_1^*} \oplus (\mathcal{D}_{T_2^*} \oplus \mathcal{D})),$$

to obtain the same conclusion as in Lemma 3.4.1.

Now, an explicit isometric dilation of a pair of commuting tuple of pure contractions can be obtained.

The section on Realization formula provided a formula for a complex valued holomorphic function on the open unit disc with the sup norm over the disc being no greater than 1. A similar argument gives such a formula for operator valued functions as well. First, let us define a *transfer function*. Let $\mathcal{H}_1$ and $\mathcal{H}_2$ be two Hilbert spaces, and

$$U = \begin{bmatrix} A & B \\ C & D \end{bmatrix} \in \mathcal{B}(\mathcal{H}_1 \oplus \mathcal{H}_2),$$

be a unitary operator. Then the $\mathcal{B}(\mathcal{H}_1)$-valued analytic function τ_U on $\mathbb{D}$ defined by

$$\tau_U(z) := A + zB(I - zD)^{-1}C \qquad (z \in \mathbb{D})$$

is called the *transfer function* of U. Using $U^*U = I$, a computation similar to the one for a complex valued transfer function yields

$$I - \tau_U(z)^*\tau_U(z) = (1 - |z|^2)C^*(I - \bar{z}D^*)^{-1}(I - zD)^{-1}C \tag{3.4.11}$$

for all $z \in \mathbb{D}$. Thus, τ_U is an operator valued contractive holomorphic function. Conversely, given any operator valued ($\mathcal{B}(\mathcal{H}_1)$ valued, say) contractive holomorphic function τ on $\mathbb{D}$, there is a Hilbert space $\mathcal{H}_2$ and a unitary operator

$$U = \begin{bmatrix} A & B \\ C & D \end{bmatrix} \in \mathcal{B}(\mathcal{H}_1 \oplus \mathcal{H}_2),$$

such that

$$\tau(z) := A + zB(I - zD)^{-1}C \qquad (z \in \mathbb{D}).$$

The proof is left as an exercise.

Now let $\mathcal{H}_1 = \mathbb{C}^m$ and $\mathcal{H}_2 = \mathbb{C}^n$, and let U be as above. Then τ_U is a contractive rational matrix-valued function on $\mathbb{D}$. The equation (3.4.11) also shows τ_U is unitary on the unit circle $\mathbb{T}$ except possibly at finite many points because D is an operator on a finite dimensional space and hence the spectrum of D is a finite set of points.

A finer analysis can be done here. Suppose D has a unimodular eigenvalue $e^{i\theta}$ with a unit eigenvector x. Then,

$$\langle D^*x, x\rangle = \langle x, Dx\rangle = e^{-i\theta}\|x\|^2$$

and $\|D^*x\|\|x\| \leq 1$ too because D is a contraction. So, equality holds in Cauchy-Schwarz inequality. So, D^*x must be a multiple of x. So, x is an eigenvector of D^* as well with eigenvalue $e^{-i\theta}$. So, the one dimensional space spanned by x is a reducing subspace for D. Thus, up to a unitary equivalence, D is a direct sum of a contraction D' and $e^{i\theta}$. Repeating the process, there is an $n \times n$ unitary matrix W such that $D = W\left(\begin{smallmatrix} \tilde{D} & 0 \\ 0 & E \end{smallmatrix}\right)W^*$ where E is diag$(e^{i\theta_1}, e^{i\theta_2}, \ldots, e^{i\theta_k})$ for some $1 \leq k \leq n$ and $\tilde{D}$ has no unimodular eigenvalue. Thus,

$$U = \begin{pmatrix} A & B \\ C & D \end{pmatrix} = \begin{pmatrix} I & 0 \\ 0 & W \end{pmatrix} \begin{pmatrix} \tilde{A} & \tilde{B} & 0 \\ \tilde{C} & \tilde{D} & 0 \\ 0 & 0 & E \end{pmatrix} \begin{pmatrix} I & 0 \\ 0 & W^* \end{pmatrix}.$$

Note that the other rows and columns corresponding to E have to be zeros to preserve contractivity. The above is equal to

$$\begin{pmatrix} \tilde{A} & (\tilde{B}, \ 0)W^* \\ & \\ W\begin{pmatrix} \tilde{C} \\ 0 \end{pmatrix} & W\begin{pmatrix} \tilde{D} & 0 \\ 0 & E \end{pmatrix}W^* \end{pmatrix}.$$

So,

$$A = \tilde{A}, B = (\tilde{B}, \ 0)W^*, C = W\begin{pmatrix} \tilde{C} \\ 0 \end{pmatrix} \text{ and } D = W\begin{pmatrix} \tilde{D} & 0 \\ 0 & E \end{pmatrix}W^*.$$

Now, putting these in the realization formula, we get

$$\begin{aligned}\tau(z) &= A + z(\tilde{B},\ 0)W^*\left(I - zW\begin{pmatrix}\tilde{D} & 0\\ 0 & E\end{pmatrix}W^*\right)^{-1}W\begin{pmatrix}\tilde{C}\\ 0\end{pmatrix}\\ &= A + z(\tilde{B},0)\begin{pmatrix}(I - z\tilde{D})^{-1} & 0\\ 0 & (I - zE)^{-1}\end{pmatrix}\begin{pmatrix}\tilde{C}\\ 0\end{pmatrix}\\ &= A + z\tilde{B}(I - z\tilde{D})^{-1}\tilde{C}\end{aligned}$$

for $z \in \mathbb{D}$. Thus, if a D with a unimodular eigenvalue occurs in the realization formula, it can be replaced by a $\tilde{D}$ which has no unimodular eigenvalue. Thus, when $\mathcal{H}_1$ and $\mathcal{H}_2$ are finite dimensional, the unitary colligation can be so chosen that the inverses on the right hand side of (3.4.11) exist for all z on the unit circle. Now, we are ready to make the following definition.

Definition 3.4.3. *A matrix valued contractive holomorphic function τ defined on the open unit disc $\mathbb{D}$ is called inner if the boundary values of the function exist everywhere on the unit circle $\mathbb{T}$ and if the boundary values are unitary matrices everywhere on the unit circle.*

Thus τ_U above is a matrix-valued rational inner function.

We now turn our attention to the study of the transfer function of the unitary matrix U^*. Set

$$\Psi(z) := \tau_{U^*}(z) = A^* + zC^*(I - zD^*)^{-1}B^* \qquad (z \in \mathbb{D}). \tag{3.4.12}$$

Then Ψ is a $\mathcal{B}(\mathcal{D}_{T_1^*})$-valued inner function on $\mathbb{D}$. Thus the multiplication operator M_Ψ on $H^2_{\mathbb{D}}(\mathcal{D}_{T_1^*})$ defined by

$$(M_\Psi f)(w) = \Psi(w)f(w) \qquad (w \in \mathbb{D}, f \in H^2_{\mathbb{D}}(\mathcal{D}_{T_1^*}),$$

is an isometry.

We can now prove the Ando dilation for a special pair.

Theorem 3.4.4. *Let (T_1, T_2) be a pair of commuting contractions on a Hilbert space $\mathcal{H}$. Let T_1 be pure and $\dim \mathcal{D}_{T_j^*} < \infty$, $j = 1, 2$. Then there exist an isometry $W : \mathcal{H} \to H^2_{\mathbb{D}}(\mathcal{D}_{T_1^*})$ and an inner function $\Psi \in H^\infty_{\mathbb{D}}(\mathcal{D}_{T_1^*})$ such that*

$$WT_1^* = M_z^* W,$$

and

$$WT_2^* = M_\Psi^* W.$$

In other words,

$$T_1 \cong P_\mathcal{Q} M_z|_\mathcal{Q}, \quad \text{and} \quad T_2 \cong P_\mathcal{Q} M_\Psi|_\mathcal{Q},$$

where $\mathcal{Q} := \operatorname{ran} W$ is a joint (M_z^*, M_Ψ^*)-invariant subspace of $H^2_{\mathbb{D}}(\mathcal{D}_{T_1^*})$.

Proof. Let $W : \mathcal{H} \to H^2_{\mathcal{D}_{T_1^*}}(\mathbb{D})$ be the isometry that embeds the space $\mathcal{H}$ into the minimal isometric dilation space of T_1 as defined in (3.4.5) and let Ψ be as in (3.4.12). Now it is enough to show that W intertwine T_2^* and M_Ψ^*. Let $h \in \mathcal{H}$, $n \geq 0$ and $\eta \in \mathcal{D}_{T_1}$. Then

$$\begin{aligned}
\langle M_\Psi^* W h, z^n\eta\rangle &= \langle Wh, \Psi(z^n\eta)\rangle \\
&= \langle D_{T_1^*}(I_{\mathcal{H}} - zT_1^*)^{-1}h, (A^* + C^*\sum_{q=0}^{\infty} D^{*q}B^*z^{q+1})(z^n\eta)\rangle \\
&= \langle D_{T_1^*}\sum_{p=0}^{\infty} z^pT_1^{*p}h, (A^* + C^*\sum_{q=0}^{\infty} D^{*q}B^*z^{q+1})(z^n\eta)\rangle \\
&= \langle D_{T_1^*}T_1^{*n}h, A^*\eta\rangle + \sum_{q=0}^{\infty}\langle D_{T_1^*}T_1^{*q+n+1}h, C^*D^{*q}B^*\eta\rangle \\
&= \langle AD_{T_1^*}T_1^{*n}h, \eta\rangle + \sum_{q=0}^{\infty}\langle BD^qCD_{T_1^*}T_1^{*q+n+1}h, \eta\rangle \\
&= \langle (AD_{T_1^*} + \sum_{q=0}^{\infty} BD^qCD_{T_1^*}T_1^{*q+1})T_1^{*n}h, \eta\rangle.
\end{aligned}$$

Then by Lemma 3.4.1, we have

$$\langle M_\Psi^* W h, z^n\eta\rangle = \langle D_{T_1^*}T_2^*(T_1^{*n}h), \eta\rangle.$$

Hence

$$\begin{aligned}
\langle WT_2^*h, z^n\eta\rangle &= \langle D_{T_1^*}(I_{\mathcal{H}} - zT_1^*)^{-1}T_2^*h, z^n\eta\rangle \\
&= \langle D_{T_1^*}\sum_{p=0}^{\infty} z^pT_1^{*p}T_2^*h, z^n\eta\rangle \\
&= \langle D_{T_1^*}T_1^{*n}T_2^*h, \eta\rangle \\
&= \langle D_{T_1^*}T_2^*(T_1^{*n}h), \eta\rangle \\
&= \langle M_\Psi^* W h, z^n\eta\rangle.
\end{aligned}$$

This implies

$$M_\Psi^* W = WT_2^*.$$

The second claim is an immediate consequence of the first part. This completes the proof of the theorem. ∎

Theorem 3.4.4 remains valid if we drop the assumption that $\dim \mathcal{D}_{T_j^*} < \infty$, $j = 1, 2$. Indeed, the only change needed in the proof of Theorem 3.4.4 is to replace the transfer function Ψ in (3.4.12) by the transfer function of U^* in Remark 3.4.2. In this case, however, the new transfer function will be a contractive multiplier. Thus we have proved the following dilation result.

Theorem 3.4.5. *Let (T_1, T_2) be a pair of commuting contractions on a Hilbert space $\mathcal{H}$ and let T_1 be a pure contraction. Then there exist an isometry $W : \mathcal{H} \to H^2_{\mathbb{D}}(\mathcal{D}_{T_1^*})$ and a contractive multiplier $\Psi \in H^\infty_{\mathbb{D}}(\mathcal{D}_{T_1^*})$ such that*

$$WT_1^* = M_z^* W,$$

and

$$WT_2^* = M_\Psi^* W.$$

In particular,

$$T_1 \cong P_{\mathcal{Q}} M_z|_{\mathcal{Q}}, \quad \text{and} \qquad T_2 \cong P_{\mathcal{Q}} M_\Psi|_{\mathcal{Q}},$$

where $\mathcal{Q} := \text{ran } W$ is a joint (M_z^*, M_Ψ^*)-invariant subspace of $H^2_{\mathbb{D}}(\mathcal{D}_{T_1^*})$.

This result is motivated by a landmark result of Agler and McCarthy where they dilate a pair of commuting contractions to two commuting isometries defined by co-ordinate multipliers on a functional Hilbert space resembling the Hardy space on the "boundary" of a distinguished variety of the bidisc, see Theorem 3.1 in [3]. In the result above by Das and Sarkar in [46], the model operators act on vector valued Hardy space of the disc. In each case, certain conditions need to be assumed. The Das and Sarkar theorem appeals for its simplicity of proof.

We would like to end this chapter by noting that the characteristic function θ_T defined for a contraction T in Chapter 1 is an example of an operator valued contractive holomorphic function. Recall that for a contraction T, the characteristic function is defined as the $\mathcal{B}(\mathcal{D}_T, \mathcal{D}_{T^*})$ valued function

$$\theta_T(z) = (-T + zD_{T^*}(I - zT^*)^{-1} D_T)|_{\mathcal{D}_T}.$$

The unitary colligation for this function is

$$U = \begin{bmatrix} -T & D_{T^*} \\ D_T & T^* \end{bmatrix} : \mathcal{D}_T \oplus \mathcal{D}_{T^*} \to \mathcal{D}_T \oplus \mathcal{D}_{T^*}.$$

CHAPTER 4

Dilation Theory in Several Variables - the Euclidean Ball

4.1. Contractive tuples

In the first and third chapter, we saw the classical Sz.-Nagy dilation theorem and its two variable generalization. We also saw that an effort of a generalization to three variables in the same vein fails. This chapter thus initiates the discussion of what is achievable in several variables. That leads us to contractive tuples of operators. We shall show that a commuting contractive tuple has a commuting dilation which is canonical in a sense to be described below. We call it the standard commuting dilation. Let d be a fixed positive integer larger than 1.

Definition 4.1.1. *Let $\mathcal{H}$ be a Hilbert space and let $\underline{T} = (T_1, T_2, \ldots, T_d)$ be a tuple of bounded operators acting on $\mathcal{H}$. Then $\underline{T}$ is called a* contractive tuple *if $\sum_{i=1}^d T_iT_i^* \leq I_{\mathcal{H}}$. The tuple is called commuting if $T_iT_j = T_jT_i$ for all $i, j = 1, 2, \ldots, d$. The positive operator $(I_{\mathcal{H}} - \sum_{i=1}^d T_iT_i^*)^{1/2}$ and the closure of its range are respectively called the* Defect operator *of $\underline{T}$ and the* Defect space *of $\underline{T}$ and are denoted by $D_{\underline{T}^*}$ and $\mathcal{D}_{\underline{T}^*}$.*

Contractivity of a tuple is a generalization of contractivity of a single operator because for $d = 1$, the contractivity condition given in the definition above boils down to the contractivity condition of a single operator. Since for a single operator T, the condition $TT^* \leq I$ is equivalent to $\|Th\| \leq \|h\|$ for all vectors h in the Hilbert space, we might ask whether a similar condition holds here in several variables. The answer is yes. For a tuple, the contractivity condition equivalent to demanding that for all $h_1, h_2, \ldots, h_d \in \mathcal{H}$,

$$\|T_1h_1 + T_2h_2 + \cdots + T_dh_d\|^2 \leq \|h_1\|^2 + \|h_2\|^2 \cdots + \|h_d\|^2.$$

This is easy to see in the following way. Denote by T the operator from $\mathcal{H}^d$ to $\mathcal{H}$ which sends a tuple $(h_1, h_2, \ldots, h_d)$ to $T_1h_1 + T_2h_2 + \cdots + T_dh_d$.

Then for $h \in \mathcal{H}$,

$$\left\langle T^*h, \begin{pmatrix} h_1 \\ h_2 \\ \vdots \\ h_d \end{pmatrix} \right\rangle = \left\langle h, T \begin{pmatrix} h_1 \\ h_2 \\ \vdots \\ h_d \end{pmatrix} \right\rangle = \langle h, T_1h_1 + T_2h_2 + \cdots + T_dh_d \rangle$$

and

$$\begin{aligned} \left\langle \begin{pmatrix} T_1^*h \\ T_2^*h \\ \vdots \\ T_d^*h \end{pmatrix}, \begin{pmatrix} h_1 \\ h_2 \\ \vdots \\ h_d \end{pmatrix} \right\rangle &= \langle T_1^*h, h_1 \rangle + \langle T_2^*h, h_2 \rangle + \cdots + \langle T_d^*h, h_d \rangle \\ &= \langle h, T_1h_1 + T_2h_2 + \cdots + T_dh_d \rangle. \end{aligned}$$

Hence the adjoint of the operator T is given by

$$T^*h = \begin{pmatrix} T_1^*h \\ T_2^*h \\ \vdots \\ T_d^*h \end{pmatrix}.$$

Thus $TT^* = \sum T_iT_i^*$. So $\underline{T}$ is a contractive tuple if and only if T as an operator is a contraction. Since a contractive tuple $\underline{T}$ acts as a row of operators from $\mathcal{H}^d$ to $\mathcal{H}$ and since this row operator is a contraction, $\underline{T}$ is sometimes called a *row contraction*.

For prototypical examples of contractive tuples, we need to discuss Fock spaces.

4.2. The full Fock space and the symmetric Fock space

The beginning of this section warrants a short discussion about the *symmetric tensor product* of Hilbert spaces because it will be used in defining the symmetric Fock space. For $k > 1$, let σ_k denote the permutation group of k symbols, i.e., σ_k is the set of all bijections of the set $\{1, 2, \ldots, k\}$. Consider the following unitary representation of the permutation group σ_k on the full tensor product space $\mathcal{H}^{\otimes k}$. It is defined on elementary tensors by

$$U_\pi(x_1 \otimes x_2 \otimes \cdots \otimes x_k) = x_{\pi^{-1}(1)} \otimes x_{\pi^{-1}(2)} \otimes \cdots \otimes x_{\pi^{-1}(k)}, \tag{4.2.1}$$

where the x_i are elements of $\mathcal{H}$ and π belongs to σ_k. It is left to the reader to verify that $\pi \to U_\pi$ is a unitary representation of σ_k. Define then the symmetric tensor product of k copies of $\mathcal{H}$ by

$$\mathcal{H}^{\circledS k} = \{\zeta \in \mathcal{H}^{\otimes k} : U_\pi\zeta = \zeta, \quad \pi \in \sigma_k\}.$$

So for $k \geq 1$, the symmetric tensor product $\mathcal{H}^{ⓢk}$ is the subspace of the full tensor product $\mathcal{H}^{\otimes k}$ consisting of all vectors fixed under the representation of the permutation group. Of course, $\mathcal{H}^{ⓢ1} = \mathcal{H}^{\otimes 1} = \mathcal{H}$. The spaces $\mathcal{H}^{\otimes 0}$ and $\mathcal{H}^{ⓢ0}$ are defined as the one dimensional vector space $\mathbb{C}$ with its usual inner product. In the following, we show that $\mathcal{H}^{ⓢk}$ is the linear span of $\{x^{\otimes k} : x \in \mathcal{H}\}$ where the notation $x^{\otimes k}$ stands for $x \otimes x \otimes \ldots \otimes x$, the k fold tensor product of the vector x with itself.

Lemma 4.2.1. *For any Hilbert space $\mathcal{H}$ and $k \in \mathbb{N}$,*

$$\mathcal{H}^{ⓢk} = \overline{\textit{span}}\ \{x^{\otimes k} : x \in \mathcal{H}\}.$$

Proof. Clearly any vector of the form $x^{\otimes k}$ is symmetric. The totality of such vectors follows easily from the following: For any $y_1, y_2, \ldots, y_k$ in $\mathcal{H}$,

$$\sum_{\pi \in \sigma_k} U_\pi(y_1 \otimes y_2 \otimes \cdots \otimes y_k)$$
$$= \frac{1}{2^k} \sum_{\{a_j = \pm 1, 1 \leq j \leq k\}} a_1 a_2 \ldots a_k (a_1 y_1 + a_2 y_2 + \cdots + a_k y_k)^{\otimes k}.$$

The proof of this statement is easy combinatorics based on the observation that in the expansion of the tensor power on the right, only those terms with distinct indices survive. ■

Definition 4.2.2. *Given a Hilbert space $\mathcal{H}$, the full Fock space over $\mathcal{H}$ is defined to be the infinite Hilbert space direct sum*

$$\Gamma(\mathcal{H}) = \mathbb{C} \oplus \mathcal{H} \oplus \mathcal{H}^{\otimes 2} \oplus \cdots \oplus \mathcal{H}^{\otimes k} \oplus \cdots .$$

By definition, this direct sum consists of all $\xi = (\xi_0, \xi_1, \xi_2, \ldots)$ such that $\sum \|\xi_k\|^2$ is finite. Note that ξ_k is in $\mathcal{H}^{\otimes k}$, the so-called *k-particle space.* The inner product on the full Fock space is defined as

$$\langle \xi, \eta \rangle = \sum \langle \xi_k, \eta_k \rangle.$$

Definition 4.2.3. *The symmetric Fock space over $\mathcal{H}$ is defined as*

$$\Gamma_s(\mathcal{H}) = \mathbb{C} \oplus \mathcal{H} \oplus \mathcal{H}^{ⓢ2} \oplus \cdots \oplus \mathcal{H}^{ⓢk} \oplus \cdots .$$

This is a closed subspace of the full Fock space because $\mathcal{H}^{ⓢk}$ is a closed subspace of $\mathcal{H}^{\otimes k}$ for $k = 1, 2, \ldots$. In both of the Fock spaces, the one dimensional subspace $\mathbb{C} \oplus \{0\} \oplus \{0\} \oplus \cdots$ is called the *vacuum space*. The unit norm element $(1, 0, 0, \ldots)$ is called the *vacuum vector* and is denoted by ω. The projection onto the vaccum space is denoted by E_0. Denote by P_s the orthogonal projection from $\Gamma(\mathcal{H})$ onto the closed subspace $\Gamma_s(\mathcal{H})$.

The action of the operator P_s on the space $\mathcal{H}^{\otimes k}$ is

$$P_s = \frac{1}{k!} \sum_{\pi \in \sigma_k} U_\pi.$$

Needless to say, this action of P_s on $\mathcal{H}^{\otimes k}$ also depends on k, but we have suppressed that to maintain brevity. Let us note that $U_\pi^* = U_{\pi^{-1}}$ since $\pi \to U_\pi$ is a unitary representation. This immediately leads to

$$P_s^* = \frac{1}{k!} \sum_{\pi \in \sigma_k} U_\pi^* = \frac{1}{k!} \sum_{\pi \in \sigma_k} U_{\pi^{-1}} = P_s.$$

Moreover, $P_s U_\pi = P_s = U_\pi P_s$. This gives

$$P_s^2 = P_s \frac{1}{k!} \sum_{\pi \in \sigma_k} U_\pi = \frac{1}{k!} \sum_{\pi \in \sigma_k} P_s U_\pi = \frac{1}{k!} \sum_{\pi \in \sigma_k} P_s = P_s.$$

So, P_s is a projection. Note that if we pick an element from $\mathcal{H}^{\circledS k}$, then P_s leaves it unchanged because each U_π leaves it unchanged. Conversely, if P_s leaves an element $\underline{x}$ of $\mathcal{H}^{\otimes k}$ unchanged, then for any π in σ_k, we have $U_\pi \underline{x} = U_\pi P_s \underline{x} = P_s \underline{x} = \underline{x}$. Thus, P_s is the projection onto $\mathcal{H}^{\circledS k}$.

It is useful to note that for x_i and y_i in $\mathcal{H}$, we have

$$\begin{aligned}
&\langle P_s(x_1 \otimes x_2 \otimes \ldots \otimes x_k), P_s(y_1 \otimes y_2 \otimes \ldots \otimes y_k)\rangle \\
=&\langle P_s(x_1 \otimes x_2 \otimes \ldots \otimes x_k), y_1 \otimes y_2 \otimes \ldots \otimes y_k\rangle \\
=&\frac{1}{k!} \sum_{\pi \in \sigma_k} \prod_{i=1}^{k} \langle x_i, y_{\pi(i)}\rangle.
\end{aligned}$$

The last quantity is called the *permanent* of the matrix $((\langle x_i, y_j\rangle))$, see page 17 of [31] and page 108 of [87] for more details on permanent and its relation to the symmetric Fock space.

We shall now consider $\Gamma(\mathbb{C}^d)$ and $\Gamma_s(\mathbb{C}^d)$, the full Fock space and the symmetric Fock space over $\mathbb{C}^d$ respectively. Let $\{e_1, e_2, \ldots, e_d\}$ be an orthonormal basis of $\mathbb{C}^d$. On the full Fock space $\Gamma(\mathbb{C}^d)$, define the *creation operator tuple* $\underline{V} = (V_1, V_2, \ldots, V_d)$ by

$$V_i \xi = e_i \otimes \xi \text{ for } i = 1, 2, \ldots, d \text{ and } \xi \in \Gamma(\mathbb{C}^d).$$

Lemma 4.2.4. *The operators V_i are isometries with orthogonal ranges.*

Proof: We compute the action of V_i^* on the orthonormal basis elements. Let $k \geq 1$ and let $i_1, i_2, \ldots, i_d$ be from $\{1, 2, \ldots, d\}$. Then,

$$\begin{aligned} \langle V_i^*(e_{i_1} \otimes e_{i_2} \otimes \cdots \otimes e_{i_k}), \xi \rangle &= \langle e_{i_1} \otimes e_{i_2} \otimes \cdots \otimes e_{i_k}, V_i \xi \rangle \\ &= \langle e_{i_1} \otimes e_{i_2} \otimes \cdots \otimes e_{i_k}, e_i \otimes \xi \rangle \\ &= \begin{cases} \langle e_{i_2} \otimes \cdots \otimes e_{i_d}, \xi \rangle & \text{if } i_1 = i \\ 0 & \text{if } i_1 \neq i. \end{cases} \end{aligned}$$

Thus

$$V_i^*(e_{i_1} \otimes e_{i_2} \otimes \cdots \otimes e_{i_k}) = \begin{cases} e_{i_2} \otimes \cdots \otimes e_{i_k} & \text{if } i_1 = i \\ 0 & \text{if } i_1 \neq i. \end{cases} \tag{4.2.2}$$

From this, it is clear that $V_i^* V_j = \delta_{ij}$ where δ_{ij} is the Kronecker delta. Hence we are done. ∎

Lemma 4.2.5. *$I_{\Gamma(\mathbb{C}^d)} - \sum_{i=1}^d V_i V_i^*$ is the one-dimensional projection onto the vacuum space.*

Proof: From the calculation of the adjoint of the V_i, it follows that $\sum V_i V_i^*(e_{i_1} \otimes e_{i_2} \otimes \cdots \otimes e_{i_k}) = e_{i_1} \otimes e_{i_2} \otimes \cdots \otimes e_{i_k}$ for any $k \geq 1$ and $1 \leq i_1, \ldots, i_k \leq d$.

Now

$$\langle V_i^* \omega, \xi \rangle = \langle \omega, e_i \otimes \xi \rangle = 0 \text{ for any } \xi \in \Gamma(\mathbb{C}^d) \text{ and any } i.$$

Thus $\sum V_i V_i^*(\omega) = 0$ and hence $I_{\Gamma(\mathbb{C}^d)} - \sum V_i V_i^*$ is the 1-dimensional projection onto the vacuum space. ∎

Define the tuple of bounded operators $\underline{S} = (S_1, S_2, \ldots, S_d)$ on $\Gamma_s(\mathbb{C}^d)$ by

$$S_i \xi = P_s(e_i \otimes \xi) \text{ for } i = 1, 2, \ldots, d \text{ and } \xi \in \Gamma_s(\mathbb{C}^d).$$

Since V_i's are isometries, the S_i's are contractions. They are symmetrized creation operators and we shall see that they commute.

An orthonormal basis for the full tensor product space $(\mathbb{C}^d)^{\otimes k}$ is $\{e_{i_1} \otimes e_{i_2} \otimes \cdots \otimes e_{i_k} : 1 \leq i_1, i_2, \ldots, i_k \leq d\}$. We wish to construct an orthonormal basis for the symmetric tensor product space $(\mathbb{C}^d)^{\circledS k}$. We need some notations for that. Any ordered d-tuple of non-negative integers $\underline{k} = (k_1, k_2, \ldots, k_d)$ will be called a $multi-index$. Let $|\underline{k}| = k_1 + k_2 + \cdots + k_d$. Given such a multi-index $\underline{k}$ with $|\underline{k}| \geq 1$, we shall, for brevity, write

$$\underline{e}^{\underline{k}} = P_s(e_1^{\otimes k_1} \otimes e_2^{\otimes k_2} \otimes \cdots \otimes e_d^{\otimes k_d}).$$

Here, we have used the notation that if $\underline{k} = (0, 0, \ldots, 0)$, then $\underline{e}^{\underline{k}}$ is the vacuum vector ω.

Lemma 4.2.6. *The norm of* $\underline{e}^{\underline{k}}$ *in the symmetric Fock space* $\Gamma_s(\mathbb{C}^d)$ *is the square root of the reciprocal of the multinomial coefficient, that is,*

$$\|\underline{e}^{\underline{k}}\| = \left(\frac{k_1!k_2!\ldots k_d!}{|\underline{k}|!}\right)^{1/2}.$$

Moreover,

$$\{\left(\frac{|\underline{k}|!}{k_1!k_2!\ldots k_d!}\right)^{1/2} \underline{e}^{\underline{k}} : \underline{k} \text{ is a multi-index}\}$$

is an orthonormal basis for $\Gamma_s(\mathbb{C}^d)$.

Proof. Let $p \geq 1$. With any a in $\mathbb{C}^d$, we associate an operator $A : (\mathbb{C}^d)^{ⓢ p-1} \to (\mathbb{C}^d)^{ⓢ p}$ as follows. For any $y_1, y_2, \ldots, y_p$ in $\mathbb{C}^d$, the action of the operator is

$$A(P_s(y_1 \otimes y_2 \otimes \cdots \otimes y_{p-1})) = P_s(a \otimes y_1 \otimes y_2 \otimes \cdots \otimes y_{p-1}).$$

We shall prove that the adjoint $A^* : (\mathbb{C}^d)^{ⓢ p} \to (\mathbb{C}^d)^{ⓢ p-1}$ is given by

$$A^*(P_s(y_1 \otimes y_2 \otimes \cdots \otimes y_p)) = \frac{1}{p}\sum_{i=1}^{p} \langle y_i, a\rangle P_s(y_1 \otimes \ldots \otimes y_{i-1} \otimes y_{i+1} \otimes \ldots \otimes y_p).$$

If ζ denotes the right side above, then for $x \in \mathbb{C}^d$, we have

$$\begin{aligned}
&\langle \zeta, x^{\otimes p-1}\rangle \\
=&\frac{1}{p}\sum_{i=1}^{p} \langle y_1, x\rangle \ldots \langle y_{i-1}, x\rangle \langle y_i, a\rangle \langle y_{i+1}, x\rangle \ldots \langle y_p, x\rangle \\
=&\langle y_1 \otimes y_2 \otimes \cdots \otimes y_p, \frac{1}{p}(a \otimes x^{\otimes p-1} + x \otimes a \otimes x^{\otimes p-2} + \cdots + x^{\otimes p-1} \otimes a)\rangle \\
=&\langle y_1 \otimes y_2 \otimes \cdots \otimes y_p, P_s(a \otimes x^{\otimes p-1})\rangle \\
=&\langle P_s(y_1 \otimes y_2 \otimes \cdots \otimes y_p), P_s(a \otimes x^{\otimes p-1})\rangle \\
=&\langle P_s(y_1 \otimes y_2 \otimes \cdots \otimes y_p), Ax^{\otimes p-1}\rangle = \langle A^* P_s(y_1 \otimes y_2 \otimes \cdots \otimes y_p), x^{\otimes p-1}\rangle.
\end{aligned}$$

Since, by Lemma 4.2.1, the vectors $x^{ⓢ p-1}$ span $(\mathbb{C}^d)^{ⓢ p-1}$, the formula for the adjoint of A is established. We used above that

$$P_s(a \otimes x^{\otimes p-1}) = \frac{1}{p}(a \otimes x^{\otimes p-1} + x \otimes a \otimes x^{\otimes p-2} + \cdots + x^{\otimes p-1} \otimes a).$$

This is a small exercise left for the reader.

It is now obvious that for any multi-index $\underline{k}$ with $|\underline{k}| \geq 1$, we have

$$S_i^* \underline{e}^{\underline{k}} = \begin{cases} 0 \text{ if } k_i = 0 \\ \frac{k_i}{|\underline{k}|} \underline{e}^{\underline{k}-\delta_i} \text{ otherwise,} \end{cases} \tag{4.2.3}$$

where $\underline{k} - \delta_i$ is the multi-index $(k_1, \dots k_{i-1}, k_i - 1, k_{i+1}, \dots, k_d)$. This directly follows by taking $a = e_i$.

Hence, we have a recursive formula as follows. If $|\underline{k}| \geq 1$, choose i so that $k_i \geq 1$. Let $\underline{k} - \delta_i$ be as above. Then,

$$\begin{aligned}\|\underline{e}^{\underline{k}}\|^2 = \langle \underline{e}^{\underline{k}}, \underline{e}^{\underline{k}} \rangle = \langle \underline{e}^{\underline{k}}, S_i \underline{e}^{\underline{k}-\delta_i} \rangle &= \langle S_i^* \underline{e}^{\underline{k}}, \underline{e}^{\underline{k}-\delta_i} \rangle \\ &= \frac{k_i}{|\underline{k}|} \langle \underline{e}^{\underline{k}-\delta_i}, \underline{e}^{\underline{k}-\delta_i} \rangle = \frac{k_i}{|\underline{k}|} \|\underline{e}^{\underline{k}-\delta_i}\|^2.\end{aligned}$$

Now, an induction argument finishes the proof since the norm formula obviously holds for $|\underline{k}| = 0$. ■

We have used above that $S_i \underline{e}^{\underline{k}-\delta_i} = \underline{e}^{\underline{k}}$. This is left as an exercise.

The formula for S_i^* obtained above immediately gives a proof of the fact that $\sum S_i S_i^* = I_{\Gamma_s(\mathbb{C}^d)} - E_0$ because $\sum S_i S_i^*(\omega) = 0$ and for any $\underline{k}$ with $|\underline{k}| \geq 1$, $\sum S_i S_i^*(\underline{e}^{\underline{k}}) = (|\underline{k}|)^{-1} \sum k_i (\underline{e}^{\underline{k}}) = \underline{e}^{\underline{k}}$.

Lemma 4.2.7. *$\underline{S}$ forms a commuting tuple.*

Proof. For $i \neq j$, by repeated applications of the formula (4.2.3), we get that for $|\underline{k}| \geq 2$,

$$S_j^* S_i^* \underline{e}^{\underline{k}} = \begin{cases} 0 & \text{if } k_i = 0 \text{ or } k_j = 0; \\ \frac{k_i k_j}{|\underline{k}|(|\underline{k}|-1)} \underline{e}^{\underline{k}-\delta_i-\delta_j} & \text{otherwise.} \end{cases}$$

Also, if $|\underline{k}| \leq 1$, then $S_j^* S_i^* \underline{e}^{\underline{k}} = 0$. From this it is clear that S_i^* and S_j^* commute. Consequently S_i and S_j commute for any i, j. ■

Definition 4.2.8. *The operator tuple $\underline{S}$ is called the commuting d-shift.*

Lemma 4.2.9. *Both $\underline{V}$ and $\underline{S}$ are contractive tuples.*

Proof. We have seen above that both $I_{\Gamma(\mathbb{C}^d)} - \sum V_i V_i^*$ and $I_{\Gamma_s(\mathbb{C}^d)} - \sum S_i S_i^*$ are projections and hence are positive operators. ■

Lemma 4.2.10. *$\sum_{i=1}^d S_i^* S_i$ is an invertible operator on $\Gamma_s(\mathbb{C}^d)$.*

Proof. We shall show that $\sum S_i^* S_i$ is a diagonal operator in the natural basis for $\Gamma_s(\mathbb{C}^d)$ mentioned above. Since we have a formula for the action of S_i^* on the basis, $\sum S_i^* S_i \omega = d\omega$ and for $|\underline{k}| \geq 1$,

$$\sum_{i=1}^{d} S_i^* S_i(\underline{e}^{\underline{k}}) = \sum S_i^*(\underline{e}^{\underline{k}+\delta_i})$$

where $\underline{k} + \delta_i$ is the multi-index $(k_1, \dots k_{i-1}, k_i + 1, k_{i+1}, \dots k_d)$. Thus,

$$\sum_{i=1}^{d} S_i^* S_i(\underline{e}^{\underline{k}}) = \sum \frac{k_i + 1}{|\underline{k}| + 1} \underline{e}^{\underline{k}} = \frac{|\underline{k}| + d}{|\underline{k}| + 1} \underline{e}^{\underline{k}}.$$

Thus $\sum S_i^* S_i$ is a diagonal operator in the basis mentioned above. Since none of the diagonal coefficients is zero and the sequence of diagonal coefficients tends to 1 as $|\underline{k}| \to \infty$, the operator is invertible. Indeed, $\sum S_i^* S_i \geq I$. ■

A good reference for the full Fock space and the symmetric Fock space is Parthasarathy [87].

4.3. The Toeplitz C^*-algebra

The C^*-subalgebra of $\mathcal{B}(\Gamma_s(\mathbb{C}^d))$ generated by $S_1, \dots, S_d$ will be denoted by $\mathcal{T}_d^c$ and called the *commutative Toeplitz C^*-algebra*. The commutative Toeplitz C^*-algebra is unital. We do not have to, a priori, include $I_{\Gamma_s(\mathbb{C}^d)}$ in the C^*-algebra $\mathcal{T}_d^c$ because the operator $\sum S_i^* S_i$ is invertible in $\mathcal{B}(\Gamma_s(\mathbb{C}^d))$. Since C^*-algebras are inverse closed, $(\sum S_i^* S_i)^{-1}$ is in the C^*-algebra generated by $S_1, \dots, S_d$ and hence $\mathcal{T}_d^c$ is unital. The subalgebra of $\mathcal{T}_d^c$ consisting of polynomials in $S_1, \dots, S_d$ and $I_{\Gamma_s(\mathbb{C}^d)}$ will be denoted by $\mathcal{A}_d^c$.

For a multi-index $\underline{k}$, the symbol $\underline{S}^{\underline{k}}$ denotes the operator $S_1^{k_1} S_2^{k_2} \dots S_d^{k_d}$.

Lemma 4.3.1. *All compact operators are in $\overline{span}\mathcal{A}_d^c(\mathcal{A}_d^c)^*$.*

Proof. Since $\sum S_i S_i^* = I_{\Gamma_s(\mathbb{C}^d)} - E_0$, the one-dimensional projection E_0 onto the vacuum space is in $\text{span}\mathcal{A}_d^c(\mathcal{A}_d^c)^*$. Now given any two multi-indices $\underline{k}$ and $\underline{l}$, the operator $\underline{S}^{\underline{k}} E_0 \underline{S}^{*\underline{l}}$ is in $\text{span}\mathcal{A}_d^c(\mathcal{A}_d^c)^*$. But note that this operator is nothing but the rank-one operator

$$\xi \to \langle \xi, \underline{e}^{\underline{l}} \rangle \underline{e}^{\underline{k}}.$$

As the set $\{\omega, \underline{e}^{\underline{k}} : \underline{k}$ is any multi-index $\}$ is a basis for $\Gamma_s(\mathbb{C}^d)$, all rank one operators are in $\overline{\text{span}}\mathcal{A}_d^c(\mathcal{A}_d^c)^*$. Thus $\overline{\text{span}}\mathcal{A}_d^c(\mathcal{A}_d^c)^*$ contains all finite rank operators and hence all compact operators. ■

Lemma 4.3.2. $\mathcal{T}_d^c = \overline{span}\mathcal{A}_d^c(\mathcal{A}_d^c)^*$.

Proof. The proof requires computing the operator $S_i^* S_j - S_j S_i^*$ for $1 \le i, j \le d$ on the canonical basis $\underline{e}^{\underline{k}}$. First, let $i = j$. Recall from (4.2.3) that for any multi-index $\underline{k}$, we have

$$S_i^* \underline{e}^{\underline{k}} = \frac{k_i}{|\underline{k}|} \underline{e}^{\underline{k} - \delta_i}$$

where $\underline{k} - \delta_i$ is the multi-index $(k_1, \dots k_{i-1}, k_i - 1, k_{i+1}, \dots, k_d)$. This immediately leads to

$$(S_i^* S_i - S_i S_i^*) \underline{e}^{\underline{k}} = \frac{|\underline{k}| - k_i}{|\underline{k}|(|\underline{k}| + 1)} \underline{e}^{\underline{k}}.$$

Thus, $S_i^* S_i - S_i S_i^*$ is a diagonal operator in this basis and the diagonal entries converge to 0 as $|\underline{k}| \to \infty$. So, it is a compact operator. By the lemma above, $S_i^* S_i - S_i S_i^*$ and hence $S_i^* S_i$ is in $\overline{span} \mathcal{A}_d^c (\mathcal{A}_d^c)^*$.

Introduce an auxilliary compact operator A which is defined as

$$\underline{e}^{\underline{k}} \to \frac{1}{|\underline{k}| + 1} \underline{e}^{\underline{k}}.$$

A direct computation now yields that for $i \neq j$, we have

$$(S_i^* S_j - S_j S_i^*) \underline{e}^{\underline{k}} = \frac{-k_i}{|\underline{k}|(|\underline{k}| + 1)} \underline{e}^{\underline{k} - \delta_i + \delta_j} = -A S_j S_i^* \underline{e}^{\underline{k} - \delta_i + \delta_j}$$

giving the operator equality

$$S_i^* S_j - S_j S_i^* = -A S_j S_i^*.$$

Now, the right hand side is compact and hence is in $\overline{span} \mathcal{A}_d^c (\mathcal{A}_d^c)^*$, so we are done. ■

We now proceed towards developing the model and dilation for a given commuting contractive tuple $\underline{T}$ on a Hilbert space $\mathcal{H}$. By an operator space we shall mean a vector subspace of $\mathcal{B}(\mathcal{L})$ where $\mathcal{L}$ is a Hilbert space. Completely positive maps can be defined on operator spaces in the same way they are defined on C^*-algebras. Given an operator space $\mathcal{E}$ and an algebra $\mathcal{A} \subseteq \mathcal{E}$, a completely positive map φ from $\mathcal{E}$ to $\mathcal{B}(\mathcal{H})$ for some Hilbert space $\mathcal{H}$ is called an $\mathcal{A} - morphism$ if

$$\varphi(AX) = \varphi(A)\varphi(X), \text{ for every } A \in \mathcal{A} \text{ and } X, AX \in \mathcal{E}.$$

Every unital $\mathcal{A}_d^c$-morphism $\varphi : \mathcal{T}_d^c \to \mathcal{B}(\mathcal{H})$ for some Hilbert space $\mathcal{H}$ gives rise to a commuting contractive tuple $(T_1, T_2, \dots, T_d)$ on $\mathcal{H}$ by way of $T_i = \varphi(S_i)$, $i = 1, 2, \dots, d$. Indeed, $\sum T_i T_i^* = \varphi(\sum S_i S_i^*) \le \varphi(I_{\Gamma_s(\mathbb{C}^d)}) = I_{\mathcal{H}}$ and

$$T_i T_j = \varphi(S_i)\varphi(S_j) = \varphi(S_i S_j) = \varphi(S_j S_i) = \varphi(S_j)\varphi(S_i) = T_j T_i$$

for all $1 \leq i, j \leq d$. Given any commuting contractive tuple $\underline{T}$ acting on $\mathcal{H}$, our aim is to produce an $\mathcal{A}_d^c$-morphism from the C^*-algebra $\mathcal{T}_d^c$ (with its subalgebra $\mathcal{A}_d^c$ as defined above) to $\mathcal{B}(\mathcal{H})$. This will be achieved by the help of the following crucial thorem. The $\mathcal{A}_d^c$-morphism is the key element in finding dilation and proving von Neumann inequality. In fact, it could be thought of as the model for $\underline{T}$.

We shall associate a completely positive map $P_{\underline{T}}$ with $\underline{T}$ which acts on $\mathcal{B}(\mathcal{H})$ by

$$P_{\underline{T}}(X) = \sum_{i=1}^{d} T_i X T_i^*.$$

Since $\underline{T}$ is a contractive tuple, the completely positive map $P_{\underline{T}}$ is contractive and hence $I_{\mathcal{H}} \geq P_{\underline{T}}(I_{\mathcal{H}}) \geq P_{\underline{T}}^2(I_{\mathcal{H}}) \geq \cdots$. This decreasing sequence of positive contractions converges strongly and T_∞ will denote the positive contraction which is the strong limit:

$$T_\infty = \lim_{m\to\infty} P_{\underline{T}}^m(I_{\mathcal{H}}).$$

A detailed analysis of such limits has been carried out later in Section 4.6. The commuting contractive tuple $\underline{T}$ will be called *pure* if $T_\infty = 0$.

Theorem 4.3.3. *Let $(T_1, T_2, \ldots, T_d)$ be a commuting contractive tuple of operators on a Hilbert space $\mathcal{H}$. Then there is a unique bounded operator $L : \Gamma_s(\mathbb{C}^d) \otimes \mathcal{D}_{\underline{T}^*} \to \mathcal{H}$ satisfying $L(\omega \otimes \xi) = D_{\underline{T}^*}\xi$ and*

$$L(\underline{e}^{\underline{k}} \otimes \xi) = \underline{T}^{\underline{k}} D_{\underline{T}^*}\xi \tag{4.3.1}$$

for every multi-index $\underline{k}$ with $|\underline{k}| = 1, 2, \ldots$. In general $\|L\| \leq 1$, and if $(T_1, T_2, \ldots, T_d)$ is a pure tuple, then L is a coisometry: $LL^* = I_{\mathcal{H}}$.

Proof. A bounded operator L satisfying (4.3.1) is obviously unique because $\Gamma_s(\mathbb{C}^d)$ is spanned by the set of vectors

$$\{\omega, \underline{e}^{\underline{k}} : \underline{k} \text{ is a multi-index with } |\underline{k}| = 1, 2, \ldots\}.$$

We define L by exhibiting its adjoint, i.e., we shall exhibit a contraction $A : \mathcal{H} \to \Gamma_s(\mathbb{C}^d) \otimes \mathcal{D}_{\underline{T}^*}$, define $L = A^*$ and then show that L has the required properties.

For every $h \in \mathcal{H}$, let Ah be the sequence of vectors $(\zeta_0, \zeta_1, \zeta_2, \ldots)$ where $\zeta_l \in (\mathbb{C}^d)^{\otimes l} \otimes \mathcal{D}_{\underline{T}^*}$ is defined by

$$\zeta_l = \sum_{i_1,\ldots,i_l=1}^{d} e_{i_1} \otimes \cdots \otimes e_{i_l} \otimes D_{\underline{T}^*} T_{i_l}^* \ldots T_{i_1}^* h$$

for $l \geq 1$ and $\zeta_0 = \omega \otimes D_{\underline{T}^*}h$. Notice that since $T_1^*, \dots, T_d^*$ commute, ζ_l actually belongs to the symmetric subspace $(\mathbb{C}^d)^{\circledS l} \otimes \mathcal{D}_{\underline{T}^*}$ so that in fact A maps into $\Gamma_s(\mathbb{C}^d) \otimes \mathcal{D}_{\underline{T}^*}$. To show that A is a contraction we claim that

$$\sum_{l=0}^{\infty} \|\zeta_l\|^2 \leq \|h\|^2.$$

Indeed,

$$\|\zeta_l\|^2 = \sum_{i_1,\dots,i_l=1}^{d} \|D_{\underline{T}^*}T_{i_l}^* \dots T_{i_1}^* h\|^2 = \sum_{i_1,\dots,i_l=1}^{d} \langle T_{i_1} \dots T_{i_l} D_{\underline{T}^*}^2 T_{i_l}^* \dots T_{i_1}^* h, h\rangle.$$

Noting that $D_{\underline{T}^*}^2 = I_{\mathcal{H}} - P_{\underline{T}}(I_{\mathcal{H}})$ we find that

$$\sum_{i_1,\dots,i_l=1}^{d} T_{i_1} \dots T_{i_l} D_{\underline{T}^*}^2 T_{i_l}^* \dots T_{i_1}^* = P_{\underline{T}}^l(I_{\mathcal{H}} - P(I_{\mathcal{H}})) = P_{\underline{T}}^l(1_{\mathcal{H}}) - P_{\underline{T}}^{l+1}(I_{\mathcal{H}}),$$

and hence

$$\|\zeta_l\|^2 = \langle P_{\underline{T}}^l(I_{\mathcal{H}})h, h\rangle - \langle P_{\underline{T}}^{l+1}(I_{\mathcal{H}})h, h\rangle.$$

The series $\|\zeta_0\|^2 + \|\zeta_1\|^2 + \dots$ therefore telescopes and we are left with

$$\sum_{l=0}^{\infty} \|\zeta_l\|^2 = \|h\|^2 - \langle T_\infty h, h\rangle \leq \|h\|^2.$$

Now let $\zeta = \underline{e}^{\underline{k}} \otimes \xi$ for some multi-index $\underline{k}$ with $|\underline{k}| = m \geq 1$ and $\xi \in \mathcal{D}_{\underline{T}^*}$. Then for any $h \in \mathcal{H}$,

$$\begin{aligned}
&\langle L(\zeta), h\rangle \\
=&\langle \underline{e}^{\underline{k}} \otimes \xi, Ah\rangle \\
=&\sum_{l=1}^{\infty} \sum_{i_1,\dots,i_l=1}^{d} \langle \underline{e}^{\underline{k}} \otimes \xi, e_{i_1} \otimes \dots \otimes e_{i_l} \otimes D_{\underline{T}^*} T_{i_l}^* \dots T_{i_1}^* h\rangle \\
=& \sum_{i_1,\dots,i_{|\underline{k}|}=1}^{d} \langle P_s(e_1^{\otimes k_1} \otimes \dots \otimes e_d^{\otimes k_d}), e_{i_1} \otimes \dots \otimes e_{i_{|\underline{k}|}}\rangle \langle \xi, D_{\underline{T}^*} T_{i_m}^* \dots T_{i_1}^* h\rangle \\
=&\langle \xi, D_{\underline{T}^*}(\underline{T}^{\underline{k}})^* h\rangle = \langle \underline{T}^{\underline{k}} D_{\underline{T}^*}\xi, h\rangle.
\end{aligned}$$

For $\zeta = \omega \otimes \xi$ with $\xi \in \mathcal{D}_{\underline{T}^*}$ we have

$$\langle \omega \otimes \xi, Ah\rangle = \langle \omega \otimes \xi, \omega \otimes D_{\underline{T}^*}h\rangle = \langle \xi, D_{\underline{T}^*}h\rangle = \langle D_{\underline{T}^*}\xi, h\rangle,$$

as required. If $(T_1, T_2, \ldots, T_d)$ is pure, then $T_\infty = 0$. Thus A is an isometry and hence L is a coisometry. ■

Given a commuting contractive tuple, the following theorem constructs an $\mathcal{A}_d^c$-morphism from the commutative Toeplitz C^*-algebra into the unital C^*-algebra generated by $T_1, T_2, \ldots, T_d$. Note that (4.3.1) implies that $L(\underline{S}^{\underline{k}} \otimes I_{\mathcal{D}_{\underline{T}^*}}) = \underline{T}^{\underline{k}} L$.

Theorem 4.3.4. *For every commuting contractive tuple $(T_1, T_2, \ldots, T_d)$ acting on a Hilbert space $\mathcal{H}$ there is a unique unital $\mathcal{A}_d^c$-morphism*

$$\varphi : \mathcal{T}_d^c \to \mathcal{B}(\mathcal{H})$$

such that $\varphi(S_i) = T_i$, $i = 1, \ldots, n$.

Proof. The uniqueness assertion is immediate by Lemma 4.3.2.

For existence, first assume that the commuting contractive tuple $\underline{T} = (T_1, T_2 \ldots, T_d)$ is pure. Recall that this means $P_{\underline{T}}^n(I_{\mathcal{H}})$ converges strongly to 0 as n tends to ∞. We first construct an $\mathcal{A}_d^c$-morphism for such a tuple. The lemma above asserts that there is a unique bounded operator $L : \Gamma_s(\mathbb{C}^d) \otimes \mathcal{D}_{\underline{T}^*} \to \mathcal{H}$ satisfying $L(\omega \otimes \xi) = D_{\underline{T}^*}\xi$ for $\xi \in \mathcal{D}_{\underline{T}^*}$ and

$$L(\underline{e}^{\underline{k}} \otimes \xi) = \underline{T}^{\underline{k}} D_{\underline{T}^*}\xi,$$

for $|\underline{k}| = 1, 2, \ldots$, $\xi \in \mathcal{D}_{\underline{T}^*}$; moreover, since $(T_1, T_2, \ldots, T_d)$ is a pure contractive tuple, L is a coisometry.

Let $\varphi : \mathcal{T}_d^c \to \mathcal{B}(\mathcal{H})$ be the completely positive map

$$\varphi(X) = L(X \otimes I_{\mathcal{D}_{\underline{T}^*}})L^*, \ X \in \mathcal{T}_d.$$

The map φ is unital because L is a co-isometry. (4.3.1) implies that for every $X \in \mathcal{T}_d^c$ we have $\varphi(\underline{S}^{\underline{k}}X) = L(\underline{S}^{\underline{k}}X \otimes I_{\mathcal{D}_{\underline{T}}})L^* = L(\underline{S}^{\underline{k}} \otimes I_{\mathcal{D}_{\underline{T}}})(X \otimes I_{\mathcal{D}_{\underline{T}}})L^* = \underline{T}^{\underline{k}} L(X \otimes I_{\mathcal{D}_{\underline{T}}})L^* = \underline{T}^{\underline{k}}\varphi(X)$. So for any polynomial f,

$$\varphi(f(\underline{S})X) = f(\underline{T})\varphi(X).$$

Hence φ is an $\mathcal{A}_d^c$-morphism having the required properties.

The general case is deduced from this by a classical device. For any commuting contractive tuple $\underline{T} = (T_1, T_2, \ldots, T_d)$ define $\underline{T}_r$ for $0 < r < 1$ to be

$$\underline{T}_r = (rT_1, \ldots, rT_d).$$

It is clear that $\underline{T}_r$ is a pure commuting contractive tuple. Thus there is an $\mathcal{A}_d^c$-morphism $\varphi_r : \mathcal{T}_d^c \to \mathcal{B}(\mathcal{H})$ satisfying

$$\varphi_r(S_i) = rT_i, \qquad i = 1, \ldots, d.$$

We have

$$\varphi_r(\underline{S}^{\underline{k}}(\underline{S}^{\underline{l}})^*) = \underline{T}_r^{\underline{k}}(\underline{T}_r^{\underline{l}})^*$$

for any multi-indices $\underline{k}$ and $\underline{l}$. Since operators of the form $\underline{S}^{\underline{k}}(\underline{S}^{\underline{l}})^*$ span $\mathcal{T}_d^c$ and since the family of maps φ_r, $0 < r < 1$ is uniformly bounded, it follows that φ_r converges point-norm to an $\mathcal{A}_d^c$-morphism φ as $r \uparrow 1$, and $\varphi(S_i) = T_i$ for all i. ∎

Corollary 4.3.5. *[von Neumann's inequality] Let $\underline{T} = (T_1, T_2, \ldots, T_d)$ be any commuting contractive tuple acting on a Hilbert space $\mathcal{H}$ and $\underline{S} = (S_1, S_2, \ldots, S_d)$ be the d-shift defined in Definition 4.2.8. Then for any polynomial f in d-variables,*

$$\|f(T_1, T_2, \ldots, T_d)\| \leq \|f(S_1, S_2, \ldots, S_d)\|.$$

Proof. Making use of the unital completely positive map φ of the last theorem which maps $f(S_1, \ldots, S_d)$ to $f(T_1, T_2, \ldots, T_d)$, we have

$$\begin{aligned}\|f(T_1, T_2, \ldots, T_d)\| &= \|\varphi(f(S_1, S_2, \ldots, S_d))\| \\ &\leq \|\varphi\|\|f(S_1, S_2, \ldots, S_d)\| = \|f(S_1, S_2, \ldots, S_d)\|.\end{aligned}$$

∎

4.4. Standard commuting dilation

The theorems above lead us to the following dilation theorem for any commuting contractive tuple $\underline{T}$ acting on some Hilbert space $\mathcal{H}$. We need some notation. If m is a positive integer or ∞ and $\mathcal{M}$ is a Hilbert space of dimension m, we shall mean by $m \cdot \underline{S}$, the operator tuple $(S_1 \otimes I_{\mathcal{M}}, \ldots, S_d \otimes I_{\mathcal{M}})$ acting on $\Gamma_s(\mathbb{C}^d) \otimes \mathcal{M}$. In the next theorem, we are going to express $\underline{T}$ as a compression of a direct sum, one of whose components might be absent. To assimilate this in a single notation, we make the convention that $m \cdot \underline{S}$ is absent if $m = 0$.

Given two operator tuples $\underline{T}$ and $\underline{R}$ on Hilbert spaces $\mathcal{H}_1$ and $\mathcal{H}_2$ respectively, we denote by $\underline{T} \oplus \underline{R}$ the operator tuple

$$(T_1 \oplus R_1, T_2 \oplus R_2, \ldots, T_d \oplus R_d) \text{ on the space } \mathcal{H}_1 \oplus \mathcal{H}_2.$$

Also, given a Hilbert space $\mathcal{N}$, and a representation β of $\mathcal{T}_d^c$ on $\mathcal{N}$, the notation $\beta(\underline{S})$ denotes the operator tuple $(\beta(S_1), \beta(S_2), \ldots, \beta(S_d))$. The operator tuple

$$\underline{A} \stackrel{\text{def}}{=} m \cdot \underline{S} \oplus \beta(\underline{S})$$

is clearly a commuting contractive tuple on $\widehat{\mathcal{H}} \stackrel{\text{def}}{=} (\Gamma_s(\mathbb{C}^d) \otimes \mathcal{M}) \oplus \mathcal{N}$. Let $\mathcal{H}$ be a subspace of $\widehat{\mathcal{H}}$ such that $A_i^* \mathcal{H} \subseteq \mathcal{H}$ for all $i = 1, \ldots, d$. Recall that such subspaces are called *co-invariant* with respect to the tuple $\underline{A}$. Consider the compression $\underline{T}$ of $\underline{A}$ to $\mathcal{H}$ as follows.

$$T_i \stackrel{\text{def}}{=} P_{\mathcal{H}} A_i|_{\mathcal{H}}.$$

This $\underline{T}$ is clearly a commuting contractive tuple on $\mathcal{H}$ and moreover, for any polynomial $f(z_1, \ldots, z_d)$, $f(\underline{T})$ is the compression of $f(\underline{A})$ due to the co-invariance of $\mathcal{H}$ with respect to $\underline{A}$. We prove that every commuting contractive tuple has such a realization with β sending all compact operators to zero. We recall that the notation $D_{\underline{T}^*}$ stands for the operator $(I - \sum T_i T_i^*)^{1/2}$.

Theorem 4.4.1. *[Dilation] Let $\underline{T}$ be any commuting contractive tuple acting on a separable Hilbert space $\mathcal{H}$ and rank$D_{\underline{T}^*} = m$ (which is a non-negative integer or ∞). Then there is a separable Hilbert space $\mathcal{M}$ of dimension m, another separable Hilbert space $\mathcal{N}$ with a commuting tuple of normal operators $\underline{Z} = (Z_1, Z_2, \ldots, Z_d)$ acting on it, satisfying $Z_1 Z_1^* + Z_2 Z_2^* + \cdots + Z_d Z_d^* = I_{\mathcal{N}}$ such that:*

(a) $\mathcal{H}$ is contained in $\widehat{\mathcal{H}} \stackrel{\text{def}}{=} (\Gamma_s(\mathbb{C}^d) \otimes \mathcal{M}) \oplus \mathcal{N}$ as a subspace and it is co-invariant under $\underline{A} \stackrel{\text{def}}{=} m \cdot \underline{S} \oplus \underline{Z}$.

(b) $\underline{T}$ is the compression of $\underline{A}$ to $\mathcal{H}$, that is, $\underline{T}^{\underline{k}} = P_{\mathcal{H}} \underline{A}^{\underline{k}}|_{\mathcal{H}}$ for every multi-index $\underline{k}$.

(c) The space $\widehat{\mathcal{H}}$ is minimal in the sense that $\widehat{\mathcal{H}} = \overline{span}\{\underline{A}^{\underline{k}} h : h \in \mathcal{H}$ and $\underline{k}$ is any multi-index$\}$.

Thus any commuting contractive tuple has a minimal commuting dilation. Moreover, this dilation is unique up to unitary equivalence.

Proof. We consider the minimal Stinespring dilation of φ. Thus we get a Hilbert space $\widehat{\mathcal{H}}$ containing $\mathcal{H}$ and a representation π of $\mathcal{T}_d^c$ on $\widehat{\mathcal{H}}$ such that

$$\varphi(X) = P_{\mathcal{H}} \pi(X) P_{\mathcal{H}} \text{ for } X \in \mathcal{T}_d^c,$$

where $P_{\mathcal{H}}$ is the projection onto $\mathcal{H}$ (We are identifying any operator $Z \in \mathcal{B}(\mathcal{H})$ with $P_{\mathcal{H}} Z P_{\mathcal{H}} \in \mathcal{B}(\hat{\mathcal{H}})$). So

$$\widehat{\mathcal{H}} = \overline{\text{span}} \ \{\pi(X)u : X \in \mathcal{T}_d^c \text{ and } u \in \mathcal{H}\}.$$

The C^*-algebra $\mathcal{T}_d^c$ is separable and hence the Hilbert space $\widehat{\mathcal{H}}$ is also separable. The tuple $(\pi(S_1), \pi(S_2), \ldots, \pi(S_d))$ is a dilation of the tuple $(T_1, T_2, \ldots, T_d)$ in the sense that for any polynomial f,

$$f(T_1, T_2, \ldots, T_d) = P_{\mathcal{H}} f(\pi(S_1), \pi(S_2), \ldots, \pi(S_d))|_{\mathcal{H}},$$

and $\mathcal{H}$ is a co-invariant subspace for $(\pi(S_1), \pi(S_2), \ldots, \pi(S_d))$.

Let us denote the set of all compact operators on $\Gamma_s(\mathbb{C}^d)$ by $\mathcal{B}_0(\Gamma_s(\mathbb{C}^d))$ (or just $\mathcal{B}_0$ when there is no chance of confusion). Since $\mathcal{T}_d^c$ contains $\mathcal{B}_0$, by standard theory of representations of C^*-algebras (see [48], Chapter I for example), the representation π decomposes as $\pi = \pi_0 \oplus \pi_1$, where $\pi_i : \mathcal{T}_d^c \to \mathcal{B}(\widehat{\mathcal{H}}_i)$ with π_0 being a non-degenerate representation of $\mathcal{B}_0$ on $\widehat{\mathcal{H}}_0$, π_1 being 0 on $\mathcal{B}_0$ and $\widehat{\mathcal{H}} = \widehat{\mathcal{H}}_0 \oplus \widehat{\mathcal{H}}_1$ (one of π_0 and π_1 could be absent too). Since the only non-degenerate representation of the C^*-algebra of compact operators is the identity representation with some multiplicity and since a represenation which is non-degenerate on an ideal, extends uniquely to the entire C^*-algebra, it follows that π_0 is just the identity representation with some multiplicity i.e., up to unitary isomorphism, $\hat{\mathcal{H}}_0 = \Gamma_s(\mathbb{C}^d) \otimes \mathcal{M}$ and $\pi_0(X) = X \otimes I_{\mathcal{M}}$ for some Hilbert space $\mathcal{M}$. So if we take $\mathcal{N} = \hat{\mathcal{H}}_1$, and $\pi_1(S_i) = Z_i$ then $(Z_1, Z_2, \ldots, Z_d)$ is a commuting contractive tuple and *(a), (b)* are satisfied. Moreover $\sum Z_i Z_i^* = I_{\mathcal{N}}$ as π_1 kills compact operators and $I_{\Gamma_s(\mathbb{C}^d)} - \sum S_i S_i^*$ is compact.

It remains to prove that the multiplicity i.e., dim $(\mathcal{M})$ is just the rank of $D_{\underline{T}^*}$. For this, note that dim $\mathcal{M}$ = dim(range $\pi_0(E)$) where E is any one-dimensional projection in $\mathcal{T}_d^c$. Taking $E = E_0$, the projection onto the vacuum space, and making use of minimality of Stinespring representation, we have

$$\text{range } \pi(E_0) = \{\pi(E_0)\xi : \xi \in \widehat{\mathcal{H}}\} = \overline{\text{span}}\{\pi(E_0)\pi(X)u : X \in \mathcal{T}_d^c, u \in \mathcal{H}\}.$$

Hence,

$$\begin{aligned}
&\text{range } \pi(E_0) \\
=&\overline{\text{span}}\{\pi(E_0)\pi(X)u : X \in \mathcal{B}_0, u \in \mathcal{H}\} \\
=&\overline{\text{span}}\{\pi(E_0)\pi(\underline{S}^{\underline{k}} E_0 (\underline{S}^{\underline{l}})^*)u : \text{all multi-indices } \underline{k}, \underline{l}, \text{ and } u \in \mathcal{H}\} \\
=&\overline{\text{span}}\{\pi(E_0)\pi((\underline{S}^{\underline{l}})^*)u : \text{all multi-indices } \underline{l}, \text{ and } u \in \mathcal{H}\}.
\end{aligned}$$

Now we define a unitary U : range $\pi(E_0) \to \overline{\text{range}} D_{\underline{T}^*}$ by setting

$$U\pi(E_0)\pi((\underline{S}^{\underline{l}})^*)u = D_{\underline{T}^*}(\underline{T}^{\underline{l}})^*u$$

and extending linearly. Then U is isometric because for $u, v \in \mathcal{H}$ and all $\underline{k}$ and $\underline{l}$,

$$\begin{aligned}\langle \pi(E_0)\pi((\underline{S}^{\underline{k}})^*)u, \pi(E_0)\pi((\underline{S}^{\underline{l}})^*)v\rangle &= \langle u, \pi(\underline{S}^{\underline{k}})\pi(E_0)\pi((\underline{S}^{\underline{l}})^*)v\rangle \\ &= \langle u, \underline{T}^{\underline{k}} D^2_{\underline{T}^*}(\underline{T}^{\underline{l}})^*v\rangle \\ &= \langle D_{\underline{T}^*}(\underline{T}^{\underline{k}})^*u, D_{\underline{T}^*}(\underline{T}^{\underline{l}})^*v\rangle.\end{aligned}$$

Taking $\underline{l} = 0$, it is clear that U is onto. This proves that $\overline{\text{range}} D_{\underline{T}^*}$ and $\mathcal{M}$ have the same dimensions. Now the uniqueness of the dilation follows from that of the minimal Stinespring dilation of a completely positive map. ∎

As remarked before in the direct sum for $\hat{\mathcal{H}}$ and $\underline{A}$ appearing in this theorem one of the summands could be absent. $\mathcal{M}$ and $n \cdot \underline{S}$ are absent iff $n = 0$, that is, iff $\sum T_i T_i^* = I_{\mathcal{H}}$. It can be shown that $\mathcal{N}$ and $\underline{Z}$ are absent if and only if $\underline{T}$ is pure.

The dilation operator $\underline{A}$ is a commuting tuple and the final section will show us that it is the *best* that can be done retaining commutativity.

The ideas of this section are from Arveson [17], Drury [59] and Arias-Popescu [10], [11], [95], [96], [97], although we closely follow the methods of Arveson. In fact, the standard commuting dilation was looked at by Drury [59] in his study of von Neumann inequality for tuples and similar ideas have been explored by Agler [2], Athavale [20] and others for different classes of operators using various reproducing kernels. Popescu concentrated mainly on the non-commutative case more of which we shall see in the next section, many commutative results fall out as special cases from his theory. The crucial operator L of Theorem 4.3.3 is actually the adjoint of the Poisson transform defined by Popescu in [97]. The proof of the dilation theorem here can be found in [27].

4.5. Standard non-commuting dilation

In this section, we concentrate on general contractive tuples, which are not necessarily commuting. The dilation will consist of isometries with orthogonal ranges. As in Section 3, we consider a model d-tuple

$\underline{V} = (V_1, V_2, \ldots, V_d)$. Recall from Section 2 that $\underline{V}$ is the tuple of creation operators on the full Fock space over $\mathbb{C}^d$. Let $\mathcal{T}_d^{nc}$ denote the C^*-subalgebra of $\mathcal{B}(\Gamma(\mathbb{C}^d))$ generated by $V_1, V_2, \ldots, V_d$. Recall from Lemma 4.2.4 that $V_1, V_2, \ldots, V_d$ are isometries with orthogonal ranges.

Thus, the C^*-algebra $\mathcal{T}_d^{nc}$ is automatically unital. Let $\mathcal{A}_d^{nc}$ be the algebra generated by $I_{\Gamma(\mathbb{C}^d)}$ and $V_1, V_2, \ldots, V_d$. This is called the *non-commutative disk algebra*, see Popescu [97]. This is the closure of the set of all operators of the form $f(\underline{V})$ where f varies over all polynomials in n non-commuting variables. There is a non-commutative Poisson transform and the $\mathcal{A}$-morphism which will be used for getting dilation. In this case too, the C^*-algebra is the same as the operator space $\overline{\text{span}}\mathcal{A}_d^{nc}(\mathcal{A}_d^{nc})^*$ because $V_i^*V_j = \delta_{ij}$.

Lemma 4.5.1. *All compact operators on $\Gamma(\mathbb{C}^d)$ are in $\mathcal{T}_d^{nc}$.*

Proof. First $E_0 = I_{\Gamma(\mathbb{C}^d)} - \sum V_iV_i^*$ is in $\mathcal{T}_d^{nc}$. Now for any $k \geq 1, l \geq 1$, $1 \leq i_1, i_2, \ldots, i_k \leq d$ and $1 \leq j_1, j_2, \ldots, j_l \leq d$, we see that the operator $V_{i_1}V_{i_2}\ldots V_{i_k}E_0(V_{j_1}V_{j_2}\ldots V_{j_l})^*$ is the same as

$$\xi \to \langle \xi, e_{i_1} \otimes e_{i_2} \otimes \cdots \otimes e_{i_k} \rangle e_{j_1} \otimes e_{j_2} \otimes \cdots \otimes e_{j_l}.$$

Since $\{\omega, e_{i_1} \otimes e_{i_2} \otimes \cdots \otimes e_{i_k} : k \geq 1 \text{ and } 1 \leq i_1, i_2, \ldots, i_k \leq d\}$ forms a basis for $\Gamma(\mathbb{C}^d)$, all rank one operators are in $\mathcal{T}_d^{nc}$ and hence all compact operators are in $\mathcal{T}_d^{nc}$. ∎

In the following we are going to describe the Poisson kernel and the Poisson transform associated with a contractive tuple following Popescu [97]. So take a contractive tuple $\underline{T} = (T_1, T_2, \ldots, T_d)$ of operators acting on $\mathcal{H}$. The *Poisson kernel* $\{K_r(\underline{T})\}_{0 \leq r \leq 1}$ associated with $\underline{T}$ is the family of bounded operators $K_r(\underline{T}) : \mathcal{H} \to \Gamma(\mathbb{C}^d) \otimes \mathcal{D}_{\underline{T}^*}$ defined by $K_r(\underline{T})h = (\zeta_0, \zeta_1, \zeta_2, \ldots)$ where $\zeta_0 = \omega \otimes D_{\underline{T}^*}h$ and for $k \geq 1$,

$$\zeta_k = r^k \sum_{i_1, \ldots, i_k = 1}^{d} e_{i_1} \otimes e_{i_2} \otimes \cdots \otimes e_{i_k} \otimes D_r(\underline{T})(T_{i_1}T_{i_2}\ldots T_{i_k})^*h,$$

where $D_r(\underline{T})$ is the defect operator $(I_{\mathcal{H}} - \sum r^2 T_iT_i^*)^{1/2}$.

Since $(rT_1, \ldots, rT_d)$ is pure for $0 < r < 1$, it is easy to check that $K_r(\underline{T})$ is an isometry for $0 < r < 1$ and $K(\underline{T}) \stackrel{\text{def}}{=} K_1(\underline{T})$ is a contraction with $K(\underline{T})^*K(\underline{T}) = I_{\mathcal{H}} - A_\infty$ where A_∞ is the operator defined in Section 3. Thus $K(\underline{T})$ is an isometry if and only if $\underline{T}$ is pure.

The *Poisson transform* associated with the contractive tuple $\underline{T}$ is the completely positive map $\alpha : \mathcal{T}_d^{nc} \to \mathcal{B}(\mathcal{H})$ defined by

$$\alpha(X) = \lim_{r \to 1} K_r(\underline{T})^*(X \otimes I_{\mathcal{D}_{\underline{T}}})K_r(\underline{T}),$$

where the limit is in the norm topology of $\mathcal{B}(\mathcal{H})$. For each $0 < r < 1$, the $K_r(\underline{T})$ satisfies

$$K_r(\underline{T})^*(V_{i_1}V_{i_2}\dots V_{i_k} \otimes I_{\mathcal{D}_{\underline{T}^*}}) = T_{i_1}T_{i_2}\dots T_{i_k}K_r(\underline{T})^*$$

for $k \geq 1$ and $1 \leq i_1, i_2, \dots, i_k \leq d$. This implies that the Poisson transform is a unital completely contractive $\mathcal{A}$-morphism. Depending on the property of the tuple $\underline{T}$, the image of $K(\underline{T})$ can be contained in a subspace, say $\mathcal{L}_{\underline{T}} \otimes \mathcal{D}_{\underline{T}}$ where $\mathcal{L}_{\underline{T}}$ is a subspace of $\Gamma(\mathbb{C}^d)$. We state the discussion above as a theorem.

Theorem 4.5.2. *If $\underline{T} = (T_1, T_2, \dots, T_d)$ is a contractive tuple acting on a Hilbert space $\mathcal{H}$, then there is a unique unital $\mathcal{A}_d^{nc}$-morphism $\alpha : \mathcal{T}_d^{nc} \to \mathcal{B}(\mathcal{H})$ such that $\alpha(V_j) = T_j$.*

Proof. The only point which needs a proof is uniqueness which is clear too. An $\mathcal{A}$-morphism which sends V_j to T_j for every j has to be unique because $\mathcal{T}_d^{nc} = \overline{\text{span}}\mathcal{A}_d^{nc}(\mathcal{A}_d^{nc})^*$. ■

Corollary 4.5.3. *[von Neumann's inequality] Let $\underline{T} = (T_1, T_2, \dots, T_d)$ be a contractive tuple acting on a Hilbert space $\mathcal{H}$. Then for any polynomial f in d non-commuting indeterminates,*

$$\|f(T_1, T_2, \dots, T_d)\| \leq \|f(V_1, V_2, \dots, V_d)\|.$$

Proof. The proof is obvious using the $\mathcal{A}_d^{nc}$-morphism α obtained above. ■

Thus for a commuting contractive tuple $\underline{T}$, there are at least two von Neumann inequalities given by Corollary 4.3.5 and Corollary 4.5.3. Arias and Popescu noted that given any pure contractive tuple $\underline{T}$, the best von Nemann inequality arises from the smallest co-invariant (with respect to $\underline{V}$) subspace, say $\mathcal{L}_{\underline{T}}$, of $\Gamma(\mathbb{C}^d)$ such that $K(\underline{T})$ takes its values in $\mathcal{L}_{\underline{T}} \otimes \mathcal{D}_{\underline{T}^*}$. If $B_i = P_{\mathcal{L}_{\underline{T}}}V_i|_{\mathcal{L}_{\underline{T}}}$ for $i = 1, 2, \dots, n$, then

$$\|f(\underline{T})\| \leq \|f(\underline{B})\|$$

for all polynomials f in n non-commuting indeterminates. See [10]. They also give a formula for $\mathcal{L}_{\underline{T}}$ in terms of $\underline{T}$.

Theorem 4.5.2 gives the following dilation theorem for any contractive tuple $\underline{T}$ acting on some Hilbert space $\mathcal{H}$. We are going to use the notation employed in the Dilation theorem of Section 4. Given a Hilbert space $\mathcal{N}$, and a representation β of $\mathcal{T}_d^{nc}$ on $\mathcal{N}$, the operator tuple

$$\underline{A} \stackrel{\text{def}}{=} m \cdot \underline{V} \oplus \beta(\underline{V})$$

is clearly a contractive tuple on $\widehat{\mathcal{H}} \stackrel{\text{def}}{=} (\mathcal{H} \otimes \mathcal{M}) \oplus \mathcal{N}$. Let $\mathcal{H}$ be a co-invariant subspace of $\widehat{\mathcal{H}}$ under the tuple $\underline{A}$. Then the compression $\underline{T}$ of $\underline{A}$ to $\mathcal{H}$

$$T_i \stackrel{\text{def}}{=} P_{\mathcal{H}} A_i|_{\mathcal{H}}$$

is a contractive tuple on $\mathcal{H}$ and moreover, for any polynomial f in d non-commuting indeterminates, $f(\underline{T})$ is the compression of $f(\underline{A})$ due to the co-invariance of $\mathcal{H}$ with respect to $\underline{A}$. The following theorem proves a converse. This theorem is to be compared with Theorem 4.4.1 which we proved in the commutative case.

Theorem 4.5.4. *[Dilation] Let $\underline{T}$ be any contractive tuple acting on a separable Hilbert space $\mathcal{H}$ and rank$D_{\underline{T}^*} = m$ (which is a non-negative integer or ∞). Then there is a separable Hilbert space $\mathcal{M}$ of dimension m, another separable Hilbert space $\mathcal{N}$ with a tuple of operators $\underline{Z} = (Z_1, Z_2, \dots, Z_d)$ acting on it, satisfying $Z_i^* Z_j = \delta_{ij}$ for $1 \le i, j \le d$ and $Z_1 Z_1^* + \cdots + Z_d Z_d^* = I_{\mathcal{N}}$ such that:*

(a) $\mathcal{H}$ is contained in $\widehat{\mathcal{H}} \stackrel{\text{def}}{=} (\mathcal{H} \otimes \mathcal{M}) \oplus \mathcal{N}$ as a subspace and it is co-invariant under $\underline{A} \stackrel{\text{def}}{=} m \cdot \underline{V} \oplus \underline{Z}$.

(b) $\underline{T}$ is the compression of $\underline{A}$ to $\mathcal{H}$, i.e.,

$$T_{i_1} T_{i_2} \dots T_{i_k} h = P_{\mathcal{H}} A_{i_1} A_{i_2} \dots A_{i_k} h$$

for every $h \in \mathcal{H}, k \ge 1$ and $1 \le i_1, i_2, \dots, i_k \le d$.

(c) The space $\widehat{\mathcal{H}}$ is minimal in the sense that $\widehat{\mathcal{H}} = \overline{span}\{A_{i_1} A_{i_2} \dots A_{i_k} h$ where $h \in \mathcal{H}, k \ge 1$ and $1 \le i_1, i_2, \dots\dots, i_k \le d\}$.

In other words, every contractive tuple $\underline{T}$ has a minimal dilation to isometries with orthogonal ranges.

For a contractive tuple $\underline{T}$, any two such minimal dilations consisting of isometries with orthogonal ranges are unitarily equivalent.

Proof. We consider a Stinespring dilation of the Poisson transform α. Thus we get a Hilbert space $\widehat{\mathcal{H}}$ containing $\mathcal{H}$ and a representation π of $\mathcal{T}_d^{nc}$ on $\widehat{\mathcal{H}}$ such that

$$\varphi(X) = P_{\mathcal{H}} \pi(X) P_{\mathcal{H}} \text{ for } X \in \mathcal{T}_d^{nc},$$

where $P_{\mathcal{H}}$ is the projection onto $\mathcal{H}$ (We are identifying any operator $Z \in \mathcal{B}(\mathcal{H})$ with $P_{\mathcal{H}} Z P_{\mathcal{H}} \in \mathcal{B}(\widehat{\mathcal{H}})$). Now

$$\begin{aligned}
&\alpha(V_i)\alpha(V_i^*)\\
&= \alpha(V_i V_i^*)\\
&= P_{\mathcal{H}}\pi(V_i V_i^*)P_{\mathcal{H}}\\
&= P_{\mathcal{H}}\pi(V_i)\pi(V_i^*)P_{\mathcal{H}}\\
&= P_{\mathcal{H}}\pi(V_i)(P_{\mathcal{H}} + P_{\mathcal{H}}^{\perp})(P_{\mathcal{H}} + P_{\mathcal{H}}^{\perp})\pi(V_i^*)P_{\mathcal{H}}\\
&= (P_{\mathcal{H}}\pi(V_i)P_{\mathcal{H}} + P_{\mathcal{H}}\pi(V_i)P_{\mathcal{H}}^{\perp})(P_{\mathcal{H}}\pi(V_i^*)P_{\mathcal{H}} + P_{\mathcal{H}}^{\perp}\pi(V_i^*)P_{\mathcal{H}})\\
&= \alpha(V_i)\alpha(V_i^*) + (P_{\mathcal{H}}\pi(V_i)P_{\mathcal{H}}^{\perp})(P_{\mathcal{H}}\pi(V_i)P_{\mathcal{H}}^{\perp})^*.
\end{aligned}$$

Thus

$$P_{\mathcal{H}}\pi(V_i)P_{\mathcal{H}}^{\perp} = 0. \tag{4.5.1}$$

Note that we may and do assume π to be a *minimal* Stinespring dilation. So

$$\widehat{\mathcal{H}} = \overline{\text{span}}\ \{\pi(X)u : X \in \mathcal{T}_d^{nc} \text{ and } u \in \mathcal{H}\}.$$

The C^*-algebra $\mathcal{T}_d^{nc}$ is separable and hence the Hilbert space $\widehat{\mathcal{H}}$ is also separable. The tuple $(\pi(V_1), \ldots, \pi(V_d))$ is a dilation of $(T_1, T_2, \ldots, T_d)$ in the sense that for any polynomial f (in d non-commuting variables),

$$f(T_1, T_2, \ldots, T_d) = P_{\mathcal{H}} f(\pi(V_1), \ldots, \pi(V_d))|_{\mathcal{H}},$$

and $\mathcal{H}$ is a co-invariant subspace for $(\pi(V_1), \pi(V_2), \ldots, \pi(V_d))$ in view of (4.5.1). The proof for obtaining the spaces $\mathcal{M}$ and $\mathcal{N}$ and the operator tuples $\underline{Z}$ and $\underline{A}$ are exactly the same as in Theorem 4.4.1 and so we omit that. We give proof of the multiplicity i.e., dim $(\mathcal{M})$ = rank of $D_{\underline{T}^*}$ which involves now non-commuting polynomials. We have

$$\begin{aligned}
&\text{range } \pi(E_0)\\
&= \{\pi(E_0)\xi : \xi \in \widehat{\mathcal{H}}\}\\
&= \overline{\text{span}}\{\pi(E_0)\pi(X)u : X \in \mathcal{T}_d^{nc}, u \in \mathcal{H}\}\\
&= \overline{\text{span}}\{\pi(E_0)\pi(E_0X)u : X \in \mathcal{T}_d^{nc}, u \in \mathcal{H}\}\\
&= \overline{\text{span}}\{\pi(E_0)\pi(X)u : X \in \mathcal{B}_0(\Gamma(\mathbb{C}^d)), u \in \mathcal{H}\}\\
&= \overline{\text{span}}\{\pi(E_0)\pi(V_{i_1}\ldots V_{i_k}E_0(V_{j_1}\ldots V_{j_l})^*)u : k, l \geq 1;\\
&1 \leq i_1, \ldots, i_k, j_1, \ldots, j_l \leq d \text{ and } u \in \mathcal{H}\}\\
&= \overline{\text{span}}\{\pi(E_0)\pi((V_{j_1}\ldots V_{j_l})^*)u : l \geq 1; 1 \leq j_1, \ldots, j_l \leq d \text{ and } u \in \mathcal{H}\}.
\end{aligned}$$

Now a unitary U : range $\pi(E_0) \to \overline{\text{range}} D_{\underline{T}^*}$ can be defined by setting

$$U\pi(E_0)\pi((V_{j_1}\ldots V_{j_l})^*)u = D_{\underline{T}^*}(T_{j_1}\ldots T_{j_l})^*u$$

and extending linearly. Then U is isometric because for $u, v \in \mathcal{H}$ and $1 \leq i_1, \ldots, i_k, j_1, \ldots, j_l \leq d$,

$$\begin{aligned}
& \langle \pi(E_0)\pi((V_{i_1} \ldots V_{i_k})^*)u, \pi(E_0)\pi((V_{j_1} \ldots V_{j_l})^*)v \rangle \\
= & \langle u, \pi(V_{i_1} \ldots V_{i_k})\pi(E_0)\pi((V_{j_1} \ldots V_{j_l})^*)v \rangle \\
= & \langle u, T_{i_1} \ldots T_{i_k} D^2_{\underline{T}^*}(T_{j_1} \ldots T_{j_l})^* v \rangle \\
= & \langle D_{\underline{T}^*}(T_{i_1} \ldots T_{i_k})^* u, D_{\underline{T}^*}(T_{j_1} \ldots T_{j_l})^* v \rangle .
\end{aligned}$$

It is clear that U is onto. This proves that $\overline{\text{range}} D_{\underline{T}^*}$ and $\mathcal{M}$ have the same dimensions.

The final contention of course follows from the fact that minimal Stinespring dilation is unique up to unitary equivalence. ■

Thus any contractive tuple can be dilated to a tuple of isometries with orthogonal ranges. The dilation obtained in the theorem above is called the *standard non-commuting dilation*. The decomposition of $\underline{A}$ into the direct sum of $m \cdot \underline{V}$ and $\underline{Z}$ can actually be thought of as the Wold decomposition, see Popescu [95]. Clearly, the $m \cdot \underline{V}$ part (the non-commutative shift) is absent if and only if $\underline{T}$ satisfies $\sum T_i T_i^* = I_{\mathcal{H}}$ and the $\underline{Z}$ part (the analogue of the unitary part when $d = 1$) is absent if and only if $\underline{T}$ is pure. Thus the dilation tuple $\underline{A}$ satisfies $\sum A_i A_i^* = I_{\widehat{\mathcal{H}}}$ if and only if $\sum T_i T_i^* = I_{\mathcal{H}}$.

Bunce [38] and Frazho [63] had explored the ideas of isometeric dilation with orthogonal ranges. Popescu was the first to study this and other related notions from one variable operator theory systematically and obtained generalizations of such notions as Wold decomposition, characteristic function and functional calculus. He extensively developed the model theory. Popescu's original arguments for the dilation was a bit different. He also treated infinite sequences of operators. Most of the results here will generalize to infinite sequences of commuting or non-commuting operators, we do not go into the infinite case here because notationally it is more difficult to present.

An alternative proof of the dilation theorem also follows from the discrete time case of dilation of quantum dynamical semigroups [24].

In the short next section, we note some interesting facts about the limiting operator T_∞.

4.6. The limiting operator T_∞

Let $\underline{T} = (T_1, T_2, \dots T_d)$ be a contractive tuple of bounded operators acting on the Hilbert space $\mathcal{H}$ and let $P_{\underline{T}}$ be the natural completely positive map associated with $\underline{T}$. We have noticed that $P^n_{\underline{T}}(I_{\mathcal{H}})$ has a strong limit which we have denoted by T_∞.

Lemma 4.6.1. *The norm of T_∞ is either 0 or 1.*

Proof. First define for any integer $m \geq 1$, the operator T^m in $\mathcal{B}(\mathcal{H}^{n^m}, \mathcal{H})$ as follows:

$$\underline{h} = \{h_{i_1 \dots i_m} : 1 \leq i_1, \dots, i_m \leq n\} \to \sum_{1 \leq i_1, \dots, i_m \leq n} T_{i_1} \dots T_{i_m} h_{i_1 \dots i_m}.$$

Its adjoint $T^{m^*} \in \mathcal{B}(\mathcal{H}, H)^{n^m}$ maps a vector h to the n^m column vector $(T^*_{i_1} \dots T^*_{i_m} h)_{1 \leq i_i, \dots, i_m \leq n}$ in $\mathcal{H}^{n^m}$. By the above definition, $T^m T^{m^*} = P^m_{\underline{T}}(1)$. Thus we find that

$$\begin{aligned} \|T_\infty^{\frac{1}{2}} h\|^2 &= \langle T_\infty h, h\rangle = \langle T^m T_\infty T^{m*} h, h\rangle = \|T_\infty^{\frac{1}{2}} T^{m*} h\|^2 \\ &\leq \|T_\infty^{\frac{1}{2}}\|^2 \|T^{m*} h\|^2 = \|T_\infty\| \|T^{m*} h\|^2 \xrightarrow{m} \|T_\infty\| \|T_\infty^{\frac{1}{2}} h\|^2. \end{aligned}$$

Hence either $T_\infty^{1/2} = 0$ or $\|T_\infty\| \geq 1$. But T_∞ being a contraction, this means that $\|T_\infty\| = 1$. ∎

This fact holds in more generality. Suppose $\mathcal{A}$ is a unital C^*-subalgebra of $\mathcal{B}(\mathcal{H})$, and $\tau : \mathcal{A} \to \mathcal{A}$ is a contractive completely positive map. Then as $\{\tau^n(I)\}_{n \geq 0}$ is a decreasing sequence of positive contractions, it converges in strong operator topology to some operator in $\mathcal{B}(\mathcal{H})$. Note that in general $\tau^n(I)$ may not converge in norm as n tends to infinity. For instance, consider the completely positive map $X \mapsto SXS^*$ in $\mathcal{B}(l^2)$, where S is the isometric unilateral shift.

Proposition 4.6.2. *Let $\mathcal{A}$ be a unital C^*-subalgebra of $\mathcal{B}(\mathcal{H})$, and let $\tau : \mathcal{A} \to \mathcal{A}$ be a contractive completely positive map. Let $D = \lim_{n \to \infty} \tau^n(I)$ where the limit is in the strong operator topology. If $D \neq 0$, then $\|\tau^n(I)\| = 1$ for every n. Further, if $0 \neq D \in \mathcal{A}$ and $\tau(D) = D$, then $\|D\| = 1$.*

Proof. Suppose for some n, $\alpha := \|\tau^n(I)\| < 1$. Then $\tau^n(\tau^n(I)) \leq \alpha.\tau^n(I) \leq \alpha^2 I$. By induction $\tau^{nm}(I) \leq \alpha^m I$ for all $m \geq 1$. As $\alpha < 1$, $\tau^{nm}(I)$ converges to 0 in norm as $m \to \infty$, and hence $D = 0$.

Now if $D \in \mathcal{A}$ and $\tau(D) = D$, take $\beta = \|D\|$. If $0 \leq \beta < 1$, from $0 \leq D \leq \beta I$, we get $0 \leq \tau(D) \leq \beta\tau(I)$, or $0 \leq D \leq \beta\tau(I)$ or more generally $0 \leq D \leq \beta\tau^n(I)$ for all n. Taking strong operator topology limit, we get $0 \leq D \leq \beta.D$. This clearly implies $\beta = 1$ if $D \neq 0$.

■

The condition here that $D \in \mathcal{A}$ and is a fixed point of τ is basically a continuity condition, as we want to interchange the action of τ and the limit, while computing $\tau(D) = \tau(\lim_{n\to\infty} \tau^n(I))$. This clearly holds true if $\tau^n(I)$ converges in norm to D. It would also happen if $\mathcal{A}$ is a von Neumann algebra (refer to Kadison and Ringrose [70]) and τ is normal (refer to Kadison and Ringrose [71]). The following two examples illustrate the possible failure of continuity.

Example 4.6.3. *Suppose $\mathcal{A}$ is the space of convergent complex sequences with supremum norm: $\mathcal{A} = \{(x_n)_{n\in\mathbb{N}} : \lim_{n\to\infty} x_n$ exists$\}$ and $\|(x_n)\| = \sup_{n\in\mathbb{N}} |x_n|$. Let $\tau : \mathcal{A} \to \mathcal{A}$ be defined by $\tau(x) = y$ where*

$$y_j = \begin{cases} x_j & \text{if } j = 2k \text{ for some } k \in \mathbb{N} \\ \frac{x_1}{2} & \text{if } j = 1 \\ x_{j-2} & \text{if } j = 2k+1, k \in \mathbb{N}. \end{cases}$$

It is easy to see that $\tau^n((1,1,\ldots))$ decreases to $(0,1,0,1,\ldots)$ which is not in the C^*-algebra $\mathcal{A}$.

Example 4.6.4. *Suppose $\mathcal{A}$ is the sequence space $l^\infty(\mathbb{N})$. Let φ be a Banach limit, that is, φ is a state on l^∞, such that*

(i) *$\varphi((x_n)) = \lim_{n\to\infty} x_n$ whenever the limit exists and*
(ii) *$\varphi((x_2, x_3, \ldots,)) = \varphi((x_1, x_2, \ldots))$ for all $(x_1, x_2, \ldots)$ in l^∞.*

The existence of such Banach limits follows from a careful application of Hahn-Banach Theorem. The details can be found in Conway [41]. The condition (ii), means that the Banach limit does not change if we insert/delete finitely many terms to a sequence. Now fix a scalar $0 < \beta < 1$ and define $\tau : \mathcal{A} \to \mathcal{A}$ by $\tau((x_n)_{n\in\mathbb{N}}) = (\beta\varphi((x_n)_{n\in\mathbb{N}}), 0, x_2, x_3, \ldots)$. Clearly, τ is a contractive completely positive map,

$$D = \lim_{n\to\infty} \tau^n((1,1,\ldots)) = (\beta, 0, 0\ldots)$$

and $\tau(D) = 0$. Therefore D is not a fixed point and $\|D\| = \beta < 1$. Actually, the completely positive map τ considered here is not normal.

We end this Section with the following converse result. The completely positive map considered here arose in the study of asymptotic roots of completely positive maps in [30].

Proposition 4.6.5. *Let $\mathcal{A}$ be a unital C^*-algebra and let D be a positive contraction in $\mathcal{A}$ such that either $\|D\| = 1$ or $D = 0$. Then D is the norm limit of $\tau^n(1)$ for some contractive completely positive map τ.*

Proof. Suppose $0 \leq D \leq 1$ and $\|D\| = 1$. As D is positive, and in particular self-adjoint, $\|D\|$ is a point in the spectrum of D. Then by spectral theory there exists a state ϕ on $\mathcal{A}$ such that $\phi(D) = \|D\| = 1$. Fix a scalar $0 < \lambda < 1$. Consider $\tau : \mathcal{A} \to \mathcal{A}$ defined by

$$\tau(X) = \lambda X + (1-\lambda)\phi(X)D, \quad \text{for all } X \in \mathcal{A}.$$

Then τ being a convex combination of two contractive CP maps it is also a contractive CP map. By direct computation

$$\tau^n(X) = \lambda^n X + (1-\lambda^n)\phi(X)D \text{ for } n \geq 1.$$

Clearly the norm limit of $\tau^n(1)$ is D. If $D = 0$, we can simply consider $\tau(X) = \lambda X$. That completes the proof. ■

4.7. Maximality of the standard commuting dilation

In this section, we would answer a natural question. If $\underline{T}$ is a commuting contractive tuple of bounded operators acting on a Hilbert space $\mathcal{H}$, then it possesses two dilations. On one hand, there is the standard commuting one which retains commutativity but the constituent operators are only a direct sum of the commutative shift and a spherical isometry. On the other hand, one has the standard non-commuting one which gives the dilation tuple as isometries with orthogonal ranges, but loses commutativity. So the natural question is what is the best dilation retaining commutativity. The standard commuting dilation is, in fact, the *maximal commuting tuple contained* in the minimal isometric dilation. We shall briefly outline the ideas here referring the reader to [28] for the details of proofs. The main theorem of this section is Theorem 4.7.7.

Since we shall deal with both the standard commuting dilation and the standard non-commuting dilation of a commuting tuple, let us fix some notations. The standard commuting dilation of a commuting contractive tuple will be denoted by $\underline{\tilde{S}}$ and the standard non-commuting dilation of a contractive tuple will be denoted by $\underline{\tilde{V}}$. If $\underline{T}$ is a commuting contractive tuple acting on $\mathcal{H}$, then we have seen that both the dilations are of such a type that $\mathcal{H}$ is a co-invariant subspace. Recall that in such a case, $\underline{T}$ is called a piece of the dilation tuple. For an arbitrary tuple of operators $\underline{R} = (R_1, R_2, \ldots, R_d)$ acting on a Hilbert space $\mathcal{L}$ and having a co-invariant subspace $\mathcal{H}$, the tuple $(P_{\mathcal{H}}R_1|_{\mathcal{H}}, P_{\mathcal{H}}R_2|_{\mathcal{H}}, \ldots, P_{\mathcal{H}}R_d|_{\mathcal{H}})$ is called a piece of $\underline{R}$. It is the same as saying that $(R_1^*|_{\mathcal{H}}, R_2^*|_{\mathcal{H}}, \ldots, R_d^*|_{\mathcal{H}})$ is a part of $\underline{R}^*$ in the sense of Halmos [65]. Let $\underline{R}$ be a d-tuple of bounded operators on a Hilbert space $\mathcal{L}$. Consider

$$\begin{aligned}\mathcal{C}(\underline{R}) = \{\mathcal{M} : \mathcal{M} \text{ is a co-invariant subspace for each } R_i,\\ \text{and } R_i^* R_j^* h = R_j^* R_i^* h, \text{ for all } h \in \mathcal{M}, \text{ and } i, j = 1, \ldots, d\}.\end{aligned}$$

So $\mathcal{C}(\underline{R})$ consists of all co-invariant subspaces of a d-tuple of operators $\underline{R}$ such that the compressions of the tuple $\underline{R}$ to the co-invariant subspace form a commuting tuple. It is a complete lattice, in the sense that arbitrary intersections and closures of spans of arbitrary unions of such spaces are again in this collection. Therefore it has a maximal element. We denote it by $\mathcal{L}^c(\underline{R})$ (or by $\mathcal{L}^c$ when the tuple under consideration is clear).

Definition 4.7.1. *Suppose $\underline{R}$ is a d-tuple of operators on a Hilbert space $\mathcal{L}$. Then the* maximal commuting piece *of $\underline{R}$ is defined as the commuting piece $\underline{R}^c = (R_1^c, \ldots, R_d^c)$ obtained by compressing $\underline{R}$ to the maximal element $\mathcal{L}^c(\underline{R})$ of $\mathcal{C}(\underline{R})$. The maximal commuting piece is said to be* trivial *if the space $\mathcal{L}^c(\underline{R})$ is just the zero space.*

The following result is quite useful in determining the maximal commuting piece. Before going into it, we introduce another notation which will be very useful in this section. As we have seen, very often we need to consider products of the form $T_{i_1}T_{i_2}\cdots T_{i_k}$, where $1 \le i_1, i_2, \ldots, i_k \le d$ and $\underline{T} = (T_1, T_2, \ldots, T_d)$ is a tuple of operators acting on $\mathcal{H}$. It will be convenient to have a notation for this. Let Λ denote the set $\{1, 2, \ldots, d\}$ and Λ^k denote the k-fold cartesian product of Λ with itself for $k \ge 1$. Given $\underline{i} = (i_1, \ldots, i_k)$ in Λ^k, $\underline{T}^{\underline{i}}$ will mean the operator $T_{i_1}T_{i_2}\cdots T_{i_k}$. Let $\tilde{\Lambda}$ denote $\cup_{n=0}^{\infty}\Lambda^n$, where Λ^0 is just the set $\{0\}$ by convention and by $\underline{T}^0$ we would mean $I_{\mathcal{H}}$. We recall that $e_1, e_2, \ldots, e_d$ stands for the standard orthonormal basis of $\mathbb{C}^d$.

Lemma 4.7.2. *Let $\underline{R}$ be a d-tuple of bounded operators on a Hilbert space $\mathcal{L}$. Let $\mathcal{K}_{ij} = \overline{span}\{\underline{R}^{\underline{i}}(R_iR_j - R_jR_i)h : h \in \mathcal{L}, \underline{i} \in \tilde{\Lambda}\}$ for all $1 \le i, j \le d$, and $\mathcal{K} = \overline{span}\{\cup_{i,j=1}^d \mathcal{K}_{ij}\}$. Then $\mathcal{L}^c(\underline{R}) = \mathcal{K}^\perp$. In other words, $\mathcal{L}^c(\underline{R}) = \{h \in \mathcal{L} : (R_i^*R_j^* - R_j^*R_i^*)(\underline{R}^{\underline{i}})^*h = 0,$ for all $i, j = 1, 2, \ldots, d$ and $\underline{i} \in \tilde{\Lambda}\}$.*

Proof. Firstly $\mathcal{K}^\perp$ is a co-invariant subspace of $\underline{R}$. This is obvious as each R_i leaves $\mathcal{K}$ invariant. Now for $i, j \in \{1, 2, \ldots, d\}$, $h_1 \in \mathcal{L}^c$, and

$h_2 \in \mathcal{L}$,

$$\langle (R_i^* R_j^* - R_j^* R_i^*) h_1, h_2 \rangle = \langle h_1, (R_j R_i - R_i R_j) h_2 \rangle = 0.$$

So we get $(R_i^* R_j^* - R_j^* R_i^*) h_1 = 0$. So $\mathcal{K}^\perp$ is in $\mathcal{C}(\underline{R})$. Now if $\mathcal{M}$ is an element of $\mathcal{C}(\underline{R})$, take $i, j \in \{1, 2, \ldots, d\}, \underline{i} \in \tilde{\Lambda}, h_1 \in \mathcal{M}$, and $h \in \mathcal{L}$. We have

$$\langle h_1, \underline{R}^{\underline{i}} (R_i R_j - R_j R_i) h \rangle = \langle (R_j^* R_i^* - R_i^* R_j^*)(\underline{R}^{\underline{i}})^* h_1, h \rangle = 0$$

as $(\underline{R}^{\underline{i}})^* h_1 \in \mathcal{M}$ and R_i^* commutes with R_j^* on $\mathcal{M}$. Hence $\mathcal{M}$ is contained in $\mathcal{K}^\perp$ Now the last statement is easy to see. ∎

Corollary 4.7.3. *Suppose $\underline{R}$, $\underline{T}$ are d-tuples of operators on two Hilbert spaces $\mathcal{L}, \mathcal{M}$. Then the maximal commuting piece of $(R_1 \oplus T_1, \ldots, R_d \oplus T_d)$ acting on $\mathcal{L} \oplus \mathcal{M}$ is $(R_1^c \oplus T_1^c, \ldots, R_d^c \oplus T_d^c)$ acting on $\mathcal{L}^c \oplus \mathcal{M}^c$. The maximal commuting piece of $(R_1 \otimes I_{\mathcal{M}}, \ldots, R_d \otimes I_{\mathcal{M}})$ acting on $\mathcal{L} \otimes \mathcal{M}$ is $(R_1^c \otimes I_{\mathcal{M}}, \ldots, R_d^c \otimes I_{\mathcal{M}})$ acting on $\mathcal{L}^c \otimes \mathcal{M}$.*

Proof. As the product $(\underline{R} \oplus \underline{T})^{\underline{i}}$ breaks into the direct sum $\underline{R}^{\underline{i}} \oplus \underline{T}^{\underline{i}}$ and the product $(\underline{R} \otimes I_{\mathcal{M}})^{\underline{i}}$ breaks into the tensor product $\underline{R}^{\underline{i}} \otimes I_{\mathcal{M}}$, so the proof is clear from the Lemma above. ∎

Lemma 4.7.4. *Let $\underline{V} = (V_1, V_2, \ldots, V_d)$ and $\underline{S} = (S_1, S_2, \ldots, S_d)$ be standard contractive tuples on full Fock space $\Gamma(\mathbb{C}^d)$ and the symmetric Fock space $\Gamma_s(\mathbb{C}^d)$ respectively. Then the maximal commuting piece of $\underline{V}$ is $\underline{S}$.*

Proof. It is clear that $\underline{S}$ is a commuting piece of $\underline{V}$. To show maximality suppose $x \in \Gamma(\mathbb{C}^d)$ and $\langle x, \underline{V}^{\underline{i}}(V_j V_j - V_j V_i) y \rangle = 0$ for all $\underline{i} \in \tilde{\Lambda}, 1 \leq i, j \leq d$ and $y \in \Gamma(\mathbb{C}^d)$. We wish to show that $x \in \Gamma_s(\mathbb{C}^d)$. Suppose $x = \oplus_{m \geq 0} x_m$ with $x_m \in (\mathbb{C}^d)^{\otimes^m}$ for $m \geq 0$. For $m \geq 2$ and any permutation σ of $\{1, 2, \ldots, m\}$ we need to show that the unitary $U_\sigma : (\mathbb{C}^d)^{\otimes^m} \to (\mathbb{C}^d)^{\otimes^m}$, defined by

$$U_\sigma(u_1 \otimes \cdots \otimes u_m) = u_{\sigma^{-1}(1)} \otimes \cdots \otimes u_{\sigma^{-1}(m)},$$

leaves x_m fixed. Since the group of permutations of $\{1, 2, \ldots, m\}$ is generated by permutations $\{(1, 2), \ldots, (m-1, m)\}$ it is enough to verify $U_\sigma(x_m) = x_m$ for permutations σ of the form $(i, i+1)$. So fix m and i with $m \geq 2$ and $1 \leq i \leq (m-1)$. We have

$$\langle \oplus_p x_p, \underline{V}^{\underline{i}}(V_k V_l - V_l V_k) y \rangle = 0,$$

for every $y \in \Gamma(\mathbb{C}^d)$, $1 \leq k, l \leq d$. As $\underline{i}$ is arbitrary, this means that

$$\langle x_m, z \otimes (e_k \otimes e_l - e_l \otimes e_k) \otimes w \rangle = 0$$

for any $z \in (\mathbb{C}^n)^{\otimes(i-1)}, w \in (\mathbb{C}^d)^{\otimes(m-i-1)}$. This clearly implies $U_\sigma(x_m) = x_m$, for $\sigma = (i, i+1)$. ■

Here is an important lemma which will be used in the proof of the main theorem.

Lemma 4.7.5. *Suppose $\underline{T}, \underline{R}$ are d-tuples of bounded operators on $\mathcal{H}$, $\mathcal{L}$, with $\mathcal{H} \subseteq \mathcal{L}$, such that $\underline{R}$ is a dilation of $\underline{T}$. Then $\mathcal{H}^c(\underline{T}) = \mathcal{L}^c(\underline{R}) \bigcap \mathcal{H}$ and $\underline{R}^c$ is a dilation of $\underline{T}^c$.*

Proof. We have $R_i^* h = T_i^* h$, for $h \in \mathcal{H}$. Therefore,

$$(R_i^* R_j^* - R_j^* R_i^*)(\underline{R}^{\underline{i}})^* h = (T_i^* T_j^* - T_j^* T_i^*)(\underline{T}^{\underline{i}})^* h$$

for $h \in \mathcal{H}$, $1 \leq i, j \leq d$, and $\underline{i} \in \tilde{\Lambda}$. Now the first part of the result is clear from Lemma 4.7.2. Further for $h \in \mathcal{L}^c(\underline{R})$, $R_i^* h = (R_i^c)^* h$ and so for $h \in \mathcal{H}^c(\underline{T}) = \mathcal{L}^c(\underline{R}) \bigcap \mathcal{H}$, $(R_i^c)^* h = R_i^* h = T_i^* h = (T_i^c)^* h$. This proves the claim. ■

Suppose $\sum T_i T_i^* = I_{\mathcal{H}}$, then it is easy to see that $P_{\underline{T}}^m(I_{\mathcal{H}}) = I_{\mathcal{H}}$ for all m and there is no-way this sequence can converge to zero. So in the pure case the defect operator and the defect spaces are non-trivial. First we restrict our attention to pure tuples (not necessarily commuting).

Theorem 4.7.6. *Let $\underline{T}$ be a pure contractive tuple on a Hilbert space $\mathcal{H}$. Then the maximal commuting piece $\underline{\tilde{V}}^c$ of the minimal isometric dilation $\underline{\tilde{V}}$ of $\underline{T}$ is a realization of the standard commuting dilation of $\underline{T}^c$ if and only if $\overline{D_{\underline{T}}(\mathcal{H})} = \overline{D_{\underline{T}^c}(\mathcal{H}^c(\underline{T}))}$. In such a case rank $(D_{\underline{T}})$ = rank $(D_{\underline{T}^c})$ = rank $(D_{\underline{\tilde{V}}})$ = rank $(D_{\underline{\tilde{V}}^c})$.*

Proof. Note that one of the realizations of the standard non-commuting dilation of $\underline{T}$ consists of the space $\Gamma(\mathbb{C}^d) \otimes \mathcal{D}_{\underline{T}}$ and the operator tuple $\tilde{V} = (V_1 \otimes I_{\mathcal{D}_{\underline{T}}}, \ldots, V_d \otimes I_{\mathcal{D}_{\underline{T}}})$. The isometric embedding of $\mathcal{H}$ into the dilation space $\Gamma(\mathbb{C}^d) \otimes \mathcal{D}_{\underline{T}}$ is given by the Poisson transform $K(\underline{T})$. Since $\underline{T}$ is purely contractive it is obvious that $\underline{T}^c$ is a pure contractive tuple. We know from Lemma 4.7.5 that $\underline{\tilde{V}}^c = \underline{S} \otimes I_{\mathcal{D}_{\underline{T}}}$ on $\Gamma_s(\mathbb{C}^d) \otimes \mathcal{D}_{\underline{T}}$ is a commuting dilation of $\underline{T}^c$. On the other hand, the standard commuting dilation of $\underline{T}^c$ consists of the space $\Gamma_s(\mathbb{C}^d) \otimes \mathcal{M}$ where $\mathcal{M} = \overline{D_{\underline{T}^c}(\mathcal{H}^c(\underline{T}))}$ and the operator tuple $\underline{S} \otimes I_{\mathcal{M}}$. Since the standard commuting dilation is unique up to unitary equivalence, these two dilations are same if and only if $\mathcal{D}_{\underline{T}} = \overline{D_{\underline{T}^c}(\mathcal{H}^c(\underline{T}))}$.

The last contention can be seen as follows. When $\underline{\tilde{V}}^c$ is a realization of the standard commuting dilation of $\underline{T}^c$, then rank $(D_{\underline{T}^c}) =$ rank $(D_{\underline{\tilde{V}}^c})$. Also as $\underline{\tilde{V}}$ is the minimal isometric dilation of $\underline{T}$, rank $(D_{\underline{T}}) =$ rank $(D_{\underline{\tilde{V}}})$. Finally as $\underline{\tilde{V}}^c = (\underline{S} \otimes I_{\mathcal{D}_{\underline{T}}})$, rank $(D_{\underline{\tilde{V}}^c}) =$ rank $(D_{\underline{T}})$. ■

The main theorem is the following.

Theorem 4.7.7. *Suppose $\underline{T}$ is a commuting contractive tuple on a Hilbert space $\mathcal{H}$. Then the maximal commuting piece of the minimal isometric dilation of $\underline{T}$ is a realization of the standard commuting dilation of $\underline{T}$.*

Our approach to prove this theorem is as follows. First we consider the standard commuting dilation of $\underline{T}$ on a Hilbert space $\mathcal{H}_1$ as described above. Now the standard tuple $\underline{S}$ is also a contractive tuple. So we have a unique unital completely positive map $\eta : \mathcal{T}_d^{nc} \to \mathcal{T}_d^c$, satisfying

$$\eta(\underline{V}^{\underline{i}}(\underline{V}^{\underline{j}})^*) = \underline{S}^{\underline{i}}(\underline{S}^{\underline{j}})^* \ \underline{i}, \underline{j} \in \tilde{\Lambda}.$$

Now clearly $\psi = \varphi \circ \eta$. Consider the minimal Stinespring dilation of the composed map $\pi_1 \circ \eta : \mathcal{T}_d^{nc} \to \mathcal{B}(\mathcal{H}_1)$. Here we obtain a Hilbert space $\mathcal{H}_2$ containing $\mathcal{H}_1$ and a unital $*$-homomorphism $\pi_2 : \mathcal{T}_d^{nc} \to \mathcal{B}(\mathcal{H}_2)$, such that

$$\pi_1 \circ \eta(X) = P_{\mathcal{H}_1}\pi_2(X)|_{\mathcal{H}_1}, \quad \text{for all } X \in \mathcal{T}_d^{nc},$$

and $\overline{\text{span}}\,\{\pi_2(X)h : X \in \mathcal{T}_d^{nc}, h \in \mathcal{H}_1\} = \mathcal{H}_2$.

Now we have a commuting diagram as follows

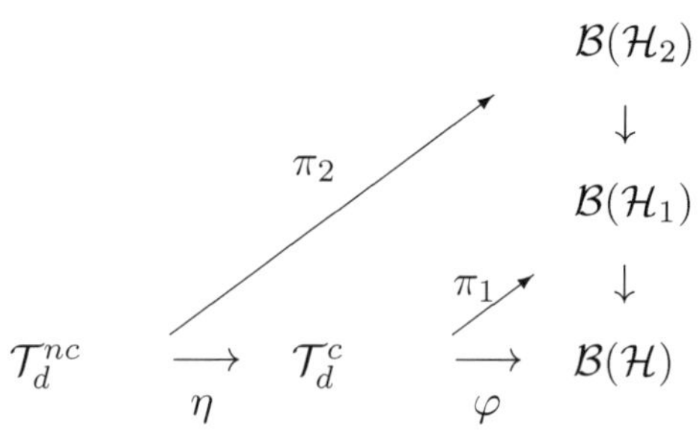

where all the down arrows are compression maps, horizontal arrows are unital completely positive maps and diagonal arrows are unital $*$-homomorphisms.

Taking $\underline{\hat{V}} = (\hat{V}_1, \ldots, \hat{V}_d) = (\pi_2(V_1), \ldots, \pi_2(V_d))$, we need to show
(i) $\underline{\hat{V}}$ is the minimal isometric dilation of $\underline{T}$ and
(ii) $\underline{\tilde{S}} = (\pi_1(S_1), \ldots, \pi_1(S_d))$ is the maximal commuting piece of $\underline{\hat{V}}$.

Due to uniqueness up to unitary equivalence of minimal Stinespring dilation, we have (i) if we can show that π_2 is a minimal dilation of $\psi = \varphi \circ \eta$. For proving this we actually make use of (ii). At first the assertion (ii) is proved in a very special case.

Definition 4.7.8. A d-tuple $\underline{T} = (T_1, T_2, \ldots, T_d)$ of operators on a Hilbert space $\mathcal{H}$ is called a spherical unitary if it is commuting, each T_i is normal, and $T_1T_1^* + \cdots + T_dT_d^* = I_{\mathcal{H}}$.

Given a d-tuple $\underline{T} = (T_1, T_2, \ldots, T_d)$ of operators satisfying $T_1T_1^* + \cdots + T_dT_d^* = I_{\mathcal{H}}$, it follows by Theorem 4.4.1 that its standard commuting dilation is a tuple of normal operators, hence each T_i^* is subnormal (or see [21] for this result). If moreover the space $\mathcal{H}$, where the operators T_i act, is a finite dimensional Hilbert space then applying the result that all finite dimensional subnormal operators are normal (see [65]), we get that normality of T_i is automatic. Thus any d-tuple $\underline{T} = (T_1, T_2, \ldots, T_d)$ of operators satisfying $T_1T_1^* + \cdots + T_dT_d^* = I_{\mathcal{H}}$ on a finite-dimensional Hilbert space is a spherical unitary.

Note that if $\underline{T}$ is a spherical unitary we have

$$\varphi(\underline{S}^{\underline{i}}(I_{\Gamma_s(\mathbb{C}^d)} - \sum S_iS_i^*)(\underline{S}^{\underline{j}})^*) = \underline{T}^{\underline{i}}(I_{\mathcal{H}} - \sum T_iT_i^*)(\underline{T}^{\underline{j}})^* = 0$$

for any $\underline{i}, \underline{j} \in \tilde{\Lambda}$. This forces that $\varphi(X) = 0$ for any compact operator X in $\mathcal{T}_d^c$. Now as the commutators $[S_i^*, S_j]$ are all compact we see that φ is a unital $*$-homomorphism. So the minimal Stinespring dilation of φ is itself. Hence the standard commuting dilation of a spherical unitary is itself. So the following result would yield Theorem 6.11.6 for spherical unitaries.

Theorem 4.7.9. *Let $\underline{T}$ be a spherical unitary on a Hilbert space $\mathcal{H}$. Then the maximal commuting piece of the minimal isometric dilation of $\underline{T}$ is $\underline{T}$.*

The proof of this theorem involves lengthy computations and we refer the reader to [28] for its proof, omitting the proof here. But assuming this, we prove the main Theorem.

Proof of Theorem 4.7.7 : As $\mathcal{T}_d^c$ contains the ideal of all compact operators by standard C^*-algebra theory we have a direct sum decomposition of π_1 as follows. Take $\mathcal{H}_1 = \mathcal{H}_{1C} \oplus \mathcal{H}_{1N}$ where $\mathcal{H}_{1C} = \overline{\text{span}}\{\pi(X)h : h \in$

$\mathcal{H}, X \in C^*(\underline{S})$ and X is compact$\}$ and $\mathcal{H}_{1N} = \mathcal{H}_1 \ominus \mathcal{H}_{1C}$, Clearly $\mathcal{H}_{1C}$ is a reducing subspace for π_1. Therefore

$$\pi_1(X) = \begin{pmatrix} \pi_{1C}(X) & \\ & \pi_{1N}(X) \end{pmatrix}$$

that is, $\pi_1 = \pi_{1C} \oplus \pi_{1N}$ where $\pi_{1C}(X) = P_{\mathcal{H}_{1C}}\pi_1(X)P_{\mathcal{H}_{1C}}$, $\pi_{1N}(X) = P_{\mathcal{H}_{1N}}\pi_1(X)P_{\mathcal{H}_{1N}}$. As observed by Arveson [17], $\pi_{1C}(X)$ is just the identity representation with some multiplicity. More precisely, $\mathcal{H}_{1C}$ can be factored as $\mathcal{H}_{1C} = \Gamma_s(\mathbb{C}^d) \otimes \overline{D_{\underline{T}}(\mathcal{H})}$, such that $\pi_{1C}(X) = X \otimes I$, in particular $\pi_{1C}(S_i) = S_i \otimes I$ where I here is the identity operator on $\overline{D_{\underline{T}}(\mathcal{H})}$. Also $\pi_{1N}(X) = 0$ for compact X. Therefore, taking $Z_i = \pi_{1N}(S_i)$, $\underline{Z} = (Z_1, \ldots, Z_d)$ is a spherical unitary.

Now as $\pi_1 \circ \eta = (\pi_{1C} \circ \eta) \oplus (\pi_{1N} \circ \eta)$ and the minimal Stinespring dilation of a direct sum of two completely positive maps is the direct sum of minimal Stinespring dilations so $\mathcal{H}_2$ decomposes as $\mathcal{H}_2 = \mathcal{H}_{2C} \oplus \mathcal{H}_{2N}$, where $\mathcal{H}_{2C}, \mathcal{H}_{2N}$ are orthogonal reducing subspaces of π_2, such that π_2 also decomposes, say $\pi_2 = \pi_{2C} \oplus \pi_{2N}$, with

$$\pi_{1C} \circ \eta(X) = P_{\mathcal{H}_{1C}}\pi_{2C}(X)|_{\mathcal{H}_{1C}}, \; \pi_{1N} \circ \eta(X) = P_{\mathcal{H}_{1N}}\pi_{2N}(X)|_{\mathcal{H}_{1N}},$$

for $X \in \mathcal{T}_d^{nc}$ with $\mathcal{H}_{2C} = \overline{\text{span}}\, \{\pi_{2C}(X)h : X \in \mathcal{T}_d^{nc}, h \in \mathcal{H}_{1C}\}$ and $\mathcal{H}_{2N} = \overline{\text{span}}\, \{\pi_{2N}(X)h : X \in \mathcal{T}_d^{nc}, h \in \mathcal{H}_{1N}\}$. It is also not difficult to see that $\mathcal{H}_{2C} = \overline{\text{span}}\, \{\pi_{2C}(X)h : X \in \mathcal{T}_d^{nc}, X \text{ compact}, h \in \mathcal{H}_{1C}\}$ and hence $\mathcal{H}_{2C}$ factors as $\mathcal{H}_{2C} = \Gamma(\mathbb{C}^d) \otimes \overline{D_{\underline{T}}(\mathcal{H})}$ with $\pi_{2C}(V_i) = V_i \otimes I$. Also $(\pi_{2N}(V_1), \ldots, \pi_{2N}(V_d))$ is a minimal isometric dilation of spherical isometry $(Z_1, \ldots, Z_d)$. Now we get that $(\pi_1(S_1), \ldots, \pi_1(S_d))$ acting on $\mathcal{H}_1$ is the maximal commuting piece of $(\pi_2(V_1), \ldots, \pi_2(V_d))$.

All that remains to show is that π_2 is the minimal Stinespring dilation of $\varphi \circ \eta$. Suppose this is not the case. Then we get a reducing subspace $\mathcal{H}_{20}$ for π_2 by taking $\mathcal{H}_{20} = \overline{\text{span}}\, \{\pi_2(X)h : X \in \mathcal{T}_d^{nc}, h \in \mathcal{H}\}$. Take $\mathcal{H}_{21} = \mathcal{H}_2 \ominus \mathcal{H}_{20}$ and correspondingly decompose π_2 as $\pi_2 = \pi_{20} \oplus \pi_{21}$,

$$\pi_2(X) = \begin{pmatrix} \pi_{20}(X) & \\ & \pi_{21}(X) \end{pmatrix}$$

Note that we already have $\mathcal{H} \subseteq \mathcal{H}_{20}$. We claim that $\mathcal{H}_2 \subseteq \mathcal{H}_{20}$. Firstly, as $\mathcal{H}_1$ is the space where the maximal commuting piece of

$$(\pi_2(V_1), \ldots, \pi_2(V_d)) = (\pi_{20}(V_1) \oplus \pi_{21}(V_1), \ldots, \pi_{20}(V_d) \oplus \pi_{21}(V_d))$$

acts, by the first part of Corollary 4.7.3, $\mathcal{H}_1$ decomposes as $\mathcal{H}_1 = \mathcal{H}_{10} \oplus \mathcal{H}_{11}$ for some subspaces $\mathcal{H}_{10} \subseteq \mathcal{H}_{20}$, and $\mathcal{H}_{11} \subseteq \mathcal{H}_{21}$. So for $X \in \mathcal{T}_d^{nc}$,

$P_{\mathcal{H}_1}\pi_2(X)P_{\mathcal{H}_1}$, has the form (see the diagram)

$$P_{\mathcal{H}_1}\pi_2(X)P_{\mathcal{H}_1} = \begin{pmatrix} \pi_{10}\circ\eta(X) & 0 & & \\ 0 & 0 & & \\ & & \pi_{11}\circ\eta(X) & 0 \\ & & 0 & 0 \end{pmatrix}$$

where π_{10}, π_{11} are compressions of π_1 to $\mathcal{H}_{10}$, $\mathcal{H}_{11}$ respectively. As the mapping η from $\mathcal{T}_d^{nc}$ to $\mathcal{T}_d^c$ is clearly surjective, it follows that $\mathcal{H}_{10}, \mathcal{H}_{11}$ are reducing subspaces for π_1. Now as $\mathcal{H}$ is contained in $\mathcal{H}_{20}$, in view of minimality of π_1 as a Stinespring dilation, $\mathcal{H}_1 \subseteq \mathcal{H}_{20}$. But then the minimality of π_2 shows that $\mathcal{H}_2 \subseteq \mathcal{H}_{20}$. Therefore, $\mathcal{H}_2 = \mathcal{H}_{20}$. ∎

This theorem puts in perspective the two standard dilations of a commuting tuple, but that is not its only merit. As an application, the article [28] goes on to find a neat classification of all representations of Cuntz algebra $\mathcal{O}_d$ coming from dilations of commuting contractive tuples. In [49], Dey showed that if $\underline{T}$ is a tuple of q-commuting operators, then the standard q-commuting dilation is the maximal q-commuting part of its standard non-commuting dilation.

We finish with the comment that the material presented here are merely the basics and starting from here one can pursue any of a number of directions. Popescu has generalized much of the Sz. Nagy - Foias theory to the case of an infinite set of contractions satisfying the contractivity property, see [95] - [94] and his papers with Arias [10], [11]. The dilation theory over the years has inspired many a branch. Some of the prominent upshots are the study of Hilbert modules which have been systematically studied by Douglas, Misra, Paulsen and Varughese [51], [53], [55] and the theory of curvature of contractive modules by Arveson [18]. The dilations of quantum dynamical semigroups by Arveson [15], [19] and Bhat [24], [26] draw their original motivation from Sz-Nagy Foias dilation. In the linked notion of dilating homomorphisms, there arise natural questions of which contractive homomorphisms are completely contractive. For a study of this question on various domains in $\mathbb{C}^d$, see the works of Misra, Pati, Paulsen and Sastry [77], [89], [79], [80]. For various generalizations of the von Neumann inequality to Banach space setting, we refer to the monograph of Pisier [92] and the references therein.

CHAPTER 5

The Euclidean Ball - The Drury Arveson Space

5.1. The space

In this chapter, we recast some of the results of the last chapter in a function theoretic setting. Let $\mathbb{B}_d$ be the open unit ball in the complex d-dimensional space $\mathbb{C}^d$ with the Euclidean norm:

$$\mathbb{B}_d = \left\{ z = (z_1, z_2, \ldots, z_d) \in \mathbb{C}^d : \|z\| < 1 \right\}.$$

Here $\|z\| = (|z_1|^2 + |z_z|^2 + \cdots + |z_d|^2)^{1/2}$.

We begin by recalling the definition of a holomorphic function in several variables. Let Ω be a domain in $\mathbb{C}^d$, i.e., an open connected subset. A function $f : \Omega \to \mathbb{C}$ is holomorphic if for each $z^0 \in \Omega$, there is an $r = r(z^0) > 0$ such that $\{z \in \mathbb{C}^d : \|z - z^0\| < r\} \subset \Omega$ and f can be written in an absolutely and uniformly convergent power series

$$f(z) = \sum_{\underline{k}} a_{\underline{k}} (z - z^0)^{\underline{k}}$$

for all z satisfying $\|z - z^0\| < r$. Here we have used the notation $\underline{k}$ for the multi-index $(k_1, k_2, \ldots, k_d)$ and $z^{\underline{k}}$ for $z_1^{k_1} z_2^{k_2} \ldots z_d^{k_d}$. For several equivalent definitions of a holomorphic function, see page 5 of [73].

Recall from Section 4.2 that for a multi-index $\underline{k} = (k_1, k_2, \ldots, k_d)$, the symbol $|\underline{k}|$ denotes the sum $k_1 + k_2 + \cdots + k_d$. Let $\gamma_{\underline{k}}$ be the multinomial coefficient

$$\gamma_{\underline{k}} = \frac{|\underline{k}|!}{k_1! k_2! \ldots k_d!}. \tag{5.1.1}$$

Consider the set H_d^2 of all functions which are holomorphic on $\mathbb{B}_d$ (and hence have absolutely and uniformly convergent power series $f(z) =$

$\sum_{\underline{k}} a_{\underline{k}} z^{\underline{k}}$ in $\mathbb{B}_d$) and satisfy

$$\sum_{\underline{k}} \frac{|a_{\underline{k}}|^2}{\gamma_{\underline{k}}} < \infty.$$

It is straightforward that H^2_d is a vector space. Moreover, it is an inner product space by declaring, for $f(z) = \sum_{\underline{k}} a_{\underline{k}} z^{\underline{k}}$ and $g(z) = \sum_{\underline{k}} b_{\underline{k}} z^{\underline{k}}$,

$$\langle f, g\rangle = \sum_{\underline{k}} \frac{a_{\underline{k}} \overline{b_{\underline{k}}}}{\gamma_{\underline{k}}}. \tag{5.1.2}$$

This inner product space, which we shall show to be complete, is called the *Drury-Arveson space.*

Introduce the following set of vectors in H^2_d:

$$k_w(z) = \frac{1}{1 - \langle z, w\rangle_{\mathbb{C}^d}} = \sum_{n=0}^{\infty} \langle z, w\rangle^k = \sum_{\underline{k}} \gamma_{\underline{k}} z^{\underline{k}} \overline{w}^{\underline{k}}.$$

Here, w varies over $\mathbb{B}_d$ and $\overline{w}$ stands for $(\overline{w}_1, \overline{w}_2, \ldots, \overline{w}_d)$.

Lemma 5.1.1. *The set $\{k_w : w \in \mathbb{B}_d\}$ is reproducing in H^2_d in the sense that for any f in H^2_d, we have*

$$\langle f, k_w\rangle_{H^2_d} = f(w).$$

Proof. Let $f(z) = \sum_{\underline{k}} a_{\underline{k}} z^{\underline{k}}$. Then, $\langle f, k_w\rangle_{H^2_d} = \sum_{k=0}^{\infty} \langle w^{\otimes k}, \xi_k\rangle = f(w)$. ∎

Lemma 5.1.2. *The set $\{k_w : w \in \mathbb{B}_d\}$ is total in H^2_d.*

Proof. Since $\langle f, k_w\rangle_{H^2_d} = f(w)$, if f is orthogonal to the set $\{k_w : w \in \mathbb{B}_d\}$, then f is identically 0 in $\mathbb{B}_d$. ∎

Lemma 5.1.3. *If a sequence $\{f_n\}$ of elements of H^2_d converges in the norm of H^2_d, then it converges uniformly on compact subsets of $\mathbb{B}_d$.*

Proof. The proof is the same as the one variable case. Let the limit be f. Let K be a compact subset of $\mathbb{B}_d$. Get an r such that $K \subset r\mathbb{B}_d$. So, for z in K, we have

$$\begin{aligned} |f_n(z) - f(z)| = |\langle f_n - f, k_z\rangle| &\le \|f_n - f\|_{H^2_d} \|k_z\|_{H^2_d} \\ &= \frac{\|f_n - f\|_{H^2_d}}{1 - \|z\|^2} \le \frac{\|f_n - f\|_{H^2_d}}{1 - r^2}. \end{aligned}$$

Consequently, the sequence $\{f_n\}$ converges uniformly to f in the compact set K. ∎

Lemma 5.1.4. *H_d^2 is the completion of the set of all polynomials in the inner product defined in* (5.1.2). *Hence, H_d^2 is a Hilbert space.*

Proof. Clearly, H_d^2 is contained the completion of the set of all polynomials in the inner product defined in (5.1.2). Let f be an element of the completion and let $\{f_n\}$ be a Cauchy sequence of polynomials converging to f. Thus, f_n converges uniformly to f on compact sets. Thus, f is holomorphic on $\mathbb{B}_d$. (The proof of this one variable result holds equally well in several variables, see for example [64].) Let $f(z) = \sum_{\underline{k}} a_{\underline{k}} z^{\underline{k}}$ be the series expansion of f. Let P_m be the projection onto the finite dimensional space of polynomials of degree up to m. Then,

$$\sum_{\underline{k}:|\underline{k}|\leq m} \gamma_{\underline{k}} |a_{\underline{k}}|^2 = \|P_m f\|^2 = \lim_{n\to\infty} \|P_m f_n\|^2 \leq \lim_{n\to\infty} \|f_n\|^2 = \|f\|^2.$$

So, f is in H_d^2. ■

Definition 5.1.5. *A multiplier φ in H_d^2 is a holomorphic function on $\mathbb{B}_d$ such that $\varphi \cdot f$ is in H_d^2 for every $f \in H_d^2$. We shall denote the algebra of all multipliers on H_d^2 by $\mathcal{M}(H_d^2)$, or simply $\mathcal{M}$ if there is no confusion.*

A straightforward application of Closed Graph Theorem implies that for every $\varphi \in \mathcal{M}$, the linear transformation $M_\varphi : H_d^2 \to H_d^2$ defined by

$$M_\varphi f = \varphi \cdot f$$

is bounded. We endow $\mathcal{M}$ with its natural norm

$$\|\varphi\| = \|M_\varphi\|.$$

The notation $H^\infty(\mathbb{B}_d)$ or just H^∞ in case there is no chance of confusion will denote the Banach algebra of all bounded holomorphic function on $\mathbb{B}_d$.

Lemma 5.1.6. *The algebra $\mathcal{M}$ is contained in H^∞ and the inclusion map is a contraction.*

Proof. First note that $M_\varphi^* k_w = \overline{\varphi(w)} k_w$ for $\varphi \in M$ and $w \in \mathbb{B}_d$. Indeed, for any $f \in H_d^2$, we have

$$\begin{aligned}\langle M_\varphi^* k_w, f\rangle &= \langle k_w, M_\varphi f\rangle = \overline{\langle \varphi \cdot f, k_w\rangle} \\ &= \overline{\varphi f(w)} = \overline{\varphi(w) f(w)} = \overline{\varphi(w)} \langle k_w, f\rangle.\end{aligned}$$

Since f is arbitrary, we have $M_\varphi^* k_w = \overline{\varphi(w)} k_w$. So, for each $w \in \mathbb{B}_d$, the number $\overline{\varphi(w)}$ is an eigenvalue of M_φ^*. Since spectral radius is not larger than operator norm, we have $|\varphi(w)| \leq \|M_\varphi\|$.

Since the inequality holds for each $w \in \mathbb{B}_d$, we have $\|\varphi\|_\infty \leq \|M_\varphi\|$. ∎

Definition 5.1.7. *We shall denote by* $M_{z_1}, M_{z_2}, \ldots, M_{z_d}$ *the multipliers by the co-ordinate functions* $z_1, z_2, \ldots, z_d$.

Lemma 5.1.8. *The tuple* $(M_{z_1}, M_{z_2}, \ldots, M_{z_d})$ *forms a commuting contractive tuple.*

Proof. It is clear that the tuple of operators is a commuting one. We need to prove that it is contractive. To that end, first note that the family of vectors $\{k_w : w \in \mathbb{B}_d\}$ forms a total set. Indeed, if $f \in H^2_d$ is orthogonal to k_w for every $w \in \mathbb{B}_d$, then $f(w) = \langle f, k_w\rangle = 0$ for all $w \in \mathbb{B}_d$ showing that f is identically zero. Now,

$$\begin{aligned}&(M_{z_1}M_{z_1}^* + M_{z_2}M_{z_2}^* + \cdots + M_{z_d}M_{z_d}^* k_w)(z)\\ =&z_1\overline{w}_1 k_w(z) + z_2\overline{w}_2 k_w(z) + \cdots + z_d\overline{w}_d k_w(z)\\ =&\frac{z_1\overline{w}_1 + z_2\overline{w}_2 + \cdots + z_d\overline{w}_d}{1 - \langle z, w\rangle}\\ =&\frac{1}{1-\langle z,w\rangle} - 1 = k_w(z) - E_0 k_w(z)\end{aligned}$$

where E_0 is the projection onto the one-dimensional space of constant functions. The above then proves the operator equality

$$\sum_{k=1}^{n} M_{z_k}M_{z_k}^* = I - E_0.$$

This proves the assertion. ∎

Corollary 5.1.9. *For any* $\underline{w} \in \overline{\mathbb{B}_d}$, *we have* $\|\sum_{i=1}^d w_i M_{z_i}\| \leq 1$.

Proof. The proof is as follows.

$$\begin{aligned}&\|\sum_{i=1}^{d} w_i M_{z_i}\|\\ =&\|(M_{z_1}, M_{z_2}, \ldots, M_{z_d})\begin{pmatrix} w_1\\ w_2\\ \vdots\\ w_d\end{pmatrix}(\overline{w_1}, \overline{w_2}, \ldots, \overline{w_d})\begin{pmatrix} M_{z_1}^*\\ M_{z_2}^*\\ \vdots\\ M_{z_d}^*\end{pmatrix}\|^{1/2}\\ \leq&\|(M_{z_1}, M_{z_2}, \ldots, M_{z_d})\begin{pmatrix} M_{z_1}^*\\ M_{z_2}^*\\ \vdots\\ M_{z_d}^*\end{pmatrix}\|^{1/2} \leq 1.\end{aligned}$$

∎

The following lemma connects the co-ordinate multipliers with the creation operators on the symmetric Fock space discussed in the previous chapter. We would like to note here that when we talk of the co-ordinates of z in $\mathbb{C}^d$, this means that we are tacitly using a fixed orthonormal basis $e_1, \ldots, e_d$, say. Usually, it is taken to be the standard orthonormal basis $e_i = (0, \ldots, 0, 1, 0, \ldots, 0)$ of $\mathbb{C}^d$ with 1 at the ith place and 0 elsewhere, but any orthonormal basis will work.

Lemma 5.1.10. *There is a unique unitary operator $W : H_d^2 \to \Gamma_s(\mathbb{C}^d)$ such that $W(1) = 1$ and $WM_{z_i} = S_i W$ for $i = 1, 2, \ldots, d$, where S_i on $\Gamma_s\left(\mathbb{C}^d\right)$ are defined by*

$$S_i(\xi) = \mathcal{P}_{\Gamma_s(\mathbb{C}^d)}(e_i \otimes \xi), \xi \in \Gamma_s(\mathbb{C}^d).$$

Proof. For $z \in \mathbb{B}_d$, let $\overline{z}$ be $(\overline{z}_1, \overline{z}_2, \ldots, \overline{z}_d)$ which is also an element of $\mathbb{B}_d$. Consider the vector

$$v_{\overline{z}} = 1 \oplus \overline{z} \oplus \overline{z}^{\otimes 2} \oplus \ldots$$

in $\Gamma_s(\mathbb{C}^d)$ and define a linear transformation W from H_d^2 to $\Gamma_s(\mathbb{C}^d)$ by first defining

$$Wk_z = v_{\overline{z}},$$

then extending it to linear combinations of k_z for $z \in \mathbb{B}_d$. It is an isometry mapping the dense subspace spanned by $\{k_z : z \in \mathbb{B}_d\}$ onto the dense subspace spanned by $\{v_z : z \in \mathbb{B}_d\}$. Indeed,

$$\begin{aligned}
\langle v_{\overline{z}}, v_{\overline{w}} \rangle &= 1 + \langle \overline{z}, \overline{w} \rangle + \langle \overline{z}^{\otimes 2}, \overline{w}^{\otimes 2} \rangle + \ldots \\
&= 1 + \langle \overline{z}, \overline{w} \rangle + \langle \overline{z}, \overline{w} \rangle^2 + \ldots \\
&= (1 - \langle \overline{z}, \overline{w} \rangle)^{-1} \\
&= (1 - \langle w, z \rangle)^{-1} = \langle k_z, k_w \rangle
\end{aligned}$$

and hence

$$\begin{aligned}
\langle \sum c_i v_{\overline{z}_i}, \sum d_j v_{\overline{w}_j} \rangle &= \sum c_i \overline{d_j} \langle v_{\overline{z}_i}, v_{\overline{w}_j} \rangle \\
&= \sum c_i \overline{d_j} \langle k_{z_i}, k_{w_j} \rangle = \langle \sum c_i k_{z_i}, \sum d_j k_{w_j} \rangle.
\end{aligned}$$

Thus, W extends as a surjective isometry from H_d^2 to $\Gamma_s(\mathbb{C}^d)$. We now need to show that the unitary W intertwines M_{z_i} on H_d^2 with S_i on $\Gamma_s(\mathbb{C}^d)$. To that end, we recall that we showed in the last chapter that if $a \in \mathbb{C}^d$ and $A : (\mathbb{C}^d)^{\circledS p-1} \to (\mathbb{C}^d)^{\circledS p}$ is defined as

$$A(P_s(y_1 \otimes y_2 \otimes \cdots \otimes y_{p-1})) = P_s(a \otimes y_1 \otimes \ldots y_{p-1}),$$

then

$$A^*(P_s(y_1 \otimes y_2 \otimes \cdots \otimes y_p)) = \frac{1}{p} \sum_{i=1}^{p} \langle y_i, a \rangle p_s(y_1 \otimes \cdots \otimes y_{i-1} \otimes y_{i+1} \otimes \cdots \otimes y_p).$$

Applying this result to $y_1 = y_2 = \cdots = y_p = \overline{z}$ and $a = e_i$, we get that

$$S_i^*(\overline{z}^{\otimes p}) = \overline{z}_i \overline{z}^{\otimes p-1} \text{ for } p = 1, 2, \ldots.$$

Thus,

$$\begin{aligned} W^* S_i^* v_{\overline{z}} = W^* \left(\overline{z}_i \cdot 1 \oplus \overline{z}_i \overline{z} \oplus \overline{z}_i \overline{z}^{\otimes 2} \oplus \ldots\right) &= W^*(\overline{z}_i v_{\overline{z}}) \\ &= \overline{z}_i W^* v_{\overline{z}} = \overline{z}_i k_z. \end{aligned}$$

And $M_{z_i}^* W^* v_{\overline{z}} = M_{z_i}^* k_z = \overline{z}_i k_z$.

Since, $W^* S_i^*$ and $M_{z_i}^* W^*$ agree on $v_{\overline{z}}$ which span the whole space, $W^* S_i^* = M_{z_i}^* W^*$ for $i = 1, 2, \ldots, d$. That completes the proof. ∎

The result above shows that the tuples $(M_{z_1}, M_{z_2}, \ldots, M_{z_d})$ on H_d^2 and $(S_1, S_2, \ldots, S_d)$ on $\Gamma_s\left(\mathbb{C}^d\right)$ are unitarily equivalent through W.

Corollary 5.1.11. *The norm of the monomial $z_1^{k_1} z_2^{k_2} \ldots z_d^{k_d}$ is the square root of the reciprocal of the multinomial coefficient, i.e.,*

$$\|z_1^{k_1} z_2^{k_2} \ldots z_d^{k_d}\| = \left(\frac{k_1! k_2! \ldots k_d!}{(k_1 + k_2 + \cdots + k_d)!}\right)^{1/2}.$$

Proof. This follows immediately from the lemma above by the action of W. We know that the constant function 1 is taken to the vacuum vector. Since W intertwines M_{z_i} and S_i, we see that the monomial $z_1^{k_1} z_2^{k_2} \ldots z_d^{k_d}$ is mapped to $e_1^{k_1} e_2^{k_2} \ldots e_d^{k_d}$ which has norm $(\frac{k_1! k_2! \ldots k_d!}{|\underline{k}|!})^{1/2}$ as obtained in Chapter 4. ∎

Theorem 5.1.12. *Let $\mathcal{B}$ be the Banach subalgebra of $\mathcal{B}(H_d^2)$ generated by $M_{z_1}.M_{z_2}, \ldots, M_{z_d}$, i.e., $\mathcal{B}$ is the closure in $\mathcal{B}(H_d^2)$ of the algebra of all polynomials in $M_{z_1}.M_{z_2}, \ldots, M_{z_d}$. Then the set of multiplicative linear functionals $\hat{\mathcal{B}}$ with the weak-* topology is homeomorphic to $\overline{\mathbb{B}_d}$.*

Proof. To any $\underline{z}$ in $\overline{\mathbb{B}_d}$, we associate a multiplicative linear functional on $\mathcal{B}$ by first defining it on polynomials $p(M_{z_1}, M_{z_2}, \ldots, M_{z_d})$ by

$$\phi_{\underline{z}}(p(M_{z_1}, M_{z_2}, \ldots, M_{z_d})) = p(\underline{z}).$$

This is clearly linear and multiplicative on the dense set

$$\{p(M_{z_1}, M_{z_2}, \ldots, M_{z_d}) : p \text{ is a polynomial } \}.$$

It is bounded there because

$$\begin{aligned}|\phi_{\underline{z}}(p(M_{z_1}, M_{z_2}, \ldots, M_{z_d}))| &= |p(\underline{z})| \\ &\leq \|p\|_{\infty,\mathbb{B}_d} \\ &\leq \|M_p\| \text{ (by Lemma 5.1.6)} \\ &= \|p(M_{z_1}, M_{z_2}, \ldots, M_{z_d})\|.\end{aligned}$$

So, it extends to the whole of $\mathcal{B}$ as a bounded linear functional. It is trivial too check that the extension remains multiplicative.

Given any ϕ in $\hat{\mathcal{B}}$, define $\underline{z} = (\phi(M_{z_1}), \phi(M_{z_2}), \ldots, \phi(M_{z_d}))$. First, we note that $\underline{z} \in \overline{\mathbb{B}_d}$ because for any $\underline{w} \in \overline{\mathbb{B}_d}$, we have

$$\begin{aligned}|\langle \underline{z}, \underline{w}\rangle| = |\sum_{i=1}^{d} z_i\overline{w_i}| &= |\sum_{i=1}^{d} \phi(M_{z_i})\overline{w_i}| \\ &= |\phi(\sum_{i=1}^{d} \overline{w_i}M_{z_i})| \leq \|\sum_{i=1}^{d} \overline{w_i}M_{z_i}\| \leq 1\end{aligned}$$

by Corollary 5.1.9.

Since ϕ is a multiplicative linear functional,

$$\begin{aligned}&\phi_{\underline{z}}(p(M_{z_1}, M_{z_2}, \ldots, M_{z_d})) \\ &\stackrel{\text{def}}{=} p(\underline{z}) \\ &= p(\phi(M_{z_1}), \phi(M_{z_2}), \ldots, \phi(M_{z_d}) \text{ by the definition of } \underline{z} \text{ above} \\ &= \phi(p(M_{z_1}, M_{z_2}, \ldots, M_{z_d}))\end{aligned}$$

and hence $\phi = \phi_{\underline{z}}$.

That $\underline{z} \to \phi_{\underline{z}}$ is a homeomorphism is easy to check. ■

We saw in the last chapter that any commuting contractive tuple $\underline{T} = (T_1, T_2, \ldots, T_d)$ acting on H has a dilation. In view of the unitary identification of $\Gamma_s(\mathbb{C}^d)$ with H_d^2, the dilation space $\hat{H}$ is $(H_d^2 \otimes D_{T^*}) \oplus N$ and the dilation operator tuple is

$$((M_{z_1} \otimes 1) \oplus Z_1, (M_{z_2} \otimes 1) \oplus Z_2, \ldots, (M_{z_d} \otimes 1) \oplus Z_d),$$

where $(Z_1, Z_2, \ldots, Z_d)$ is a tuple of commuting normal operators with

$$\sum_{i=1}^{d} Z_i Z_i^* = 1.$$

The dilation space and the dilation tuple is unique only up to unitary equivalence. This is why $\Gamma_s(\mathbb{C}^d)$ and H^2_d along with their canonical tuple $(S_1, S_2, \ldots, S_d)$ or $(M_{z_1}, M_{z_2}, \ldots, M_{z_d})$ can be used interchangeably.

Conventionally, given a tuple of commuting bounded operators, a dilation tuple is usually a tuple of commuting normal operators, like the unitary dilation of a contraction or the commuting unitary dilation of a pair of commuting contractions. More on this theme will be expounded in the next chapter. It is a natural question why we are not looking for such a dilation for a commuting contractive tuple. We recall that the dilation that has been produced for a commuting contractive tuple does have such a part $(Z_1, Z_2, \ldots, Z_d)$ which is indeed a tuple of commuting normal operators. So, one would be done if the other part

$$(M_{z_1} \otimes I, M_{z_2} \otimes I, \ldots, M_{z_d} \otimes I)$$

on $(H^2_d \otimes D_{T^*})$ could be dilated to a tuple of commuting normal operators.

Given a tuple of commuting normal operators $\underline{N} = (N_1, N_2, \ldots, N_d)$, consider the unital abelian C^*-algebra $C^*(\underline{N})$ generated by $N_1, N_2, \ldots, N_d$ and the identity operator. Recall from Chapter 2 that the set of multiplicative linear functionals (or the character space) $\widehat{C^*(\underline{N})}$ of $C^*(\underline{N})$ is the set of complex homomorphisms on $C^*(\underline{N})$. It is compact in the weak-* topology. Recall also that for X in $C^*(\underline{N})$, the Gelfand transform which is a continuous function on $\widehat{C^*(\underline{N})}$ is denoted by $\hat{X}$. Since continuous image of a compact set is compact, the set

$$\begin{aligned}\sigma(\underline{N}) &= \{(w(N_1), w(N_2), \ldots, w(N_d)) : w \in \widehat{C^*(\underline{N})}\} \\ &= \{(\hat{N}_1(w), \hat{N}_2(w), \ldots, \hat{N}_d(w)) : w \in \widehat{C^*(\underline{N})}\}\end{aligned}$$

is a compact subset of $\mathbb{C}^d$. This is called the *(algebraic) joint spectrum* of the tuple $\underline{N}$. In the next chapter, we shall see a general concept of joint spectrum for an arbitrary tuple of commuting bounded operators on a Hilbert space. For the purpose of this chapter, when we refer to the joint spectrum, we shall only refer to commuting bounded normal operators and then we shall always mean the algebraic joint spectrum.

Lemma 5.1.13. *Consider a tuple of commuting bounded normal operators $\underline{N} = (N_1, N_2, \ldots, N_d)$. Let p be a polynomial in d variables. Then,*

$$\|p(N_1, N_2, \ldots, N_d)\| = \sup\{|p(\lambda_1, \lambda_2, \ldots, \lambda_d)| : (\lambda_1, \lambda_2, \ldots, \lambda_d) \in \sigma(\underline{N})\}.$$

Proof. If p is any polynomial in d variables, then by Gelfand theory,

$$\begin{aligned}\sigma(p(N_1,N_2,\ldots,N_d)) &= \{w(p(N_1,N_2,\ldots,N_d)) : w \in \widehat{C^*(\underline{N})}\}\\ &= \{p(w(N_1),w(N_2),\ldots,w(N_d)) : w \in \widehat{C^*(\underline{N})}\}\\ &= p(\sigma(\underline{N})).\end{aligned}$$

Since $p(N_1,N_2,\ldots,N_d)$ is a normal operator, we have

$$\begin{aligned}&\|p(N_1,N_2,\ldots,N_d)\|\\ =&r(p(N_1,N_2,\ldots,N_d))\\ =&\sup\{|p(\lambda_1,\lambda_2,\ldots,\lambda_d)| : (\lambda_1,\lambda_2,\ldots,\lambda_d) \in \sigma(\underline{N})\}.\end{aligned}$$

■

Corollary 5.1.14. *If $\underline{N}$ is a tuple of commuting bounded normal operators whose joint spectrum is contained in the closed unit ball, then*

$$\|p(N_1,N_2,\ldots,N_d)\| \le \|p\|_{\infty,\overline{\mathbb{B}}_d}.$$

If $\underline{T} = (T_1,T_2,\ldots,T_d)$ is a commuting tuple of bounded operators acting on $\mathcal{H}$ which has a commuting normal dilation $\underline{N} = (N_1,N_2,\ldots,N_d)$ on a space $\mathcal{K} \supseteq \mathcal{H}$ with $\sigma(\underline{N}) \subseteq \overline{\mathbb{B}}_d$, then

$$\|p(T_1,T_2,\ldots,T_d)\| = \|P_H p(N_1,N_2,\ldots,N_d)|_H\| \le \|p\|_{\infty,\overline{\mathbb{B}}_d}$$

for any polynomial p in d variables. We shall end this section by showing that the tuple $(M_{z_1},M_{z_2},\ldots,M_{z_d})$ has no commuting normal dilation whose joint spectrum is contained in the closed unit Euclidean ball. The following lemma is the major step for that.

Lemma 5.1.15. *Let $d \ge 2$. Let $c_0,c_1,\ldots$ be a sequence of complex numbers such that*

(i) $\displaystyle\sum_{n=0}^{\infty}|c_n| = 1$ *(ii)* $\displaystyle\sum_{n=0}^{\infty}|c_n|^2 n^{\frac{d-1}{2}} = \infty.$

Let

$$s = \sup\{|z_1z_2\ldots z_d| : \underline{z} = (z_1,z_2,\ldots,z_d) \in \mathbb{B}_d\}.$$

Consider the polynomials

$$f_N(z_1,z_2,\ldots,z_d) = \sum_{n=0}^{N}\frac{c_n}{s^n}(z_1z_2\ldots z_d)^n.$$

Then $\|f_N\| \le 1$ for every N and

$$\lim_{N\to\infty}\|f_N(M_{z_1},M_{z_2},\ldots,M_{z_d})\| = \infty.$$

Proof. First note that arithmetic-geometric mean inequality implies that

$$\frac{1}{d}\left(|z_1|^2 + |z_2|^2 + \cdots + |z_d|^2\right) \geq \left(|z_1|^2|z_2|^2 \ldots |z_d|^2\right)^{1/d},$$

the equality being attained at $|z_1| = |z_2| = \cdots = |z_d|$. Hence $s = \sqrt{\frac{1}{d^d}}$.

Now, for $(z_1, z_2, \ldots, z_d) \in \mathbb{B}_d$.

$$|f_N(z_1, z_2, \ldots, z_d)| \leq \sum_{n=0}^{N} \frac{|c_n|}{s^n}|z_1, z_2, \ldots, z_d|^n \leq \sum_{n=0}^{\infty} |c_n| = 1.$$

But,

$$\|f_N\left(M_{z_1}, M_{z_2}, \ldots, M_{z_d}\right)\| = \|M_{f_N}\| \geq \|M_{f_N} \cdot 1\|_{H_d^2} = \|f_N\|_{H_d^2}.$$

Denote by p the polynomial $p(z_1, z_2, \ldots, z_d) = z_1 z_2 \ldots z_d$. Then

$$f_N(z_1, z_2, \ldots, z_d) = \sum_{n=0}^{N} \frac{c_n}{s^n} p^n.$$

According to the definition of the Drury-Arveson space, p^n and p^m are orthogonal to each other if $n \neq m$. Thus,

$$\|f_N\|_{H_d^2}^2 = \sum_{n=0}^{\infty} \frac{|c_n|^2}{s^{2n}} \|p^n\|_{H_d^2}^2.$$

As for the norm of the monomial p^n, invoke Corollary 5.1.11 to get that $\|p^n\|^2 = \frac{(n!)^d}{(nd)!}$.

Using Stirling's formula, $n! \sim \sqrt{2\pi}\frac{n^{n+\frac{1}{2}}}{e^n}$, we get

$$\|p^n\|^2 \sim (2\pi)^{\frac{d-1}{2}} n^{\frac{d-1}{2}} d^{-(nd+\frac{1}{2})} = \left(\frac{2\pi^{d-1}}{2}\right)^{1/2} n^{\frac{d-1}{2}} d^{-nd}.$$

Thus,

$$\|f_N\|_{H_d^2}^2 \sim \sum_{n=0}^{N} |c_n|^2 n^{\frac{d-1}{2}}.$$

Because of the way c_n have been chosen this diverges. That completes the proof. ■

Corollary 5.1.16. *The tuple* $(M_{z_1}, M_{z_2}, \ldots, M_{z_d})$ *is not subnormal.*

Proof. Let f be any polynomial in d variables. Then, by Theorem 5.1.12, we have

$$\|f(M_{z_1}, M_{z_2}, \ldots, M_{z_d})\| = \sup_{|z_1|^2+|z_2|^2+\cdots+|z_d|^2\leq 1} |f(z_1, z_2, \ldots, z_d)|.$$

If the tuple $(M_{z_1}, M_{z_2}, \ldots, M_{z_d})$ is subnormal, then so is the operator $f(M_{z_1}, M_{z_2}, \ldots, M_{z_d})$. For a subnormal operator, the norm is the same as the spectral radius. This, now, contradicts Lemma 5.1.15. ∎

Corollary 5.1.17. *The tuple* $(M_{z_1}, M_{z_2}, \ldots, M_{z_d})$ *has no dilation to a commuting tuple* $\underline{N} = (N_1, N_2, \ldots, N_d)$ *of bounded normal operators with joint spectrum contained in the closed unit ball.*

Proof. This conclusion is straightforward from Corollary 5.1.14, the discussion following it and Lemma 5.1.15. ∎

The material of this section was from Arveson [17], Drury [60] and Muller and Vasilescu [82].

5.2. The characteristic function

Now, we move in a direction parallel to a significant result about a complete unitary invariant that we proved in the first chapter. The invariant is the characteristic function. Recall that in Chapter 1, we showed that the characteristic function is a complete unitary invariant for a pure (or $C_{\cdot 0}$) contraction. We shall now define such a function for a commuting contractive tuple and will show its complete unitary invariance for a pure commuting contractive tuple. Enroute, we shall provide a functional model too.

We briefly recall the result that is true for a single contraction. A key ingredient for studying contraction operators on Hilbert spaces is the following analytic operator-valued function, called the characteristic function of T:

$$\theta_{\underline{T}}(z) = -T + zD_{T^*}(I_{\mathcal{H}} - zT^*)^{-1}D_T, \; z \in \mathbb{D}. \tag{5.2.1}$$

The operators D_T and D_{T^*} are the so-called defect operators $(I_{\mathcal{H}} - T^*T)^{1/2}$ and $(I_{\mathcal{H}} - TT^*)^{1/2}$ of T and T^*, respectively. By virtue of the relation $TD_T = D_{T^*}T$ (see (1.10.1)), the values $\theta_{\underline{T}}(z)$ of the characteristic function can be regarded as bounded operators from $\mathcal{D}_{\underline{T}} = \overline{\text{Ran}} D_T$ into $\mathcal{D}_{\underline{T}^*} = \overline{\text{Ran}} D_{T^*}$.

We showed in Chapter 1 that $\theta_{\underline{T}}(z)$ is contraction valued and that $\| \theta_{\underline{T}}(0)\xi \| < \| \xi \|$ for all $\xi \in \mathcal{D}_{\underline{T}}$. The characteristic functions $\theta_{\underline{T}}$ and θ_R of two contractions T and R are said to coincide if there are unitary operators $\sigma_1 : \mathcal{D}_{\underline{T}} \to \mathcal{D}_R$ and $\sigma_2 : \mathcal{D}_{\underline{T}^*} \to \mathcal{D}_{R^*}$ such that

$$\theta_{\underline{T}}(z) = \sigma_2^{-1}\theta_R(z)\sigma_1 \text{ for all } z \in \mathbb{D}. \tag{5.2.2}$$

It is easy to see that if T and R are two unitarily equivalent contractions, i.e., if there is a unitary operator U such that $T = URU^*$, then the characteristic functions $\theta_{\underline{T}}$ and θ_R coincide. One can easily construct examples to show that the converse of this is not true in this generality. However, we proved in Theorem 1.11.4 that the converse is true if both T and R are pure contractions.

Theorem 5.2.1. *Two pure contractions are unitarily equivalent if and only if their characteristic functions coincide.*

The route to prove the theorem is via constructing a functional model.

Now we embark on proving such a theorem for a commuting contractive tuple. If $\mathcal{E}$ is a complex Hilbert space, define $\mathcal{O}(\mathbb{B}_d, \mathcal{E})$ to be the class of all $\mathcal{E}$-valued analytic functions on $\mathbb{B}_d$. Then consider the Hilbert space

$$H_d^2(\mathcal{E}) = \{f \in \mathcal{O}(\mathbb{B}_d, \mathcal{E}) : f = \sum_{k \in \mathbb{N}^d} a_k z^k \tag{5.2.3}$$

$$\text{with } a_k \in \mathcal{E} \text{ and } \|f\|^2 = \sum_{k \in \mathbb{N}^d} \frac{\|a_k\|^2}{\gamma_k} < \infty\}$$

where γ_k is the multinomial coefficient defined in (5.1.1). When $\mathcal{E} = \mathbb{C}$, the space is the same as the one considered in the last section. Of course, when $d = 1$ and $\mathcal{E} = \mathbb{C}$, this is the usual Hardy space on the unit disk. Given complex Hilbert spaces $\mathcal{E}$ and $\mathcal{E}_*$, the multiplier space $M(\mathcal{E}, \mathcal{E}_*)$ consists of all $\varphi \in \mathcal{O}(\mathbb{B}_d, \mathcal{B}(\mathcal{E}, \mathcal{E}_*))$ such that $\varphi H_d^2(\mathcal{E}) \subset H_d^2(\mathcal{E}_*)$. By the closed graph theorem, for each function $\varphi \in M(\mathcal{E}, \mathcal{E}_*)$, the induced multiplication operator $M_\varphi : H_d^2(\mathcal{E}) \to H_d^2(\mathcal{E}_*), f \mapsto \varphi f$ is continuous.

Recall that a contraction T is said to be of class $C_{\cdot 0}$ if T^{*m} converges strongly to 0 as $m \to \infty$ and the characteristic function θ_T of a single contraction T is a multiplier from the Hardy space $H_{\mathbb{D}}^2(\mathcal{D}_{\underline{T}})$ to the Hardy space $H_{\mathbb{D}}^2(D_{T^*})$. If T is a $C_{\cdot 0}$ contraction acting on a Hilbert space $\mathcal{H}$, then there is a unitary operator U from $\mathcal{H}$ onto $\mathbb{H} = H_{\mathbb{D}}^2(\mathcal{D}_{T^*}) \ominus M_{\theta_T} H_{\mathbb{D}}^2(\mathcal{D}_{\underline{T}})$ such that $UTU^* = P_{\mathbb{H}} M_z|_{\mathbb{H}}$ where M_z is the multiplication operator with the independent variable z on $H_{\mathbb{D}}^2(\mathcal{D}_{T^*})$. Thus any $C_{\cdot 0}$ contraction can be realized as $P_{\mathbb{H}} M_z|_{\mathbb{H}}$ where the model space $\mathbb{H}$ is the orthocomplement of the range of M_{θ_T}.

For a commuting contractive tuple $\underline{T} = (T_1, T_2, \ldots, T_d)$, we shall show below that a similar model can be constructed. The positive operator $(I_{\mathcal{H}} - \sum_{i=1}^{d} T_i T_i^*)^{1/2}$ and the closure of its range are called the *defect operator* $D_{\underline{T}^*}$ and the *defect space* $\mathcal{D}_{\underline{T}^*}$ of T^*.

We denote by T the bounded operator from $\mathcal{H}^d$ to $\mathcal{H}$ which maps

$$(h_1, h_2, \ldots, h_d) \text{ to } T_1 h_1 + T_2 h_2 + \cdots + T_d h_d.$$

The action of the adjoint operator $T^* : \mathcal{H} \to \mathcal{H}^d$ maps h to the column vector $(T_1^* h, T_2^* h, \ldots, T_d^* h)$ and, in fact, the tuple $\underline{T}$ is a contractive tuple if and only if the operator T is a contraction. Thus for a contractive tuple $\underline{T}$, one can also consider the defect operator $D_{\underline{T}} = (I_{\mathcal{H}_d} - T^*T)^{1/2} = ((\delta_{ij} I_{\mathcal{H}} - T_i^* T_j))^{1/2}$ in $\mathcal{B}(\mathcal{H}^d)$ and the associated defect space $\mathcal{D}_{\underline{T}} = \overline{\operatorname{Ran} D_{\underline{T}}} \subset \mathcal{H}^d$. We start with a generalization of (1.10.1).

Lemma 5.2.2. *For any commuting contractive tuple $\underline{T}$, we obtain the identity*

$$T D_{\underline{T}} = D_{\underline{T}^*} T.$$

Proof. Let R be a contraction from a Hilbert space $\mathcal{H}'$ into a Hilbert space $\mathcal{H}$. Then

$$R D_R^2 = R(I - R^*R) = (I - RR^*)R = D_{R^*}^2 R.$$

Thus, for any polynomial p, we have $Rp(D_R^2) = p(D_{R^*}^2)R$. Since the square root is a limit of polynomials, we have $RD_R = D_{R^*}R$. Here, we have the special case of the operator $R = T$ defined above from $\mathcal{H}' = \mathcal{H}^d$ into $\mathcal{H}$. ∎

Note that, for $z = (z_1, z_2, \ldots, z_d) \in \mathbb{B}_d$, the operator Z from $\mathcal{H}^d$ to $\mathcal{H}$ which maps $(h_1, h_2, \ldots, h_d)$ to $z_1 h_1 + z_2 h_2 + \cdots + z_d h_d$ is a contraction because $ZZ^* = \sum |z_i|^2 I_{\mathcal{H}}$. Thus $Z = (z_1 I_{\mathcal{H}}, z_2 I_{\mathcal{H}}, \ldots, z_d I_{\mathcal{H}})$ is a commuting contractive tuple with $\|Z\| = (\sum |z_i|^2)^{1/2} < 1$. Hence, given a commuting contractive tuple T, the operator ZT^* is a strict contraction for $z \in \mathbb{B}_d$ and hence $I_{\mathcal{H}} - ZT^*$ is invertible. We define the characteristic function of T to be the analytic operator-valued function $\theta_{\underline{T}} : \mathbb{B}_d \to \mathcal{B}(\mathcal{D}_{\underline{T}}, T_\infty \mathcal{D}_{\underline{T}^*})$ with

$$\theta_{\underline{T}}(z) = -T + D_{\underline{T}^*}(I_{\mathcal{H}} - ZT^*)^{-1} Z D_{\underline{T}}, \ z \in \mathbb{B}_d \tag{5.2.4}$$

Lemma 5.2.3. *Given a commuting contractive tuple $\underline{T}$, its characteristic function $\theta_{\underline{T}}$ is a multiplier, that is $\theta_{\underline{T}} \in M(\mathcal{D}_{\underline{T}}, \mathcal{D}_{\underline{T}^*})$, with $\|M_{\theta_{\underline{T}}}\| \leq 1$. For $z, w \in \mathbb{B}_d$, the identity*

$$1 - \theta_{\underline{T}}(z)\theta_{\underline{T}}(w)^* = (1 - ZW^*) D_{\underline{T}^*}(1 - ZT^*)^{-1}(1 - TW^*)^{-1} D_{\underline{T}} \tag{5.2.5}$$

holds.

Proof. Since $\theta_{\underline{T}}(z) = -T + D_{\underline{T}^*}(I_{\mathcal{H}} - ZT^*)^{-1}ZD_{\underline{T}}$ and $\theta_{\underline{T}}(w)^* = -T^* + D_{\underline{T}}W^*(I_{\mathcal{H}} - TW^*)^{-1}D_{\underline{T}^*}$, we have

$$
\begin{aligned}
&\theta_{\underline{T}}(z)\theta_{\underline{T}}(w)^* \\
=&TT^* - TD_{\underline{T}}W^*(I_{\mathcal{H}} - TW^*)^{-1}D_{\underline{T}^*} - D_{\underline{T}^*}(I_{\mathcal{H}} - ZT^*)^{-1}ZD_{\underline{T}}T^* \\
&+D_{\underline{T}^*}(I_{\mathcal{H}} - ZT^*)^{-1}Z(I - T^*T)W^*(I_{\mathcal{H}} - TW^*)^{-1}D_{\underline{T}^*}
\end{aligned}
$$

and hence

$$
\begin{aligned}
&I - \theta_{\underline{T}}(z)\theta_{\underline{T}}(w)^* \\
&= D_{\underline{T}^*}^2 + D_{\underline{T}^*}TW^*(I - TW^*)^{-1}D_{\underline{T}^*} + D_{\underline{T}^*}(I - ZT^*)^{-1}ZT^*D_{\underline{T}^*} \\
&- D_{\underline{T}^*}(I - ZT^*)^{-1}ZW^*(I - TW^*)^{-1}D_{\underline{T}^*} \\
&+ D_{\underline{T}^*}(I - ZT^*)^{-1}ZT^*TW^*(I - TW^*)^{-1}D_{\underline{T}^*} \\
&= D_{\underline{T}^*}[I - I + (I - TW^*)^{-1} \\
&- I + (I - ZT^*)^{-1} - \|z\|^2(I - ZT^*)^{-1}(I - TW^*)^{-1} \\
&+ (-I + (I - ZT^*)^{-1})(-I + (I - TW^*)^{-1})]D_{\underline{T}^*} \\
&= D_{\underline{T}^*}[(I - TW^*)^{-1} \\
&- I + (I - ZT^*)^{-1} - ZW^*(I - ZT^*)^{-1}(I - TW^*)^{-1} \\
&+ (I - (I - ZT^*)^{-1} - (I - TW^*)^{-1} + (I - ZT^*)^{-1}(I - TW^*)^{-1})]D_{\underline{T}^*} \\
&= (1 - ZW^*)D_{\underline{T}^*}(I - ZT^*)^{-1}(I - TW^*)^{-1})D_{\underline{T}^*}
\end{aligned}
$$

■

For $z = w$, the right-hand side of formula (5.2.5) defines a positive operator. Thus we obtain the following theorem.

Theorem 5.2.4. *Given a commuting contractive tuple $\underline{T}$, its characteristic function $\theta_{\underline{T}}$ is a bounded analytic function on $\mathbb{B}_d$ with $\sup_{z\in\mathbb{B}_d} \|\theta_{\underline{T}}(z)\| \leq 1$.*

5.3. Functional model of a pure commuting contractive tuple

This section will bring out very clearly the important role that the characteristic function plays for commuting contractive tuples.

For the rest of this chapter, we call the d-shift the shift because the dimension d is fixed. By definition, this is the commuting tuple $M_z = (M_{z_1}, M_{z_2}, \ldots, M_{z_d})$ on the scalar-valued functional Hilbert space H_d^2 consisting of the multiplication operators M_{z_i} with the coordinate functions z_i. It is not hard to show that

$$\text{SOT} - \lim_{k\to\infty} \sum_{1\leq i_1, i_2,\ldots i_k\leq n} M_{z_{i_1}} M_{z_{i_2}} \ldots M_{z_{i_k}} M_{z_{ik}}^* \ldots M_{z_{i_2}}^* M_{z_{i_1}}^* = 0.$$

Thus the shift is an example of a *pure* commuting contractive tuple in the sense of the following definition.

For a commuting contractive tuple $\underline{T}$ on a Hilbert space $\mathcal{H}$, recall the completely positive map $P_{\underline{T}} : \mathcal{B}(\mathcal{H})$ by $P_{\underline{T}}(X) = \sum_{i=1}^{d} T_i X T_i^*$ discussed in Chapter 4. Also recall that $T_\infty \in \mathcal{B}(\mathcal{H})$ denotes the strong limit of the decreasing sequence of positive operators $I \geq P_{\underline{T}}(I) \geq P_{\underline{T}}^2(I) \geq \cdots \geq 0$. The commuting contractive tuple $\underline{T}$ is called pure if $T_\infty = 0$.

Remark 5.3.1. *In the case $d = 1$, a contraction $T \in \mathcal{B}(\mathcal{H})$ is pure in the above sense if and only if it is of class $C_{\cdot 0}$.*

The operator L in the theorem below is the precursor of the functional model that we are going to construct. We have discussed the following result in Theorem 4.3.3. We repeat it here because now, instead of the Fock space, the lanmguage is that of a function space.

Theorem 5.3.2. *Let $\underline{T}$ be a commuting contractive tuple of operators on some Hilbert space $\mathcal{H}$. Then there exists a unique bounded linear operator $L : H_d^2 \otimes \mathcal{D}_{\underline{T}^*} \to \mathcal{H}$ satisfying*

$$L(f \otimes \xi) = f(T_1, T_2, \ldots, T_d) D_{\underline{T}^*} \xi$$

for all f in $\mathbb{C}[z_1, z_2, \ldots, z_d]$ and ξ in $\mathcal{D}_{\underline{T}^*}$. Furthermore, we have $LL^* = I_{\mathcal{H}} - T_\infty$ and the identity

$$L(f(M_{z_1} \otimes I_{\mathcal{D}_{\underline{T}^*}}, M_{z_2} \otimes I_{\mathcal{D}_{\underline{T}^*}}, \ldots, M_{z_d} \otimes I_{\mathcal{D}_{\underline{T}^*}})) = f(T_1, T_2, \ldots, T_d) L$$

holds for all f in $\mathbb{C}[z_1, z_2, \ldots, z_d]$ where $\mathbb{C}[z_1, z_2, \ldots, z_d]$ is the algebra of all polynomials in d complex variables.

Proof. The uniqueness is clear because polynomials are dense in H_d^2.

As usual, the operator L is constructed through its adjoint $A : \mathcal{H} \to H_d^2 \otimes \mathcal{D}_{\underline{T}^*}$ defined by

$$Ah = \sum_{\underline{k}} \frac{\underline{z}^{\underline{k}}}{\|\underline{z}^{\underline{k}}\|^2} \otimes D_{\underline{T}^*} (T_1^*)^{k_1} (T_2^*)^{k_2} \ldots (T_d^*)^{k_d} h.$$

To check that Ah indeed inside inside $H_d^2 \otimes \mathcal{D}_{\underline{T}^*}$, we do a norm computation.

$$\|Ah\|^2$$

$$= \sum_{\underline{k}} \sum_{\underline{l}} \langle \frac{\underline{z}^{\underline{k}}}{\|\underline{z}^{\underline{k}}\|^2}, \frac{\underline{z}^{\underline{l}}}{\|\underline{z}^{\underline{l}}\|^2} \rangle \langle D_{T^*}(T_1^{k_1}T_2^{k_2}\dots T_d^{k_d})^* h, D_{T^*}(T_1^{k_1}T_2^{k_2}\dots T_d^{k_d})^* h \rangle$$

$$= \sum_{\underline{k}} \frac{1}{\|\underline{z}^{\underline{k}}\|^2} \langle T_1^{k_1}T_2^{k_2}\dots T_d^{k_d} D_{T^*}^2 (T_1^{k_1}T_2^{k_2}\dots T_d^{k_d})^* h, h \rangle$$

(because the $\frac{\underline{z}^{\underline{k}}}{\|\underline{z}^{\underline{k}}\|}$ are orthonormal)

$$= \sum_{\underline{k}} \gamma_{\underline{k}} \langle T_1^{k_1}T_2^{k_2}\dots T_d^{k_d} (I - \sum_{i=1}^d T_i T_i^*)(T_1^{k_1}T_2^{k_2}\dots T_d^{k_d})^* h, h \rangle$$

$$= \sum_{n=0}^{\infty} \sum_{\underline{k}:|\underline{k}|=n} \gamma_{\underline{k}} \langle T_1^{k_1}T_2^{k_2}\dots T_d^{k_d} (I - \sum_{i=1}^d T_i T_i^*)(T_1^{k_1}T_2^{k_2}\dots T_d^{k_d})^* h, h \rangle$$

$$= \sum_{n=0}^{\infty} (\langle P_T^n(I)h, h\rangle - \langle P_T^{n+1}(I)h, h\rangle)$$

where $\gamma_{\underline{k}}$ is as defined in (5.1.1) and P_T stands for the completely positive map $P_T(X) = \sum_{i=1}^d T_i X T_i^*$.

The last sum is telescopic and hence equals $\|h\|^2 - \|T_\infty^{1/2} h\|^2$. That shows that Ah is indeed in $H_d^2 \otimes \mathcal{D}_{\underline{T}^*}$. Moreover, the operator A is a contraction and is an isometry if $T_\infty = 0$.

Now, put $L = A^*$. This L satisfies $LL^* = I_{\mathcal{H}} - T_\infty$. Let us see how L acts on elements of the form $\underline{z}^{\underline{k}} \otimes \xi$ for $\xi \in \mathcal{D}_{T^*}$.

$$\begin{aligned}\langle L(\underline{z}^{\underline{k}} \otimes \xi), h\rangle &= \langle \underline{z}^{\underline{k}} \otimes \xi, Ah\rangle \\ &= \langle \underline{z}^{\underline{k}} \otimes \xi, \sum_{\underline{l}} \frac{\underline{z}^{\underline{l}}}{\|\underline{z}^{\underline{l}}\|^2} \otimes D_{T^*}(T^*)^{\underline{l}} h\rangle \\ &= \langle \underline{z}^{\underline{k}} \otimes \xi, \frac{\underline{z}^{\underline{k}}}{\|\underline{z}^{\underline{k}}\|^2} \otimes D_{T^*}(T^*)^{\underline{k}} h\rangle = \langle T^{\underline{k}} D_{T^*}\xi, h\rangle.\end{aligned}$$

From here, chasing the powers of each z_i appropriately, we get,

$$L(M_{z_i} \otimes I)(\underline{z}^{\underline{k}} \otimes \xi) = T_i L(\underline{z}^{\underline{k}} \otimes \xi)$$

and that leads us to

$$\begin{aligned}&L(f(M_{z_1} \otimes I_{\mathcal{D}_{T^*}}, M_{z_2} \otimes I_{\mathcal{D}_{\underline{T}^*}}, \dots, M_{z_d} \otimes I_{\mathcal{D}_{\underline{T}^*}})) \\ &= f(T_1, T_2, \dots, T_d) L\end{aligned}$$

for all f in $\mathbb{C}[z_1, z_2, \dots, z_d]$.

■

Remark 5.3.3. *The tuple $\underline{T}$ is pure if and only if L is a co-isometry.*

Given a Hilbert space $\mathcal{E}$, there is a natural identification between $H_d^2 \otimes \mathcal{E}$ and $H_d^2(\mathcal{E})$. There is a canonical unitary operator $U_{\mathcal{E}} : H_d^2 \otimes \mathcal{E} \to H_d^2(\mathcal{E})$ with $U_{\mathcal{E}}(f \otimes x) = fx$ for $f \in H_d^2$ and $x \in \mathcal{E}$. In the following, we shall identify the space $H_d^2 \otimes \mathcal{E}$ and $H_d^2(\mathcal{E})$ via this unitary operator $U_{\mathcal{E}}$. We denote by $M_z^{\mathcal{E}} = (M_{z_1}^{\mathcal{E}}, M_{z_2}^{\mathcal{E}}, \dots, M_{z_d}^{\mathcal{E}}) \in \mathcal{B}(H_d^2(\mathcal{E}))^d$ the tuple of multiplication operators induced by the coordinate functions z_i. Via the unitary above, the tuple $M_z^{\mathcal{E}}$ is unitarily equivalent to $(M_{z_1} \otimes I_{\mathcal{E}}, M_{z_2} \otimes I_{\mathcal{E}}, \dots, M_{z_d} \otimes I_{\mathcal{E}})$. In this way, each multiplier $\varphi \in M(\mathcal{E}, \mathcal{E}_*)$ induces a bounded operator $M_\varphi : H_d^2 \otimes \mathcal{E} \to H_d^2 \otimes \mathcal{E}_*$.

Recall from the last section the vectors k_w defined by $k_w(z) = (1 - \langle z, w\rangle)^{-1}$ for z in $\mathbb{B}_d$.

Lemma 5.3.4. *Let $\varphi \in M(\mathcal{E}, \mathcal{E}_*)$ be a multiplier. Then the identity*

$$M_{\varphi^*}(k_z \otimes x) = k_z \otimes \varphi(z)^* x$$

holds for all $z \in \mathbb{B}_d$ and $x \in \mathcal{E}_*$.

Proof. Fix $z \in \mathbb{B}_d$ and $x \in \mathcal{E}_*$. Note first that

$$\langle f \otimes y, k_z \otimes x\rangle = f(z)\langle y, x\rangle = \langle f(z)y, x\rangle.$$

holds for all $f \in H_d^2$ and $y \in \mathcal{E}_*$. Hence it follows that $\langle f, k_z \otimes x\rangle = \langle f(z), x\rangle$ for each function $f \in H_d^2(\mathcal{E}_*)$. Using this identity twice (for $\mathcal{E}$ and $\mathcal{E}_*$-valued functions), we obtain that

$$\langle f, M_\varphi^*(k_z \otimes x)\rangle = \langle \varphi(z) f(z), x\rangle = \langle f, k_z \otimes \varphi(z)^* x\rangle$$

■

Next we relate the operator L with the characteristic function.

Lemma 5.3.5. *Given a commuting contractive tuple T, we obtain the identity*

$$L^*L + M_{\theta_{\underline{T}}} M_{\theta_{\underline{T}}}^* = I_{H_d^2 \otimes \mathcal{D}_{\underline{T}^*}}$$

Proof. For an element $z = (z_1, z_2, \dots, z_d)$ of the ball $\mathbb{B}_d$, the operator L satisfies the identity

$$L(k_z \otimes \xi) = (1 - TZ^*)^{-1} D_{T^*} \xi \quad (z \in \mathbb{B}_d, \xi \in \mathcal{D}_{\underline{T}^*}).$$

This is a straightforward inner product computation and we leave it as an exercise. Therefore, for z, w in $\mathbb{B}_d$ and ξ, η in $\mathcal{D}_{\underline{T}^*}$, we obtain that

$$\begin{aligned}
&\langle (L^*L + M_{\theta_{\underline{T}}} M_{\theta_{\underline{T}}}^*) k_z \otimes \xi, k_w \otimes \eta \rangle \\
=&\langle L(k_z \otimes \xi), L(k_w \otimes \eta) \rangle + \langle M_{\theta_{\underline{T}}}^*(k_z \otimes \xi), M_{\theta_{\underline{T}}}^*(k_w \otimes \eta) \rangle \\
=&\langle (1 - TZ^*)^{-1} D_{T^*}\xi, (1 - TW^*)^{-1} D_{T^*}\eta \rangle \\
+&\langle k_z \otimes \theta_{\underline{T}}(z)^*\xi, k_w \otimes \theta_{\underline{T}}(w)^*\eta \rangle \\
=&\langle D_{T^*}(1 - WT^*)^{-1}(1 - TZ^*)^{-1} D_T^*\xi, \eta \rangle + \langle k_z, k_w \rangle \langle \theta_{\underline{T}}(w)\theta_{\underline{T}}(z)^*\xi, \eta \rangle \\
=&\langle k_z \otimes \xi, k_w \otimes \eta \rangle
\end{aligned}$$

Using the fact that the vectors k_z form a total set in H_d^2, the assertion follows. ■

In [17], Arveson used abstract factorisation results to prove the existence of an auxiliary space, say $\mathcal{D}$, and a multiplier $\varphi \in M(\mathcal{D}, \mathcal{D}_{\underline{T}^*})$ such that

$$I_{H_d^2 \otimes \mathcal{D}_{\underline{T}^*}} - L^*L = M_\varphi M_\varphi^*.$$

The lemma above shows that $\mathcal{D}$ can be chosen to be $\mathcal{D}_T$ and φ can be chosen to be the characteristic function of T.

As usual, we call two commuting tuples $T = (T_1, T_2, \ldots, T_d)$ and $R = (R_1, R_2, \ldots, R_d)$ of bounded operators on Hilbert spaces $\mathcal{H}$ and $\mathcal{K}$ unitarily equivalent if there exists a unitary operator U from $\mathcal{H}$ to $\mathcal{K}$ such that $R_i = UT_iU^*$ holds for all $i = 1, 2, \ldots, d$. Now we have arrived at the main theorem of this section.

Theorem 5.3.6. *Every pure commuting contractive tuple T on a Hilbert space $\mathcal{H}$ is unitarily equivalent to the commuting tuple $\mathbb{T} = (\hat{T}_1, \hat{T}_2, \ldots, \hat{T}_d)$ on the functional space $\mathbb{H}_T = (H_d^2 \otimes \mathcal{D}_{\underline{T}^*}) \ominus M_{\theta_{\underline{T}}}(H_d^2 \otimes \mathcal{D}_{\underline{T}})$ defined by $\hat{T}_i = P_{\mathbb{H}_T}(M_{z_i} \otimes I_{\mathcal{D}_{\underline{T}^*}})|\mathbb{H}_T$ for $1 \leq i \leq n$.*

Proof. Since T is pure, the map

$$L^* : \mathcal{H} \to H_d^2 \otimes \mathcal{D}_{\underline{T}^*}$$

is an isometry. Thus $\mathcal{H}$ is isometrically embedded into $H_d^2 \otimes \mathcal{D}_{\underline{T}^*}$ via the identification of $\mathcal{H}$ with the closed subspace $L^*\mathcal{H}$. Now L^*L is the projection of $H_d^2 \otimes \mathcal{D}_{\underline{T}^*}$ onto the closed subspace $L^*\mathcal{H}$. But then by Lemma 3.6, the operators L^*L and $M_{\theta_{\underline{T}}} M_{\theta_{\underline{T}}}^*$ are mutually orthogonal projections which add up to identity. Therefore the subspace $L^*\mathcal{H}$ is the orthocomplement of the range of $M_{\theta_{\underline{T}}}$:

$$L^*\mathcal{H} = (H_d^2 \otimes \mathcal{D}_{\underline{T}^*}) \ominus M_{\theta_{\underline{T}}}(H_d^2 \otimes \mathcal{D}_{\underline{T}}).$$

Now by Theorem 3.3, $L^*T_i^* = (M_{z_i} \otimes I_{\mathcal{D}_{\underline{T}^*}})^* L^*$. Thus the subspace $L^*\mathcal{H}$ is co-invariant for the shift and, via the identification of $\mathcal{H}$ with $L^*\mathcal{H}$, the operators T_i in $\mathcal{B}(\mathcal{H})$ are unitarily equivalent with the compressions of the operators $M_{z_i} \otimes I_{\mathcal{D}_{\underline{T}^*}}$ to the space $\mathbb{H}_T$. ■

So every pure commuting contractive tuple $\underline{T}$ on a Hilbert space $\mathcal{H}$ is unitarily equivalent to the commuting tuple $P_{\mathbb{H}_T}(M_z \otimes I_{\mathcal{D}_{\underline{T}^*}})|\mathbb{H}_T$, where we have used M_z for the tuple $(M_{z_1}, M_{z_2}, \dots, M_{z_d})$ and $\mathbb{H}_T$ is the M_z^*-invariant subspace $(H_d^2 \otimes \mathcal{D}_{\underline{T}^*}) \ominus M_{\theta_{\underline{T}}}(H_d^2 \otimes \mathcal{D}_{\underline{T}})$ of $H_d^2 \otimes \mathcal{D}_{\underline{T}^*}$, i.e., simultaneously invariant under all M_{z_i}.

5.4. The characteristic function as a complete unitary invariant

Definition 5.4.1. *Given two commuting contractive tuples $\underline{T}$ and $\underline{R}$ on Hilbert spaces $\mathcal{H}$ and $\mathcal{K}$, the characteristic functions of $\underline{T}$ and $\underline{R}$ are said to coincide if there exist unitary operators $\tau : \mathcal{D}_{\underline{T}} \to \mathcal{D}_{\underline{R}}$ and $\tau_* : \mathcal{D}_{\underline{T}^*} \to \mathcal{D}_{\underline{R}^*}$ such that the following diagram commutes for all z in $\mathbb{B}_d$:*

$$\begin{array}{ccc} \mathcal{D}_{\underline{T}} & \xrightarrow{\theta_{\underline{T}}(z)} & \mathcal{D}_{\underline{T}^*} \\ \tau\Big\downarrow & & \Big\downarrow\tau_* \\ \mathcal{D}_{\underline{R}} & \xrightarrow[\theta_{\underline{R}}(z)]{} & \mathcal{D}_{\underline{R}^*} \end{array}$$

In this section, we prove that the characteristic function of a pure commuting contractive tuple is a complete unitary invariant.

Proposition 5.4.2. *The characteristic functions of two unitarily equivalent commuting contractive tuples coincide.*

Proof. Let $\underline{T}$ and $\underline{R}$ be two commuting contractive tuples on $\mathcal{H}$ and $\mathcal{K}$, respectively, such that there is a unitary operator $\sigma : \mathcal{H} \to \mathcal{K}$ satisfying $\sigma T_i \sigma^* = R_i$ for all i. Denote by $\underline{\sigma}$ and $\underline{\sigma}^*$ the operators

$$\oplus_{i=1}^d \sigma : \mathcal{H}^d \to \mathcal{K}^d \text{ and } \oplus_{i=1}^d \sigma^* : \mathcal{K}^d \to \mathcal{H}^d$$

Then it is easy to see that $\underline{\sigma} D_{\underline{T}}^2 \underline{\sigma}^* = D_{\underline{R}}^2$ and $\sigma D_{\underline{T}^*}^2 \sigma^* = D_{\underline{R}^*}^2$. Thus $\underline{\sigma} D_{\underline{T}} \underline{\sigma}^* = D_{\underline{R}}$ and $\sigma D_{\underline{T}^*} \sigma^* = D_{\underline{R}^*}$. The restriction $\tau = \sigma|_{\mathcal{D}_{\underline{T}}}$ is a unitary operator between $\mathcal{D}_{\underline{T}}$ and $\mathcal{D}_{\underline{R}}$. Similarly, the restriction $\tau_* = \sigma|_{\mathcal{D}_{\underline{T}^*}}$ defines a unitary operator from $\mathcal{D}_{\underline{T}^*}$ to $\mathcal{D}_{\underline{R}^*}$. Finally, note that for $h \in \mathcal{D}_{\underline{T}}$, we have

$$\begin{aligned}\theta_{\underline{R}}(z)\underline{\tau}h &= (-R + D_{R^*}(1 - ZR^*)^{-1}ZD_{\underline{R}})\sigma h\\ &= (-\sigma T + D_{\underline{R}^*}(1 - ZR^*)^{-1}Z\underline{\sigma}D_{\underline{T}})h\\ &= (-\sigma T + D_{\underline{R}^*}(1 - ZR^*)^{-1}\sigma ZD_{\underline{T}})h\\ &= (-\sigma T + \sigma D_{\underline{T}^*}(1 - ZT^*)ZD_{\underline{T}})h\\ &= \tau_*\theta_{\underline{T}}(z)h\end{aligned}$$

for all $z \in \mathbb{B}_d$. Hence the two characteristic functions $\theta_{\underline{T}}$ and θ_R coincide. ■

Next we prove the converse of the above proposition for the case of pure tuples.

Proposition 5.4.3. *Let $\underline{T}$ and $\underline{R}$ be two pure commuting contractive tuples on $\mathcal{H}$ and $\mathcal{K}$, respectively. If their characteristic functions $\theta_{\underline{T}}$ and θ_R coincide, then the tuples $\underline{T}$ and $\underline{R}$ are unitarily equivalent.*

Proof. Let $\tau' : \mathcal{D}_{\underline{T}} \to \mathcal{D}_{\underline{R}}$ and $\tau'_* : \mathcal{D}_{\underline{T}^*} \to \mathcal{D}_{\underline{R}^*}$ be two unitary operators such that the diagram

$$\begin{array}{ccc}\mathcal{D}_{\underline{T}} & \xrightarrow{\theta_{\underline{T}}(z)} & \mathcal{D}_{\underline{T}^*}\\ {\scriptstyle\tau'}\big\downarrow & & \big\downarrow{\scriptstyle\tau'_*}\\ \mathcal{D}_{\underline{R}} & \xrightarrow[\theta_{\underline{R}}(z)]{} & \mathcal{D}_{\underline{R}^*}\end{array}$$

commutes for all z in . The operators τ' and τ'_* give rise to unitary operators $\tau = 1 \otimes \tau' : H_d^2 \otimes \mathcal{D}_{\underline{T}} \to H_d^2 \otimes \mathcal{D}_{\underline{R}}$ and $\tau_* = 1 \otimes \tau'_* : H_d^2 \otimes \mathcal{D}_{\underline{T}^*} \to H_d^2 \otimes \mathcal{D}_{\underline{R}^*}$ which satisfy the intertwining relation

$$M_{\theta_{\underline{R}}}\tau = \tau_* M_{\theta_{\underline{T}}}$$

We conclude that

$$\tau_*(\mathbb{H}_T) = \tau_*((\mathrm{Ran}\, M_{\theta_{\underline{T}}})^{\perp}) = \tau_*(\mathrm{Ran}\, M_{\theta_{\underline{T}}})^{\perp} = (\mathrm{Ran}\, M_{\theta_{\underline{R}}})^{\perp} = \mathbb{H}_R$$

where $\mathbb{H}_T$ and $\mathbb{H}_R$ are the model spaces for T and R as in Theorem 5.3.6. Since the operator τ_* intertwines the tuples $(M_z \otimes I_{\mathcal{D}_{\underline{T}^*}})^*$ and $(M_z \otimes I_{\mathcal{D}_{\underline{R}^*}})^*$ component wise, the induced unitary operator $\tau_* : \mathbb{H}_T \to \mathbb{H}_R$ intertwines the adjoint of the restrictions of these tuples, which are precisely the model tuples $P_{\mathbb{H}_T}(M_z \otimes I_{\mathcal{D}_{\underline{T}^*}})|\mathbb{H}_T$ and $P_{\mathbb{H}_R}(M_z \otimes I_{\mathcal{D}_{\underline{R}^*}})|\mathbb{H}_R$. But then Theorem 5.3.6 shows that $\underline{T}$ and $\underline{R}$ are unitarily equivalent. ■

Summarizing the last two propositions we obtain the main result.

Theorem 5.4.4. *Two pure commuting contractive tuples* $\underline{T}$ *and* $\underline{R}$ *on Hilbert spaces* $\mathcal{H}$ *and* $\mathcal{K}$ *are unitarily equivalent if and only if their characteristic functions coincide.*

This theorem which appeared in [32] has been generalized to completely non-coisometric contractive tuples. The interested reader may see [33].

CHAPTER 6

Dilation Theory in Several Variables - The Symmetrized Bidisc

6.1. The joint spectrum

We begin with a short discussion on the definition of joint spectrum of a tuple of commuting bounded operators on a Hilbert space as introduced by J. L. Taylor [115]. This is known as Taylor spectrum. Here we need the concept of anti-symmetric (or Fermion) Fock space. This is very similar to the notion of symmetric (or Boson) Fock space introduced in Chapter 4.

Let $\mathcal{H}$ be a Hilbert space. Fix a natural number $k > 1$. Consider the representation $\pi \mapsto U_\pi$ of the permutation group σ_k of $\{1, 2, \ldots, k\}$ on $\mathcal{H}^{\otimes^k}$ as defined in (4.2.1). Now the k-fold anti-symmetric tensor product of $\mathcal{H}$ is defined as the subspace

$$\mathcal{H}^{@^k} = \{x \in \mathcal{H}^{\otimes^k} : U_\pi x = \epsilon(\pi)x\},$$

where $\epsilon(\pi)$ is the signature of the permutation π (It is $+1$ for even permutations and -1 for odd permutations).

The projection P_a of $\mathcal{H}^{\otimes^k}$ to $\mathcal{H}^{@^k}$ is given by

$$P_a = \frac{1}{k!} \sum_{\pi \in \sigma_k} \epsilon(\pi) U_\pi.$$

For any permutation π, $P_a U_\pi = \epsilon(\pi) P_a$. For $x_i, y_i \in \mathcal{H}$, for $i = 1, 2, \ldots, k$,

$$\langle P_a(\otimes_{i=1}^k x_i), \otimes_{i=1}^k y_i \rangle = \frac{1}{k!} \det[\langle x_i, y_j \rangle]_{1 \le i,j \le k}.$$

One may recall that similar computation for symmetric Fock spaces required the permanent instead of the determinant. We denote $P_a(y_1 \otimes y_2 \otimes \cdots \otimes y_k)$ by $y_1 \wedge y_2 \wedge \cdots \wedge y_k$. Observe that

$$y_{\pi(1)} \wedge y_{\pi(2)} \wedge \cdots \wedge y_{\pi(k)} = \epsilon(\pi) y_1 \wedge y_2 \wedge \cdots \wedge y_k$$

for any permutation π. It is then easy to see that $y_1 \wedge y_2 \wedge \cdots \wedge y_k = 0$ if $y_i = y_j$ for some $i \neq j$.

Now the anti-symmetric Fock space over $\mathcal{H}$ is defined as

$$\Gamma_a(\mathcal{H}) := \mathbb{C} \oplus \mathcal{H} \oplus \mathcal{H}^{@^2} \oplus \cdots .$$

We are especially interested in the case $\mathcal{H} = \mathbb{C}^d$, for some natural number d. As before, $\omega = 1 \oplus 0 \oplus \cdots$ will be called the vacuum vector.

We denote $\Gamma_a(\mathbb{C}^d)$ by Λ_d. Let $\{e_1, e_2, \ldots, e_d\}$ be the standard ortho-normal basis of $\mathbb{C}^d$. Then $\{P_a(e_{j_1} \otimes e_{j_2} \otimes \cdots \otimes e_{j_k}) : 1 \leq j_1 < j_2 < \cdots < j_k \leq d\}$ is a basis for $(\mathbb{C}^d)^{@^k}$. In particular, $(\mathbb{C}^d)^{@^k}$ has dimension $\begin{pmatrix} d \\ k \end{pmatrix}$ (It is equal to zero if $k > d$). So we get

$$\Lambda_d = \mathbb{C} \oplus \mathbb{C}^d \oplus (\mathbb{C}^d)^{@^2} \oplus \cdots \oplus (\mathbb{C}^d)^{@^d},$$

and it has dimension $\sum_{k=0}^{d} \begin{pmatrix} d \\ k \end{pmatrix} = 2^d$. Readers familiar with the notion of 'exterior algebra' may note that the anti-symmetric Fock space over $\mathbb{C}^d$ is just the exterior algebra in d-generators along with a natural Hilbert space structure imposed on it.

For $j \in \{1, 2, \ldots, d\}$ we define the 'left-creation operators'

$$A_j : \Lambda_d \to \Lambda_d$$

by

$$A_j(x) = P_a(e_j \otimes x), \; x \in \Lambda_d.$$

Then A_j is a partial isometry for every j, and the collection $\{A_j : 1 \leq j \leq d\}$ satisfies *canonical anti-commutation relations*} (CAR):

$$A_j A_k + A_k A_j = 0 \text{ and } A_k^* A_j + A_j A_k^* = \delta_{j,k} I, \text{ for all } j, k.$$

For more information on Boson and Fermion Fock space we refer to [87].

For a tuple $\underline{T} = (T_1, T_2, \ldots, T_d)$ of commuting bounded operators on $\mathcal{H}$ define $\Delta_{\underline{T}} : \mathcal{H} \otimes \Lambda_d \to \mathcal{H} \otimes \Lambda_d$ by

$$\Delta_{\underline{T}} = \sum_{i=1}^{d} T_i \otimes A_i.$$

Then,

$$\Delta_{\underline{T}}^2 = \sum_{i,j=1}^{d} T_i T_j \otimes A_i A_j = \sum_{i<j} T_i T_j \otimes (A_i A_j + A_j A_i) = 0$$

and hence $\operatorname{ran} \Delta_{\underline{T}} \subseteq \ker(\Delta_{\underline{T}})$. We say that $\underline{T}$ is *(Taylor) non-singular* or *(Taylor) invertible* on $\mathcal{H}$ if $\operatorname{ran} \Delta_{\underline{T}} = \ker(\Delta_{\underline{T}})$. Given $\lambda = (\lambda_1, \lambda_2, \ldots, \lambda_d)$, consider the tuple of operators $\underline{T}-\lambda = (T_1-\lambda_1 I, T_2-\lambda_2 I, \ldots, T_d-\lambda_d I)$ on $\mathcal{H}$. The *Taylor spectrum* of $\underline{T}$ on $\mathcal{H}$ is

$$\sigma_T(\underline{T}, \mathcal{H}) = \{(\lambda_1, \lambda_2, \ldots, \lambda_d) \in \mathbb{C}^d : \operatorname{ran} \Delta_{\underline{T}-\lambda} \neq \ker(\Delta_{\underline{T}-\lambda})\}.$$

When $d = 1$, the operator $\Delta_{\underline{T}}$ has the following 2×2 block operator matrix form with respect to the decomposition $\mathcal{H} \otimes \Lambda_1 = (\mathcal{H} \otimes \omega) \oplus (\mathcal{H} \otimes e_1)$:

$$\Delta_{\underline{T}} = \begin{pmatrix} 0 & 0 \\ T & 0 \end{pmatrix}.$$

So,

$$\ker(\Delta_{\underline{T}}) = \ker(T) \oplus \mathcal{H} \text{ and } \operatorname{ran} \Delta_{\underline{T}} = \{0\} \oplus \operatorname{ran} T.$$

Hence, $\ker(\Delta_{\underline{T}})/\operatorname{ran} \Delta_{\underline{T}} = \ker(T) \oplus (\mathcal{H}/\operatorname{ran} T)$ showing that T is non-singular if and only if it is one-to-one and onto. For $d = 2$, the block operator matrix form of $\Delta_{\underline{T}}$ with respect to the decomposition $\mathcal{H} \otimes \Lambda_2 = (\mathcal{H} \otimes \omega) \oplus (\mathcal{H} \otimes e_1) \oplus (\mathcal{H} \otimes e_2) \oplus (\mathcal{H} \otimes e_1 \wedge e_2)$ is

$$\begin{pmatrix} 0 & 0 & 0 & 0 \\ T_1 & 0 & 0 & 0 \\ T_2 & 0 & 0 & 0 \\ 0 & -T_2 & T_1 & 0 \end{pmatrix}.$$

So,

$$\ker(\Delta_{\underline{T}}) = \{\ker(T_1) \cap \ker(T_2)\} \oplus \{(h_1, h_2) : T_2 h_1 = T_1 h_2\} \oplus \mathcal{H}$$

and

$$\operatorname{ran}(\Delta_{\underline{T}}) = \{0\} \oplus \{(T_1 h, T_2 h) : h \in \mathcal{H}\} \oplus \operatorname{ran} T_1 + \operatorname{ran} T_2\}.$$

Thus, the commuting pair (T_1, T_2) is non-singular if and only if

(1) $\ker(T_1) \cap \ker(T_2) = \{0\}$,
(2) $\{(h_1, h_2) : T_2 h_1 = T_1 h_2\} = \{(T_1 h, T_2 h) : h \in \mathcal{H}\}$,
(3) $\mathcal{H}$ is the span of $\operatorname{ran} T_1$ and $\operatorname{ran} T_2$.

On can also consider the Koszul complex $K(\underline{T}, \mathcal{H})$:

(6.1.1)

$$0 \to \mathcal{H} \otimes \Lambda_d^0 \overset{\Delta_{\underline{T}}^0}{\to} \mathcal{H} \otimes \Lambda_d^1 \overset{\Delta_{\underline{T}}^1}{\to} \cdots \overset{\Delta_{\underline{T}}^{k-1}}{\to} \mathcal{H} \otimes \Lambda_d^k \overset{\Delta_{\underline{T}}^k}{\to} \cdots \overset{\Delta_{\underline{T}}^{d-1}}{\to} \mathcal{H} \otimes \Lambda_d^d \to 0$$

where $\Delta_{\underline{T}}^k$ is $\Delta_{\underline{T}}|_{\mathcal{H} \otimes \Lambda_d^k}$. This complex is said to be *exact* if

$$\operatorname{ran} \Delta_{\underline{T}}^{k-1} = \ker(\Delta_{\underline{T}}^k)$$

for every $k = 1, 2, \ldots, d$. Then it is easy to see that

$$\sigma_T(\underline{T}, \mathcal{H}) = \{(\lambda_1, \lambda_2, \ldots, \lambda_d) : K(\underline{T} - \lambda, \mathcal{H}) \text{ is not exact }\}.$$

For $d = 1$ and $d = 2$, this complex takes the forms

$$0 \to \mathcal{H} \xrightarrow{T} \mathcal{H} \to 0$$

and

$$0 \to \mathcal{H} \stackrel{h \to (T_1 h, T_2 h)}{\longrightarrow} \mathcal{H} \oplus \mathcal{H} \stackrel{(h_1,h_2) \to T_2 h_1 - T_1 h_2}{\longrightarrow} \mathcal{H} \to 0$$

respectively. While it is clear that for $d = 1$, the complex is exact if and only if T is one-to-one and onto, the exactness of the complex for $d = 2$ is the same as the conditions (1), (2) and (3) listed above.

The Taylor joint spectrum carries a holomorphic functional calculus. We shall only state this because the proof is beyond the scope of this book. For a bounded open subset Ω of $\mathbb{C}^d$, let $\mathcal{O}(\Omega)$ denote the algebra of all $\mathbb{C}$-valued holomorphic functions defined on Ω.

Theorem 6.1.1. *Let $\underline{T} = (T_1, T_2, \ldots, T_d)$ be a tuple of commuting bounded operators on a Hilbert space $\mathcal{H}$. Let Ω be a bounded open subset of $\mathbb{C}^d$ containing $\sigma_T(\underline{T}, \mathcal{H})$. Then there is a unique unital continuous homomorphism from $\mathcal{O}(\Omega)$ into $\mathcal{B}(\mathcal{H})$ mapping f to $f(\underline{T})$ and satisfying*

(1) *$z_i(\underline{T}) = T_i$, for $i = 1, 2, \ldots, d$,*
(2) *$f(\underline{T})$ is in the double commutant of $\{T_1, T_2, \ldots, T_d\}$, i.e., $f(\underline{T})$ commutes with any $X \in \mathcal{B}(\mathcal{H})$ that commutes with all T_i,*
(3) *$f(\underline{T}) = 0$ if $f = 0$ in a neighbourhood of $\sigma_T(\underline{T}, \mathcal{H})$,*
(4) *$\sigma(f(\underline{T})) = f(\sigma_T(\underline{T}, \mathcal{H})$.*

A detailed construction of the functional calculus can be found in [116]. The uniqueness result is due to a short and elegant paper by Putinar [99].

6.2. Normal operator tuples

Recall that in Chapter 5, we defined a notion of an algebraic joint spectrum for a tuple $\underline{N} = (N_1, N_2, \ldots, N_d)$ of commuting bounded normal operators. It was

$$\sigma(\underline{N}) = \{(w(N_1), w(N_2), \ldots, w(N_d)) : w \in \widehat{C^*(\underline{N})}\}.$$

A natural question is whether the Taylor joint spectrum $\sigma_T(\underline{N})$ is the same as the algebraic joint spectrum $\sigma(\underline{N})$. The answer is yes. We need some lemmas to prepare for the proof.

Let X and Y be Banach spaces. A bounded operator $T : X \to Y$ is said to be *bounded below* if $\|Tx\| \geq c\|x\|$ for some c and for all $x \in X$. A related concept is that of an *approximate point spectrum* of a bounded linear operator $T : X \to X$. The approximate point spectrum $\sigma_{\text{app}}(T)$ of T is the set of all complex numbers λ for which there is a sequence $\{x_n\}$ of unit vectors such that the sequence $(T-\lambda)x_n$ converges to 0 as $n \to \infty$. Since $T-\lambda$ is then not bounded below, clearly $\sigma_{\text{app}}(T) \subseteq \sigma(T)$. For details on the approximate point spectrum for operators on Hilbert spaces, we refer the reader to Halmos [66].

Lemma 6.2.1. *Let $T : \mathcal{H} \to \mathcal{K}$ be a bounded operator from a Hilbert space $\mathcal{H}$ to another Hilbert space $\mathcal{K}$. Then T is surjective if and only if T^* is bounded below.*

Proof. This is a very well-known result. The proof goes as follows.

$$\begin{aligned} T^* \text{ is bounded below } &\Leftrightarrow \text{ There is a left inverse of } T, \text{ say } L^* \\ &\Leftrightarrow TL = I (\text{ where } L \text{ is the adjoint of } L^*) \\ &\Leftrightarrow T \text{ is surjective.} \end{aligned}$$

∎

Lemma 6.2.2. *For a normal operator, the spectrum is the same as the approximate point spectrum.*

Proof. Let N be a normal operator on a Hilbert space $\mathcal{H}$. Let $\lambda \in \sigma(N)$. If the kernel of $N-\lambda$ is non-trivial, then there is nothing to prove. Hence, we assume that $\ker(N-\lambda) = \{0\}$. Since N is normal, this implies that $\ker(N-\lambda)^* = \{0\}$. Hence, $\operatorname{ran}(N-\lambda)$ is dense in $\mathcal{H}$. Since λ is in the spectrum, this implies that $N-\lambda$ is not bounded below. We are done. ∎

Lemma 6.2.3. *Let $T_1, T_2, \ldots, T_d, S_1, S_2, \ldots, S_d$ be mutually commuting bounded operators on $\mathcal{H}$ such that $\sum T_i S_i = I$. Then the tuple $\underline{T}$ is Taylor invertible.*

Proof. Note that

$$(\sum_{j=1}^{d} S_j \otimes A_j^*)\Delta_{\underline{T}} + \Delta_{\underline{T}} \sum_{j=1}^{d} S_j \otimes A_j^* = \sum_{j=1}^{d}\sum_{i=1}^{d}(S_j T_i \otimes A_j^* A_i + T_i S_j \otimes A_i A_j^*).$$

In view of the canonical anti-commutation relations, the last quantity is $\sum_{j=1}^{d} T_j S_j$ which is the same as I. Thus, if $x \in \ker\Delta_{\underline{T}}$, then

$$x = (\sum_{j=1}^{d} S_j \otimes A_j^* \Delta_{\underline{T}} + \Delta_{\underline{T}} \sum_{j=1}^{d} S_j \otimes A_j^*)x = \Delta_{\underline{T}}(\sum_{j=1}^{d} S_j \otimes A_j^*)x$$

and hence $x \in \operatorname{ran}\Delta_{\underline{T}}$. We are done. ∎

Just like the approximate point spectrum of a single bounded operator, we can consider the *joint approximate point spectrum* of a commuting tuple $\underline{T} = (T_1, T_2, \dots, T_d)$. We say that $(\lambda_1, \lambda_2, \dots, \lambda_d)$ is in the joint approximate point spectrum of $\underline{T}$ if there is a sequence $\{x_n\}$ of unit vectors such that $(T_j - \lambda_j)x_n \to 0$ as $n \to \infty$ for each j. For more on joint approximate point spectrum, see [37] and [40].

Theorem 6.2.4. *Let $\underline{N} = (N_1, N_2, \dots, N_d)$ be a tuple of commuting bounded normal operators. Then*

$$\sigma_T(\underline{N}) = \{(w(N_1), w(N_2), \dots, w(N_d)) : w \in \widehat{C^*(\underline{N})}\}$$

where $C^*(\underline{N})$ is the unital C^*-algebra generated by $N_1, N_2, \dots, N_d$.

Proof. Let $(\lambda_1, \lambda_2, \dots, \lambda_d) \in \sigma_T(\underline{N})$. Since the tuple $(N_1 - \lambda_1, N_2 - \lambda_2, \dots, N_d - \lambda_d)$ is not Taylor invertible, so there are no S_i commuting with $N_j - \lambda_j$ such that $\sum S_i(N_i - \lambda_i) = I$. Thus, the ideal generated by $N_1 - \lambda_1, N_2 - \lambda_2, \dots, N_d - \lambda_d$ in $C^*(\underline{N})$ is a proper ideal. Hence, there is a $w \in \widehat{C^*(\underline{N})}$ such that $(\lambda_1, \lambda_2, \dots, \lambda_d) = (w(N_1), w(N_2), \dots, w(N_d))$.

Conversely, if $(\lambda_1, \lambda_2, \dots, \lambda_d) = (w(N_1), w(N_2), \dots, w(N_d))$ for some $w \in \widehat{C^*(\underline{N})}$, then

$$w(\sum_{i=1}^{d}(N_i - \lambda_i)(N_i - \lambda_i)^*) = 0.$$

Thus, $\sum_{i=1}^{d}(N_i - \lambda_i)(N_i - \lambda_i)^*$ is not invertible. By Lemma 6.2.2, we know that there is a sequence $\{x_n\}$ of unit vectors such that $(\sum_{i=1}^{d}(N_i - \lambda_i)(N_i - \lambda_i)^*)x_n \to 0$ as $n \to \infty$. This means that $(\overline{\lambda_1}, \overline{\lambda_2}, \dots, \overline{\lambda_d})$ is in the joint approximate point spectrum of $(N_1^*, N_2^*, \dots, N_d^*)$ or in other words, the column operator

$$\begin{pmatrix} (N_1 - \lambda_1)^* \\ (N_2 - \lambda_2)^* \\ \vdots \\ (N_d - \lambda_d)^* \end{pmatrix} : \mathcal{H} \to \begin{pmatrix} \mathcal{H} \\ \mathcal{H} \\ \vdots \\ \mathcal{H} \end{pmatrix}$$

is not bounded below. By Lemma 6.2.1, we have that the range of the operator

$$(N_1 - \lambda_1, N_2 - \lambda_2, \dots, N_d - \lambda_d) : \begin{pmatrix} \mathcal{H} \\ \mathcal{H} \\ \vdots \\ \mathcal{H} \end{pmatrix} \to \mathcal{H}$$

is a proper subspace of $\mathcal{H}$. Hence, the exactness in the Koszul complex is broken at the last stage. So, $(\lambda_1, \lambda_2, \dots, \lambda_d) \in \sigma_T(\underline{N})$. ∎

Note that Lemma 6.2.2 also holds for a tuple of commuting bounded normal operators where the spectrum is replaced by the Taylor spectrum and the approximate point spectrum is replaced by the joint approximate point spectrum.

6.3. Spectral sets

In his seminal paper [16], Arveson introduced the novel concept of a spectral set. Let K be a subset of $\mathbb{C}^d$. Consider the algebra rat(K), the set of rational functions with poles off K, i,e., the elements of rat(K) are quotients of polynomials in d variables and the polynomial in the denominator does not have any zeros in K. If $\underline{T} = (T_1, T_2, \ldots, T_d)$ is a tuple of commuting bounded operators acting on a Hilbert space $\mathcal{H}$ with the Talor joint spectrum $\sigma_T(\underline{T}) \subseteq K$ and

$$r(z_1, z_2, \ldots, z_d) = \frac{p(z_1, z_2, \ldots, z_d)}{q(z_1, z_2, \ldots, z_d)}$$

is in rat(K), then by the functional calculus mentioned above $r(\underline{T}) = p(\underline{T})q(\underline{T})^{-1}$ is in $\mathcal{B}(\mathcal{H})$.

A compact subset K of $\mathbb{C}^d$ is called a *spectral set* for a commuting n-tuple of operators $\underline{T} = (T_1, \ldots, T_d)$ acting on a Hilbert space $\mathcal{H}$ if the Taylor joint spectrum of $\underline{T}$ is contained in K and

(6.3.1)
$$\begin{aligned}\|r(T_1, T_2, \ldots, T_d)\| \leq& \|r\|_{\infty,K}\\ =& \sup\{|r(z_1, z_2, \ldots, z_d)| : (z_1, z_2, \ldots, z_d) \in K\}\end{aligned}$$

for any rational function r in rat(K).

An $m \times m$ matrix valued rational function r is a matrix $[r_{ij}]_{i,j=1}^m$ of scalar valued rational functions. In this case too, we can make sense of $r(T_1, T_2, \ldots T_d)$, viz.,

$$r(T_1, T_2, \ldots T_d) = [r_{ij}(T_1, T_2, \ldots T_d)]_{i,j=1}^m.$$

This is then a block operator matrix acting as an operator on the direct sum of m copies of $\mathcal{H}$.

Moreover, an $m \times m$ matrix valued rational function r has the natural norm

$$\|r\|_{\infty,K} = \sup\{\|r(z_1, z_2, \ldots, z_d)\| : (z_1, z_2, \ldots, z_d) \in K\}$$

where $\|r(z_1, z_2, \ldots, z_d)\|$ denotes the operator norm of the $m \times m$ matrix $r(z_1, z_2, \ldots, z_d)$.

A set K is called a *complete spectral set* if the von Neumann type inequality (6.3.1) above holds for all rational functions taking values in matrices of any order with the norm on r being as above.

Theorem 1.2.2 and the remark following it in Arveson [16] have been revolutionary in the subject of Hilbert space operator theory. Here we shall try to understand why. We already know that a simultaneous dilation of $\underline{T}$ consists of n commuting bounded operators $N_1, N_2, \ldots, N_d$ on a Hilbert space $\mathcal{K}$ containing $\mathcal{H}$ in such a way that

$$P_{\mathcal{H}} N_1^{k_1} N_2^{k_2} \ldots N_d^{k_d}|_{\mathcal{H}} = T_1^{k_1} T_2^{k_2} \ldots T_d^{k_d} \tag{6.3.2}$$

for all non-negative integers $k_1, k_2, \ldots, k_d$. Recall that the dilation is called minimal if the whole space $\mathcal{K}$ is the closure of span of the set

$\{T_1^{k_1} T_2^{k_2} \ldots T_d^{k_d} h : k_1, k_2, \ldots, k_d$ are non-negative integers and $h \in \mathcal{H}\}$.

Naturally, one wants the dilation operators to be nicer than the given ones. The dilation $(N_1, N_2, \ldots, N_d)$ is called a *normal dilation* if the N_i are commuting normal operators. It is called a *boundary normal dilation* if it is a normal dilation with its Taylor joint spectrum contained in the Šilov boundary of K (relative to $A(K)$). The remarkable result that Arveson proved says that K is a complete spectral set for $\underline{T}$ if and only if $\underline{T}$ has a boundary normal dilation.

Say that K has the *complete spectral property* if the following holds. "If K is a spectral set for a commuting tuple $\underline{T}$, then it is a complete spectral set for $\underline{T}$." It has historically intrigued Hilbert space operator theorists which subsets of the complex plane or of higher dimensional Euclidean space have this property.

In dimension one, the unit disk has this property. Indeed, because of von Neumann's inequality, a contraction is precisely a bounded operator that has the closed unit disc as a spectral set. Sz.-Nagy dilation then tells us that if the closed unit disc is a spectral set for a bounded operator set, then it is a complete spectral set of that operator. Berger, Foias and Lebow extended that result to any simply connected domain. Whether a domain in the plane with a single hole enjoyed the complete spectral property was open for a long time till Agler showed in [1] that the annulus has this property. Dritschel and McCullough proved in [57] that a multiply connected domain in general does not. They use Gel'fand-Naimark-Segal construction combined with a cone-separation argument. Agler, Harland and Raphael in [68] have shown that there exists a planar domain with two holes and an operator T on a Hilbert space of dimension 4 such that the domain is a spectral set, but not a complete spectral set, for T. Thus, this branch of generalization from the disc to multiply connected domains is now complete.

The other direction of generalization is to higher dimensions. In dimension two, the bidisc has this property because of Ando's theorem.

As of now, no subset of $\mathbb{C}^3$ with complete spectral property is known.

If K is the closed unit ball of some norm in $\mathbb{C}^d$ where $n > 2$, then K cannot have this property. See Paulsen [88] and Pisier [91].

6.4. The symmetrized bidisc

Since it is not very common to come across sets that have the complete spectral property mentioned above, the search for such a set led to the discovery of a subset of $\mathbb{C}^2$.

This set was introduced by Agler and Young [4] and has since then been of interest to both complex analysts and functional analysts, see for example [72] and [35]. This set is the losed symmetrized bidisc

$$\Gamma = \{(z_1 + z_2, z_1 z_2) : |z_1| \leq 1, |z_2| \leq 1\} \subseteq \mathbb{C}^2.$$

The open symmetrized bidisc $\{(z_1 + z_2, z_1 z_2) : |z_1| < 1, |z_2| < 1\}$ is denoted by $\mathbb{G}$.

Definition 6.4.1. *A commuting pair of bounded operators (S, P) defined on a Hilbert space $\mathcal{H}$ is called a Γ-contraction if it has Γ as a spectral set, i.e., the Taylor joint spectrum of (S, P) is contained in Γ and (S, P) satisfies*

$$\|f(S, P)\| \leq \sup_{(z_1, z_2) \in \Gamma} |f(z_1, z_2)|$$

where f is a holomorphic function in two variables in an open set containing Γ and the supremum is over Γ. A Γ- contraction (S, P) is said to be a pure *Γ-contraction if P is a pure contraction, i.e, $P^{*n} \to 0$ as $n \to \infty$.*

The remarkably smooth theory that Agler and Young developed for Γ-contractions parallels the highly successful theory of dilation of a single contraction because they showed, in ([5]), the existence of a Γ-isometric dilation for any Γ-contraction.

In this chapter, we construct an explicit Γ-isometric dilation of a Γ-contraction, i.e., given a Γ-contraction (S, P) on a Hilbert space $\mathcal{H}$, we construct a space $\mathcal{K}$ containing $\mathcal{H}$ as a subspace and a Γ-isometry (T, V) on $\mathcal{K}$ such that $T^*|_{\mathcal{H}} = S^*$ and $V^*|_{\mathcal{H}} = P^*$. In other words, a Γ-contraction is the compression of a Γ-isometry to a co-invariant subspace. What is remarkable here is that the space $\mathcal{K}$ need not be any bigger than the minimal isometric dilation space of the contraction P and V is in fact the minimal isometric dilation of P. Moreover, T, in such a case, is uniquely determined. This chapter uses material from many papers of Agler and Young as well as from [35].

There are several ways to describe a Γ-contraction. A new way of characterizing Γ-contractions is exhibited here. To that end, we define the fundamental equation of a pair of bounded operators (S, P) with $\|P\| \leq 1$, to be the operator equation

$$S - S^*P = D_P X D_P\,, \quad X \in \mathcal{B}(\mathcal{D}_P).$$

We show the existence and uniqueness of solution to the fundamental equation for a Γ-contraction (S, P) and call the solution, the fundamental operator for (S, P). Uniqueness of minimal Γ-isometric dilation of a Γ-contraction follows from the uniqueness of the solution. This relates the theory of Γ-contractions beautifully to solving operator equations.

6.5. Polynomial convexity

This section will briefly recall *polynomial convexity*. It is a vast topic and we shall only sketch here the basic properties that are relevant for understanding the symmetrized bidisc.

Definition 6.5.1. *Let K be a compact subset of the n-dimensional complex space $\mathbb{C}^d$. Then the polynomially convex hull of K, denoted by $\hat{K}$, is defined as*

$$\hat{K} = \{z \in \mathbb{C}^d : |p(z)| \leq \|p\|_{\infty,K} \text{ for all polynomials } p\}.$$

A compact set K is said to be *polynomially convex* if $K = \hat{K}$. Obviously $K \subseteq \hat{K}$. The intuitive idea behind this definition if $n = 1$ is that the polynomially convex hull of K fills in any "holes" that may exist in K. The following proposition makes that precise.

Proposition 6.5.2. *A set $K \subset \mathbb{C}$ is a polynomially convex set if and only if $\mathbb{C} \setminus K$ is connected.*

Proof. Let $\mathbb{C} \setminus K$ be connected. Let z_0 be a point in $\mathbb{C} \setminus K$. Consider the function

$$z \mapsto \frac{1}{z - z_0}$$

which is holomorphic in a neighbourhood of K which has the property that its complement in $\mathbb{C}$ is connected. Thus, by Runge's Theorem, page

198 of [42], one can find a rational function with pole at infinity, in other words a polynomial p_n, such that

$$|p_n(z) - \frac{1}{z - z_0}| < \frac{1}{n} \text{ for all } z \in K \cup \{z_0\}.$$

We would like to draw the reader's attention to the fact this is where the connectedness of $\mathbb{C} \setminus K$ is being used, for if it were not connected, our rational function would have a pole in every component of $\mathbb{C} \setminus K$ and hence would indeed be just a rational function and not a polynomial. Now, consider the polynomial q_n given by

$$q_n(z) = (z - z_0)p_n(z) - 1.$$

By the above, $\|q_n\|_{\infty,K} \leq \sup_K |z - z_0|/n$. For a large enough n, q_n will have norm less than 1 whereas the value of $|q_n(z_0)|$ is 1 for all n. That shows that K is polynomially convex.

If $\mathbb{C} \setminus K$ is not connected, then any point in the bounded component of $\mathbb{C} \setminus K$ is in the polynomially convex hull of K by the maximum modulus principle. ■

We saw above that polynomial approximation of a holomorphic function can be done when K is a polynomially convex subset of $\mathbb{C}$. Its multivariable generalization is the famous *Oka-Weil* Theorem.

Theorem 6.5.3. *Let K be a compact polynomially convex subset of $\mathbb{C}^d$. Then for every function f holomorphic in some neighbourhood of K, we can find a sequence $\{p_j\}$ of polynomials converging to f uniformly on K.*

The proof of this theorem is outside the scope of this text. We refer the interested reader to Chapter 7 of Alexander and Wermer [7].

6.6. The Fejér-Riesz Theorem

Given a polynomial $p(z) = \sum_{n=0}^{N} c_n z^n$, where the c_n are complex scalars, the function $\tau(e^{i\theta}) = |p(e^{i\theta})|^2$ on the unit circle is clearly a non-negative function and is strictly positive if p does not have any zero on the unit circle. Note that τ, as a function, is a linear combination of

$$z^{-N}, z^{-N+1}, \ldots, z^{N-1}, z^N.$$

Such a function is called a *trigonometric polynomial* or a Laurent polynomial. A classical result says that all non-negative Laurent polynomials are squares of analytic polynomials.

Theorem 6.6.1. *[Fejér-Riesz] Let* $\tau(e^{i\theta}) = \sum_{n=-N}^{N} a_n e^{in\theta}$ *be a strictly positive function on the unit circle* $\mathbb{T}$. *Then there is a polynomial* $p(z) = \sum_{n=0}^{N} c_n z^n$ *such that* $\tau(e^{i\theta}) = |p(e^{i\theta})|^2$

Proof. Since τ is real valued, $a_{-n} = \bar{a}_n$ and a_0 is real. Without loss of generality, we can assume a_{-N} to be non-zero (Why?). Consider the polynomial $g(z) = z^N \tau(z)$. This is a polynomial of degree $2N$. Moreover, $g(0) \neq 0$. Also,

$$\overline{g\left(\frac{1}{\bar{z}}\right)} = \frac{1}{z^N} \sum_{n=-N}^{N} \overline{a_n} \frac{1}{z^n} = z^{-N} \sum_{n=-N}^{N} a_{-n} z^{-n} = z^{-N}\tau(z) = z^{-2N} g(z).$$

Thus, if z is a zero of g, so is $1/\bar{z}$. Hence, the $2N$ zeroes of g are

$$z_1, z_2, \ldots, z_N \text{ and } 1/\bar{z}_1, 1/\bar{z}_2, \ldots, 1/\bar{z}_N$$

for some $z_1, z_2, \ldots, z_N$. We can choose $z_1, z_2, \ldots, z_N$ to be outside the closed unit disc. If we set

$$q(z) = (z - z_1)(z - z_2)\ldots(z - z_N),$$

then

$$\overline{\left(z - \frac{1}{\bar{z}_1}\right)\left(z - \frac{1}{\bar{z}_2}\right)\cdots\left(z - \frac{1}{\bar{z}_N}\right)} = \frac{(-1)^N \bar{z}^N q(1/\bar{z})}{z_1 z_2 \ldots z_N}.$$

Hence, for z on the unit circle,

$$\left|\left(z - \frac{1}{\bar{z}_1}\right)\left(z - \frac{1}{\bar{z}_2}\right)\cdots\left(z - \frac{1}{\bar{z}_N}\right)\right| = \frac{1}{|z_1 z_2 \ldots z_N|}|q(z)|.$$

Now we are ready to complete the proof. Since τ is positive on the unit circle, $\tau(e^{i\theta}) = |\tau(e^{i\theta})|$. But, $\tau(e^{i\theta}) = e^{-iN\theta} g(e^{i\theta})$ and hence

$$\tau(e^{i\theta}) = |g(e^{i\theta})| = \left|\frac{a_N}{z_1 z_2 \ldots z_N}\right| |q(e^{i\theta})|^2.$$

Now, take $p(z) = |\frac{a_N}{z_1 z_2 \ldots z_N}|^{1/2} q(z)$.

■

Exercise 6.6.2. *The above remains true even if τ is positive but not strictly positive. This can be proved by a careful handling of the zeroes of τ on the circle.*

We shall now see a beautiful generalization of the Fejér-Riesz Theorem for Laurent polynomials whose values lie in the space of bounded operators. Since the positive operators form a cone, one can ask when a positive operator valued Laurent polynomial can be factorized in a way similar to the result above. The following theorem, proved first in full generality by Rosenblum, answers that question. The proof given here is due to Rosenblum. For a surprising connection to Schur complements and thereby a different proof, see the survey article by Dritschel and Rovnyak [58]. Very recently Dritschel has generalized the result to factorization of a positive Laurent polynomial in two variables, see [56].

Before we can get into this generalized Fejér-Riesz Theorem, we need some background preparation. Let σ be the normalized arc length measure on the unit circle $\mathbb{T}$. Consider the Hilbert space

$$L^2(\mathcal{E}) = \{f : f \text{ is an } \mathcal{E}\text{-valued measurable function on } \mathbb{T} \text{ such that } \int_{\mathbb{T}} \|f(z)\|^2 d\sigma(z) < \infty\}.$$

Although we do not include the symbol $\mathbb{T}$ in the notation $L^2(\mathcal{E})$ for the sake of brevity of writing, the underlying set for these functions are the unit circle. If ψ is in $L^\infty(\mathcal{E})$, then ψ induces a linear operator M_ψ on $L^2(\mathcal{E})$ by

$$(M_\psi f)(z) = \psi(z)f(z).$$

Clearly, this is a bounded operator because

$$\|M_\psi f\|^2 = \int_{\mathbb{T}} \|\psi(z)f(z)\|^2 d\sigma(z) \leq \|\psi\|_\infty^2 \|f\|^2. \tag{6.6.1}$$

It will be good to remember that $L^\infty(\mathcal{E})$ is contained in $L^2(\mathcal{E})$. Note that an arbitrary function in $L^2(\mathcal{E})$ has the form $\sum_{n=-\infty}^{\infty} e_n e^{in\theta}$ for some $\ldots, e_{-2}, e_{-1}, e_0, e_1, e_2, \ldots$ in $\mathcal{E}$ with $\sum \|e_n\|^2 < \infty$.

The Hardy space $H^2(\mathcal{E})$ of $\mathcal{E}$ valued holomorphic functions on the unit disc has a natural isometric image inside $L^2(\mathcal{E})$. Indeed, any such Hardy space function is of the form $\sum_{n=0}^{\infty} e_n z^n$ for $e_0, e_1, e_2, \ldots$ coming from $\mathcal{E}$. Map this function to $\sum_{n=0}^{\infty} e_n e^{in\theta}$. This is an isometry. Define $\mathcal{H} = \mathcal{E} \oplus \mathcal{E} \oplus \cdots$. Obviously, this is the Hardy space in disguise. Embed it in $L^2(\mathcal{E})$ isometrically by sending

$$e_0 \oplus e_1 \oplus e_2 \oplus \cdots \text{ to } \sum_{n=0}^{\infty} e_n e^{in\theta}.$$

By the embedding above, The Hardy space as well as $\mathcal{H}$ (which are to begin with isomorphic copies of one another) will be treated as a closed subspace of $L^2(\mathcal{E})$ and we shall denote the projection onto the Hardy space (or $\mathcal{H}$) from $L^2(\mathcal{E})$ by P. For ψ as above, define T_ψ on the Hardy space (or on $\mathcal{H}$) by $T_\psi = PM_\psi|_{\mathcal{H}}$. We note here that, by virtue of (6.6.1), we have

$$\|T_\psi\| \leq \|\psi\|_\infty. \tag{6.6.2}$$

If $p(z) = \sum_0^n C_n z^n$ is a polynomial with C_n coming from $\mathcal{B}(\mathcal{E})$, define T_p in the same way as T_τ, i.e., $T_p = PM_p|_{\mathcal{H}}$ (note that the projection P is redundant now).

Let us recall that our space $\mathcal{H} = \mathcal{E} \oplus \mathcal{E} \oplus \cdots$ is entitled to have a *shift operator*. It is the isometry S on $\mathcal{H}$ acting as

$$S(e_0 \oplus e_1 \oplus e_2 \oplus \cdots) = (0 \oplus e_0 \oplus e_1 \oplus \cdots).$$

In the Hardy space picture, the shift is multiplication M_z by z, i.e., it sends a function $\sum_{n=0}^{\infty} e_n z^n$ to $\sum_{n=1}^{\infty} e_{n-1} z^n$. If $T, C \in \mathcal{B}(\mathcal{H})$, then T is called a *Toeplitz operator* if $S^*TS = T$ ($M_z^* T M_z = T$ in the Hardy space picture) and C is called an *analytic operator* if $CS = SC$. It is straightforward that T_p is an analytic operator. We do a simple calculation below to show that T_ψ is a Toeplitz operator.

$$\begin{aligned}
\langle M_z^* T_\psi M_z f, g\rangle &= \langle T_\psi M_z f, M_z g\rangle \\
&= \langle PM_\psi M_z f, M_z g\rangle \\
&= \langle M_\psi M_z f, PM_z g\rangle \\
&= \langle M_\psi M_z f, M_z g\rangle \\
&= \int_{\mathbb{T}} \psi(z) z f(z) \overline{z g(z)} d\sigma(z) \\
&= \int_{\mathbb{T}} \psi(z) f(z) \overline{g(z)} d\sigma(z) = \langle T_\psi f, g\rangle.
\end{aligned}$$

It is well-known that the converse of this fact is also true, i.e., any bounded operator T on the Hardy space which satisfies $M_z^* T M_z = T$ is of the form T_ψ for some $\psi \in L^\infty(\mathcal{E})$. For more on Toeplitz operators, we refer the reader to [76].

When considered as operators on $\mathcal{H}$, a Toeplitz operator and an analytic operator will have block operator matrices. Toeplitz and analytic operators respectively have the forms

$$T = \begin{pmatrix} T_0 & T_{-1} & T_{-2} & \cdots \\ T_1 & T_0 & T_{-1} & \ddots \\ T_2 & T_1 & T_0 & \ddots \\ \vdots & \ddots & \ddots & \ddots \end{pmatrix}, \qquad C = \begin{pmatrix} C_0 & 0 & 0 & \cdots \\ C_1 & C_0 & 0 & \ddots \\ C_2 & C_1 & C_0 & \ddots \\ \vdots & \ddots & \ddots & \ddots \end{pmatrix} \tag{6.6.3}$$

as can be easily seen by first considering the matrix representations $((\, T_{ij} \,))$ and $((\, C_{ij} \,))$ of T and C respectively with respect to the decomposition $\mathcal{H} = \mathcal{E} \oplus \mathcal{E} \oplus \cdots$ and then using the relations $S^*TS = T$ and $CS = SC$.

In the scalar case, the polynomial p can be chosen not to have any zeros in $\mathbb{D}$. An analytic operator C is said to be *outer* if $\overline{\operatorname{ran}}\, C$ is a subspace of $\mathcal{H}$ of the form $\mathcal{F} \oplus \mathcal{F} \oplus \cdots$ for some closed subspace $\mathcal{F}$ of $\mathcal{E}$.

Definition 6.6.3. *A $\mathcal{B}(\mathcal{E})$ valued polynomial $p(z)$ is outer if the analytic Toeplitz operator $C = T_p$ is outer.*

We leave it as an exercise to the reader to prove that a scalar polynomial, which is not identically zero, is outer if and only if it does not have zeroes in $\mathbb{D}$.

Theorem 6.6.4. *[The Fejér-Riesz Theorem for Operator Valued Functions] Let $\tau(z) = \sum_{k=-N}^{N} A_k z^k$ be a Laurent polynomial with coefficients in $\mathcal{B}(\mathcal{E})$ for some Hilbert space $\mathcal{E}$. If $\tau(\zeta) \geq 0$ for all $\zeta \in \mathbb{T}$, then*

$$\tau(\zeta) = p(\zeta)^* p(\zeta), \qquad \zeta \in \mathbb{T}, \tag{6.6.4}$$

for some polynomial $p(z) = C_0 + C_1 z + \cdots + C_N z^N$ with coefficients in $\mathcal{B}(\mathcal{E})$. The polynomial $p(z)$ can be chosen to be outer.

The rest of this section will prove this theorem.

The first thing to note is that the Laurent polynomial τ satistifes

$$\|\tau\|_\infty \stackrel{\text{def}}{=} \sup_{z \in \mathbb{T}} \|\tau(z)\| < \infty,$$

thus defining an M_τ on $L^2(\mathcal{E})$. Moreover, for the kind of τ given in the statement of the theorem, i.e., $\tau(z)$ defining a positive operator for every $z \in \mathbb{T}$, we have

$$\langle M_\tau f, f\rangle = \int_{\mathbb{T}} \langle \tau(z) f(z), f(z)\rangle d\sigma(z) \geq 0$$

showing that M_τ is a positive operator. This implies that T_τ a positive operator too.

Note that in the block matrix forms of T_τ and T_p are as follows where $A_j = 0$ for $|j| > m$ and $C_j = 0$ for $j > m$:

$$T_\tau = \begin{pmatrix} A_0 & A_{-1} & A_{-2} & \cdots \\ A_1 & A_0 & A_{-1} & \ddots \\ A_2 & A_1 & A_0 & \ddots \\ \vdots & \ddots & \ddots & \ddots \end{pmatrix}, \qquad T_p = \begin{pmatrix} C_0 & 0 & 0 & \cdots \\ C_1 & C_0 & 0 & \ddots \\ C_2 & C_1 & C_0 & \ddots \\ \vdots & \ddots & \ddots & \ddots \end{pmatrix}. \tag{6.6.5}$$

It is easy to see that if $T_\tau = T_p^* T_p$, then $\tau(\zeta) = p(\zeta)^* p(\zeta)$ for all $\zeta \in \mathbb{T}$. Thus, the problem is to write a given nonnegative Toeplitz operator T, i.e., T_τ in the form $T = C^*C$, where C is T_p for an outer analytic polynomial of the same degree. Lemmas 6.6.5 and 6.6.6 reduce the problem to showing that a certain isometry is a shift operator.

Lemma 6.6.5 (Lowdenslager's Criterion). *Let $\mathcal{H}$ be a Hilbert space, and let $S \in \mathcal{B}(\mathcal{H})$ be a shift operator. Let $T \in \mathcal{B}(\mathcal{H})$ be Toeplitz relative to S as defined above, and suppose that $T \geq 0$. Let $\mathcal{H}_T$ be the closure of the range of $T^{1/2}$ in the inner product of $\mathcal{H}$. Then there is an isometry S_T mapping $\mathcal{H}_T$ into itself such that*

$$S_T T^{1/2} f = T^{1/2} S f, \qquad f \in \mathcal{H}.$$

In order that $T = C^*C$ for some analytic operator $C \in \mathcal{B}(\mathcal{H})$, it is necessary and sufficient that S_T is a shift operator. In this case, C can be chosen to be outer.

Proof. The identity $S^*TS = T$ implies that $T^{1/2}Sf$ and $T^{1/2}f$ have the same norms for any $f \in \mathcal{H}$. Hence, we can define the isometry S_T as $S_T T^{1/2} f = T^{1/2} S f$.

If S_T is a shift operator, we can view $\mathcal{H}_T$ as a direct sum $\mathcal{H}_T = \mathcal{E}_T \oplus \mathcal{E}_T \oplus \cdots$ with $S_T(h_0, h_1, \ldots) = (0, h_0, h_1, \ldots)$. Here $\dim \mathcal{E}_T \leq \dim \mathcal{E}$. To see this, notice first that $S_T T^{1/2}$ equals $T^{1/2}S$ as operators from $\mathcal{H}$ into $\mathcal{H}_T$. Hence, $T^{1/2}S_T^*$ and $S^*T^{1/2}$ agree on $\mathcal{H}_T$, and therefore $T^{1/2}(\ker S_T^*) \subseteq \ker S^*$. The dimension inequality then follows because $T^{1/2}$ is one-to-one on the closure of its range. Therefore we may choose an isometry V from $\mathcal{E}_T$ into $\mathcal{E}$. Define an isometry W on $\mathcal{H}_T$ into $\mathcal{H}$ by

$$W(h_0, h_1, \ldots) = (Vh_0, Vh_1, \ldots).$$

Define $C \in \mathcal{B}(\mathcal{H})$ by mapping $\mathcal{H}$ into $\mathcal{H}_T$ via $T^{1/2}$ and then $\mathcal{H}_T$ into $\mathcal{H}$ via W:

$$Cf = WT^{1/2}f, \qquad f \in \mathcal{H}.$$

The way W has been cooked up, it is clear that $WS_T = SW$. This translates into $CS = SC$, proving that C is analytic. It is clearly outer. And $C^*C = T^{1/2}W^*WT^{1/2} = T$.

Conversely, suppose that $T = C^*C$ where $C \in \mathcal{B}(\mathcal{H})$ is analytic. Define an isometry W from $\mathcal{H}_T$ into $\mathcal{H}$ by $WT^{1/2}f = Cf$, $f \in \mathcal{H}$. Then $WS_T = SW$, and hence $S_T^{*n} = W^*S^{*n}W$ for all $n \geq 1$. Since the powers of S^* tend strongly to zero, so do the powers of S_T^*, and therefore S_T is a shift operator. ∎

Lemma 6.6.6. *In Lemma 6.6.5, let $T = T_\tau$ be given by (6.6.5) for a Laurent polynomial $\tau(z)$ of degree N. If $T = C^*C$ where $C \in \mathcal{B}(\mathcal{H})$ is analytic and outer, then $C = T_p$ for some outer analytic polynomial $p(z)$ of degree N.*

Proof. Let $\tau(z) = \sum_{k=-N}^{N} A_k z^k$. We appeal to (6.6.5) keeping in mind that $A_j = 0$ for $|j| > N$. Note that we have to show that $S^{*j}CE_0 = 0$ for all $j > N$ where E_0 is the projection from $\mathcal{H}$ onto $\mathcal{E} \oplus 0 \oplus 0 \oplus \cdots$. It is sufficient to show that $S^{*N+1}CE_0 = 0$. Since $T = C^*C$ and C is analytic (i.e., C commutes with S), we have

$$E_0^* C^* S^{*j} C E_0 = E_0^* S^{*j} T E_0 = T_j = 0, \qquad j > N.$$

It follows that $\operatorname{ran} S^{*N+1}CE_0 \perp \operatorname{ran} CS^k E_0$ for all $k \geq 0$, and therefore

$$\operatorname{ran} S^{*N+1} C E_0 \perp \operatorname{ran} C. \tag{6.6.6}$$

Since C is outer, $\overline{\operatorname{ran}}\, C$ reduces S, and so $\operatorname{ran} S^{*N+1}CE_0 \subseteq \overline{\operatorname{ran}}\, C$. Therefore $S^{*N+1}CE_0 = 0$ by (6.6.6), and the result follows. ■

The proof of the operator Fejér-Riesz theorem is now easily completed.

Proof of Theorem 6.6.4. Define $T = T_\tau$ as in (6.6.5). Lemmas 6.6.5 and 6.6.6 reduce the problem to showing that the isometry S_T is a shift operator. It is sufficient to show that $\|S_T^{*n} f\| \to 0$ for every f in $\mathcal{H}_T$.

Claim: If $f = T^{1/2}h$ where $h \in \mathcal{H}$ has the form $h = (h_0, \dots, h_r, 0, \dots)$, then $S_T^{*n} f = 0$ for all sufficiently large n.

For if $u \in \mathcal{H}$ and n is any positive integer, then

$$\begin{aligned} \left\langle S_T^{*n} f, T^{1/2} u \right\rangle_{\mathcal{H}_T} &= \left\langle f, S_T^{n} T^{1/2} u \right\rangle_{\mathcal{H}_T} \\ &= \left\langle T^{1/2} h, T^{1/2} S^n u \right\rangle_{\mathcal{H}} = \langle Th, S^n u \rangle_{\mathcal{H}}. \end{aligned}$$

By the definition of $T = T_\tau$, Th has only a finite number of nonzero entries (depending on N and r), and the first n entries of $S^n u$ are zero (irrespective of u). The claim follows from the arbitrariness of u.

In view of the claim, $\|S_T^{*n} f\| \to 0$ for a dense set of vectors in $\mathcal{H}_T$, and hence by approximation this holds for all f in $\mathcal{H}_T$. Thus S_T is a shift operator, and the result follows. ■

A more general result is proved in the original version of Theorem 6.6.4 in [102]. There it is only required that $\tau(z)g$ is a Laurent polynomial for a dense set of g in $\mathcal{E}$ (the degrees of these polynomials need not be bounded). There is also an accompanying uniqueness statement: the outer polynomial $p(z)$ in Theorem 6.6.4 can be chosen such that $p(0) \geq 0$, and then it is unique. See [22] and [103].

6.7. The set Γ

Lemma 6.7.1. *Γ is polynomially convex.*

Proof. We need to show that given a point $(\lambda_1 + \lambda_2, \lambda_1\lambda_2)$ outside Γ, there always is a polynomial p such that $\|p\|_{\infty,\Gamma} \leq 1$ and $|p(\lambda_1 + \lambda_2, \lambda_1\lambda_2)| > 1$. This is the same as producing a symmetric polynomial p of (z_1, z_2) such that $\|p\|_{\infty,\mathbb{D}^2} \leq 1$ and $|p(\lambda_1, \lambda_2)| > 1$. If both λ_1 and λ_2 have moduli bigger than 1, then $p(z_1, z_2) = z_1 z_2$ suffices. If one of them has modulus bigger than 1 and the other has modulus less than or equal to 1, then $p(z_1, z_2) = (z_1^n + z_2^n)/2$ works for large n. ■

Although Γ is polynomially convex, it is not convex. In fact, Γ is not biholomorphic to any convex domain, see Costara [43]. Although the definition of Γ is pretty simple - it is indeed the symmetrization of the bidisc - there are several characterizations that are very useful in practice. The following theorem is from Agler and Young [5]. For two commuting variables S and P, define

$$\begin{aligned}\rho(S,P) &= 2(I - P^*P) - S + S^*P - S^* + P^*S \\ &= \frac{1}{2}\left((2-S)^*(2-S) - (2P-S)^*(2P-S)\right).\end{aligned}$$

Theorem 6.7.2. *Let $(s,p) \in \mathbb{C}^2$. The following are equivalent:*

(a) $(s,p) \in \Gamma$;
(b) $|s - \bar{s}p| + |p^2| \leq 1$ *and* $|s| \leq 2$;
(c) $\rho(\alpha s, \alpha^2 p) \geq 0$, *for all* $\alpha \in \mathbb{T}$, *the unit circle in* $\mathbb{C}$;
(d) $|p| \leq 1$ *and there exists* $\beta \in \mathbb{C}$ *such that* $|\beta| \leq 1$ *and* $s = \beta + \bar{\beta}p$;
(e) $|s| \leq 2$ *and* $(2\alpha p - s)(2\overline{\alpha p} - \bar{s}) \leq (2 - \alpha s)(2 - \overline{\alpha s})$ *for all* $\alpha \in \mathbb{T}$.

Proof. (a) $\Leftrightarrow$ (b): We start by choosing z_1 and z_2 in the complex plane such that $s = z_1 + z_2$ and $p = z_1 z_2$. This means that z_1 and z_2 are the roots of the polynomial $\xi(z) = z^2 - sz + p = 0$. Let $A = \begin{pmatrix} 0 & 1 \\ 0 & 0 \end{pmatrix}$ and $q(z) = \bar{p}z^2 - \bar{s}z + 1$. Form the matrix $q(A)^*q(A) - \xi(A)^*\xi(A)$. Since $A^2 = 0$, both $p(A)$ and $q(A)$ are linear functions of A and hence $p(A) = \begin{pmatrix} p & -s \\ 0 & p \end{pmatrix}$ and $q(A) = \begin{pmatrix} 1 & -\bar{s} \\ 0 & 1 \end{pmatrix}$. After a short computation, we get

$$q(A)^*q(A) - \xi(A)^*\xi(A) = \begin{pmatrix} 1 - |p|^2 & -\overline{(s - \bar{s}p)} \\ -(s - \bar{s}p) & 1 - |p|^2 \end{pmatrix}.$$

Note that $\xi(A) = (A - z_1)(A - z_2)$ and $q(z) = (1 - \overline{z_1}z)(1 - \overline{z_2}z)$ and $q(A) = (1 - \overline{z_1}A)(1 - \overline{z_2}A)$. Thus,

$$\begin{aligned}
&q(A)^*q(A) - \xi(A)^*\xi(A)\\
&= (1 - \overline{z_1}A)^*(1 - \overline{z_2}A)^*(1 - \overline{z_1}A)(1 - \overline{z_2}A)\\
&- (A - z_1)^*(A - z_2)^*(A - z_1)(A - z_2)\\
&= (1 - \overline{z_1}A)^*\left((1 - \overline{z_2}A)^*(1 - \overline{z_2}A) - (A - z_2)^*(A - z_2)\right)(1 - \overline{z_1}A)\\
&+ (A - z_2)^*\left((1 - \overline{z_1}A)^*(1 - \overline{z_1}A) - (A - z_1)^*(A - z_1)\right)(A - z_2)\\
&= (1 - |z_2|^2)(1 - \overline{z_1}A)^*(I - A^*A)(1 - \overline{z_1}A)\\
&+ (1 - |z_1|^2)(A - z_2)^*(I - A^*A)(A - z_2).
\end{aligned}$$

Let $(I - A^*A)^{1/2}(1 - \overline{z_1}A) = A_1$ and $(I - A^*A)^{1/2}(A - z_2) = A_2$. Hence we now have

$$(1 - |z_2|^2)A_1^*A_1 + (1 - |z_1|^2)A_2^*A_2 = \begin{pmatrix} 1 - |p|^2 & -\overline{(s - \overline{s}p)} \\ -(s - \overline{s}p) & 1 - |p|^2 \end{pmatrix}.$$

The statement (i) holds if and only if z_1 and z_2 have moduli no larger than 1 which by virtue of the equality above holds if and only if the matrix $\begin{pmatrix} 1 - |p|^2 & -\overline{(s - \overline{s}p)} \\ -(s - \overline{s}p) & 1 - |p|^2 \end{pmatrix}$ is non-negative definite which happens if and only if (ii) holds.

The technique used above can be used to prove a general result called Schur test for a polynomial to have all its zeroes inside the closed unit disc. This test, first proposed by Schur in 1917, states that given a polynomial ξ of degree n, its zeroes are inside the closed unit disc if and only if $q(S)^*q(S) - \xi(S)^*\xi(S)$ is a positive semidefinite matrix where q is the polynomial $q(z) = z^n\overline{\xi(1/\overline{z})}$ and S is the $n \times n$ matrix with 1 at the first super diagonal and 0 elsewhere. For a detailed discussion of this general result, see Ptak and Young [98].

(b) $\Leftrightarrow$ (e): The condition $|s| \le 2$ is part of both (b) and (e). Hence, we keep it aside. Let $s - \overline{s}p = \sigma$. Note that for α on the unit circle,

$$
\begin{aligned}
& (2\alpha p - s)(2\overline{\alpha p} - \overline{s}) \le (2 - \alpha s)(2 - \overline{\alpha s}) \\
\Leftrightarrow\ & 4|p|^2 - 2\overline{\alpha} s\overline{p} - 2\alpha\overline{s}p + |s|^2 \le 4 - 2\alpha s - 2\overline{\alpha s} + |s|^2 \\
\Leftrightarrow\ & (1 - |p|^2) - \text{ Re } \alpha(s - \overline{s}p) \ge 0 \\
\Leftrightarrow\ & \text{the trigonometric polynomial } 2(1 - |p|^2) - \alpha\sigma - \overline{\alpha\sigma} \ge 0 \\
\Leftrightarrow\ & 2(1 - |p|^2) - \alpha\sigma - \overline{\alpha\sigma} = (a + b\alpha)\overline{(a + b\alpha)} \\
& \text{for some } a \text{ and } b \text{ by Fejer-Riesz Theorem} \\
\Leftrightarrow\ & \text{there are } a \text{ and } b \text{ such that } 2(1 - |p|^2) = |a|^2 + |b|^2 \text{ and } \overline{a}b = -\sigma \\
\Leftrightarrow\ & |\sigma| \le 1 - |p|^2.
\end{aligned}
$$

(b) $\Leftrightarrow$ (d): If (d) holds, a straightforward computation shows that $s - \overline{s}p = \beta(1 - |p|^2)$. Since $|\beta| \le 1$, we have $|s - \overline{s}p| \le |1 - |p|^2|$. Also, $|s| \le 2$ because $|\beta| \le 1$ and $|p| \le 1$.

If (b) holds and $|p| < 1$, then choose $\beta = (s - \overline{s}p)/(1 - |p|^2)$. If $|p| = 1$, then $s = \overline{s}p$ and with $p = e^{i\theta}$, this gives that $e^{-i\theta/2}s$ is real (either $|s|$ or $-|s|$). Since $|s| \le 2$, we have $e^{-i\theta/2}s = 2\cos\phi$ for some ϕ. This translates to $s = \beta + \overline{\beta}e^{i\theta}$ with $\beta = e^{i(-(\phi - \theta/2))}$.

The equivalence of (c) and (e) is left as an exercise. ■

We would like to introduce a parametrized family of functions. For $\alpha \in \mathbb{T}$, the unit circle, and (s, p) satisfying $2 - \alpha s \ne 0$, let

$$\varphi(\alpha, s, p) = \frac{2\alpha p - s}{2 - \alpha s}.$$

Definition 6.7.3. *The functions $\varphi(\alpha, \cdot)$ will be called the parametrized coordinate functions, parametrized by $\alpha \in \mathbb{T}$.*

Since $|s| < 2$ for all $(s, p) \in \mathbb{G}$, this function is well–defined on $\mathbb{T} \times \mathbb{G}$. The notation $\varphi(\alpha, \cdot)$ will mean that for a fixed α, we are considering it as a function on $\mathbb{G}$. If we write λ for the pair (s, p), we shall sometimes write $\varphi(\alpha, \lambda)$. We note that for a fixed $\alpha \in \mathbb{T}$, the function $\varphi(\alpha, \cdot)$ is in $H^\infty(\mathbb{G})$. Using the functions φ, one can write down a version of the condition (e) above for $\mathbb{G}$ instead of Γ. Indeed, following the lines of arguments in the proof of the theorem above, it can be shown that $(s, p) \in \mathbb{G}$ if and only if $|s| < 2$ and $|\varphi(\alpha, s, p)| < 1$ for all α on the unit circle. This parametrized family of functions was introduced by Agler and Young in [6] and plays an important role in the study of function theory on the symmetrized bidisc, see [36].

6.8. The distinguished boundary of Γ

For a compact subset K of $\mathbb{C}^d$, and an algebra $\mathcal{F}$ of continuous complex-valued functions on K which separates the points of K, a boundary for $\mathcal{F}$ is a closed subset S of K such that

$$|f(x)| \leq \max\{|f(t)| : t \in S\}$$

for all f in $\mathcal{F}$ and x in K. The Šilov boundary of K for $\mathcal{F}$ is defined as the intersection of all boundaries S of K for $\mathcal{F}$. It was proved by Šilov that this intersection is non-empty and is a boundary for $\mathcal{F}$.

If the algebra $\mathcal{F}$ of functions is $C(K) \cap \mathcal{O}(\text{int}K)$, where $\text{int}K$ denotes the interior of K and $O(\text{int}K)$ denotes the algebra of functions which are holomorphic on $\text{int}K$, then the Shilov boundary of K for $\mathcal{F}$ is called the Bergman–Šilov boundary of K or the *distinguished boundary* of K, usually denoted by bK. By the maximum principle, bK is always a subset of the topological boundary of K. When $d \geq 2$, it is often a proper subset of the boundary of K. For example, it is easy to see that if

$$K = \{(z, w) \in \mathbb{C}^2 : |z| \leq 1, |w| \leq 1\},$$

then $bK = \{(z, w) : |z| = |w| = 1\}$. The aim of this section is to find the distinguished boundary of Γ. A finite Blaschke product is a complex function of the form

$$B(z) = \lambda \prod_{i=1}^{m} \left(\frac{z - a_i}{1 - \overline{a_i} z} \right)$$

where m is a natural number, $a_1, a_2, \ldots, a_m$ are points from the open unit disc $\mathbb{D}$ and λ is a point on the unit circle. A single Blaschke factor $\exp i\theta \frac{z-a}{1-\bar{a}z}$ is in fact a general element of the group of biholomorphic automorphisms of the unit disc.

Lemma 6.8.1. *Given λ_1 and λ_2 on the unit circle with $\lambda_1 \neq \lambda_2$, there is an automorphism φ of the unit disc such that $\varphi(\lambda_1) = 1$ and $\varphi(\lambda_2) = -1$.*

Proof. Let $\lambda_1 = \exp i\theta_1$ and $\lambda_2 = \exp i\theta_2$. Consider the automorphism

$$\varphi_1(z) = (\exp -i\frac{\theta_1 + \theta_2}{2})z.$$

Clearly, $\varphi_1(\lambda_1) = \overline{\varphi_1(\lambda_2)}$ and $\text{Re}\varphi_1(\lambda_1)$ is in the open interval $(-1, 1)$. Since $t \to 2t/(1+t^2)$ is a bijection of $(-1, 1)$ onto itself, there is a unique a in $(-1, 1)$ satisfying $\text{Re}\varphi_1(\lambda_1) = 2a/(1 + a^2)$. Thus,

$$\varphi_1(\lambda_1) + \overline{\varphi_1(\lambda_2)} = \varphi_1(\lambda_1) + \overline{\varphi_1(\lambda_1)} = \frac{4a}{1 + a^2}$$

which leads to

$$\frac{\varphi_1(\lambda_1) - a}{1 - a\varphi_1(\lambda_1)} = \frac{\varphi_1(\lambda_2) - a}{1 - a\varphi_1(\lambda_2)}.$$

So, if φ_2 is the automorphism $\varphi_2(z) = (z - a)/(1 - az)$, then we have $(\varphi_2 \circ \varphi_1)(\lambda_1) = -(\varphi_2 \circ \varphi_1)(\lambda_2)$. A composition of automorphisms is again an automorphism and every automorphism maps the unit circle into the unit circle. Let $(\varphi_2 \circ \varphi_1)(\lambda_1) = \exp i\theta$ and $\varphi_3(z) = (\exp -i\theta)z$. The required automorphism φ is $\varphi_3 \circ \varphi_2 \circ \varphi_1$. ■

We are now ready for the distinguished boundary of the symetrized bidisc.

Theorem 6.8.2. *The distinguished boundary of the set* Γ, *denoted by* $b\Gamma$ *is the set*

$$b\Gamma = \{(\lambda_1 + \lambda_2\ , \lambda_1\lambda_2) : \ |\lambda_1| = |\lambda_2| = 1\}.$$

Proof. Any function in $C(\Gamma) \cap \mathcal{O}(\mathbb{G})$ attains its maximum on the set $\{(\lambda_1 + \lambda_2\ , \lambda_1\lambda_2) :\ |\lambda_1| = |\lambda_2| = 1\}$. Indeed, any such function is the symmetric function in $C(\overline{\mathbb{D}^2}) \cap \mathcal{O}(\mathbb{D}^2)$ and that attains its maximum on $b\mathbb{D}^2 = \{(\lambda_1, \lambda_2) : |\lambda_1| = |\lambda_2| = 1\}$.

To conclude that $\{(\lambda_1 + \lambda_2\ , \lambda_1\lambda_2) :\ |\lambda_1| = |\lambda_2| = 1\}$ is indeed the smallest boundary, we shall show that for every point in this set, there is a function in $C(\Gamma) \cap \mathcal{O}(\mathbb{G})$ which attains its maximum only at that point. To that end, consider a point of the form $(2\lambda, \lambda^2)$ with $|\lambda| = 1$. It is easy to see that the function $f(s, p) = \overline{\lambda}s + 2$ attains its maximum 4 only at this point.

On the other hand, if we consider a point $(\lambda_1 + \lambda_2, \lambda_1\lambda_2)$ with $|\lambda_1| = |\lambda_2| = 1$ and $\lambda_1 \neq \lambda_2$, then by Lemma 6.8.1. we get an automorphism φ of the unit disc such that $\varphi(\lambda_1) = 1$ and $\varphi(\lambda_2) = -1$. Consider the map H_φ defined on Γ by

$$H_\varphi(z_1 + z_2, z_1z_2) = (\varphi(z_1) + \varphi(z_2), \varphi(z_1)\varphi(z_2))$$

for $z_1, z_2 \in \overline{\mathbb{D}}$. It is easy to check that H_φ is a biholomorphic automorphism of $\mathbb{G}$ (in fact, a bijection of Γ onto itself) which sends $(\lambda_1 + \lambda_2, \lambda_1\lambda_2)$ to $(0, -1)$. Let $f(s, p) = s^2 - 4p + 2$ and $F = f \circ H_\varphi$. A straightforward computation shows that f attains its maximum only at $(0, -1)$ and hence F attains its maximum only at $(\lambda_1 + \lambda_2, \lambda_1\lambda_2)$. ■

We find below some equivalent descriptions of $b\Gamma$ due to Agler and Young.

Theorem 6.8.3. *Let* $(s, p) \in \mathbb{C}^2$. *Then the following are equivalent:*

(1) $(s, p) \in b\Gamma$;

(2) $(s,p) = (2xe^{i\frac{\theta}{2}}, e^{i\theta})$ *for some* $\theta \in \mathbb{R}$, *and* $x \in [-1,1]$.
(3) $|p| = 1,\ s = \bar{s}p$ *and* $|s| \leq 2$.
(4) $|p| = 1,\ s = \beta + \bar{\beta}p$ *for some* $\beta \in \mathbb{C}$ *of modulus* 1.

Proof. (1) ⇒ (2): Let $s = \lambda_1 + \lambda_2$ and $p = \lambda_1\lambda_2$ for some λ_1 and λ_2 both of modulus 1. Let $\lambda_1\lambda_2 = e^{i\theta}$. Then $s = \lambda_1 + \overline{\lambda_1}e^{i\theta} = e^{i\theta/2}(e^{-i\theta/2}\lambda_1 + e^{i\theta/2}\overline{\lambda_1})$. Now, taking x to be the real part of $e^{i\theta/2}\overline{\lambda_1}$, we are done.

(2) ⇒ (3) is clear.

(3) ⇒ (1): From part (b) of Theorem 6.7.2, we already know that (s,p) is in Γ. Hence, there are z_1 and z_2 in the closed unit disc so that $s = z_1 + z_2$ and $p = z_1z_2$. Using the fact that $|z_1z_2| = 1$, we get that $|z_1|$ and $|z_2|$ both have to be 1.

(1) holds if and only if (4) holds: Let $|p| = 1$ and $s = \beta + \bar{\beta}p$ for some $\beta \in \mathbb{C}$ of modulus 1. Taking $\lambda_1 = \beta$ and $\lambda_2 = \bar{\beta}p$ we see that $s = \lambda_1 + \lambda_2$ and $p = \lambda_1\lambda_2$ where clearly $|\lambda_1| = |\lambda_2| = 1$. Hence $(s,p) \in b\Gamma$. Conversely, let $(s,p) \in b\Gamma$. Then $s = \lambda_1 + \lambda_2$ and $p = \lambda_1\lambda_2$ for some λ_1, λ_2 of modulus 1. Clearly $|p| = 1$ and $\lambda_2 = \bar{\lambda}_1 p$. Thus we have $s = \lambda_1 + \bar{\lambda}_1 p = \beta + \bar{\beta}p$, where $\beta = \lambda_1$. ∎

6.9. Structure theorems for Γ−isometries and Γ−unitaries

We begin with an easy observation.

Lemma 6.9.1. *A commuting pair of bounded operators* (S,P) *is a* Γ *contraction if and only if*

$$\|f(S,P)\| \leq \|f\|_{\infty,\Gamma} = \sup\{|f(x_1,x_2)| : (x_1,x_2) \in \Gamma\} \tag{6.9.1}$$

for any holomorphic polynomial f *in two variables.*

Proof. If (S,P) is a Γ contraction, then of course (6.9.1) just follows from definition.

The converse proof can be easily done by using polynomial convexity of Γ. Indeed, if the Taylor spectrum $\sigma(S,P)$ is not contained in Γ, then there is a point (λ_1,λ_2) in $\sigma(S,P)$ that is not in Γ. By polynomial convexity of Γ, there is a polynomial f such that $|f(\lambda_1,\lambda_2)| > \|f\|_{\infty,\Gamma}$. By polynomial spectral mapping theorem,

$$\sigma(f(S,P)) = \{f(x_1,x_2) : (x_1,x_2) \in \sigma(S,P)\}$$

and hence the spectral radius of $f(S, P)$ is bigger than $\|f\|_{\infty,\Gamma}$. But then $\|f(S, P)\| > \|f\|_{\infty,\Gamma}$, contradicting the fact that Γ is a spectral set for (S, P).

By polynomial convexity of Γ, a pair satisfying (6.9.1) will also satisfy

$$\|f(S, P)\| \leq \|f\|_{\infty,\Gamma}$$

for any function holomorphic in a neighbourhood of Γ. Indeed, Oka-Weil Theorem allows us to approximate f uniformly by polynomials. That completes the proof. ■

The restriction of a Γ-contraction to a joint invariant subspace is a Γ-contraction. Indeed, if $\mathcal{M}$ is a such a subspace, and f is any polynomial in two variables, then

$$\|f(S|_{\mathcal{M}}, P|_{\mathcal{M}})\| = \|f(S, P)|_{\mathcal{M}}\| \leq \|f(S, P)\| \leq \|f\|_{\infty,\Gamma}.$$

The other thing that follows immediately from the lemma above is that the adjoint pair (S^*, P^*) is a Γ-contraction too.

Taking cue from such classically beautiful concepts that have been explained in the beginning of this chapter, Agler and Young introduced the following definitions.

Definition 6.9.2. *A commuting pair (S, P) is called a* Γ-unitary *if S and P are normal operators and the joint spectrum $\sigma(S, P)$ of (S, P) is contained in the distinguished boundary of Γ.*

Definition 6.9.3. *A commuting pair (S, P) is called a Γ-isometry if there exist a Hilbert space $\mathcal{N}$ containing $\mathcal{H}$ and a Γ-unitary $(\tilde{S}, \tilde{P})$ on $\mathcal{N}$ such that $\mathcal{H}$ is left invariant by both $\tilde{S}$ and $\tilde{P}$, and*

$$S = \tilde{S}|_{\mathcal{H}} \text{ and } P = \tilde{P}|_{\mathcal{H}}.$$

In other words, $(\tilde{S}, \tilde{P})$ is a Γ-unitary extension of (S, P). A commuting pair (S, P) is a Γ-co-isometry *if (S^*, P^*) is a Γ-isometry. Moreover, a Γ-isometry (S, P) is said to be a pure Γ-isometry if P is a pure isometry, i.e, there is no non trivial subspace of $\mathcal{H}$ on which P acts as a unitary operator.*

Here and henceforth, when we say joint spectrum, we shall mean the Taylor joint spectrum unless otherwise mentioned. Recall that

$$\begin{aligned}\rho(S, P) &= 2(I - P^*P) - (S - S^*P) - (S^* - P^*S)\\ &= \frac{1}{2}\{(2 - S)^*(2 - S) - (2P - S)^*(2P - S)\}.\end{aligned}$$

The following result was proved in [4]. There, in fact, it was proved that positivity of $\rho(S,P)$ is a necessary and sufficient condition for (S,P) to be a Γ-contraction. A straightforward proof of one direction is given below using joint spectral theory.

Proposition 6.9.4. *Let (S,P) be a Γ-contraction. Then $\rho(\alpha S,\alpha^2 P)\ \geq 0$, for all $\alpha \in \overline{\mathbb{D}}$.*

Proof. Using the definition of a Γ-contraction and the function

$$f(s,p) = (2\alpha^2 p - \alpha s)/(2 - \alpha s)$$

which is holomorphic in a neighbourhood of Γ for $\alpha \in \mathbb{D}$, we get

$$\|\,(2\alpha^2 P - \alpha S)(2 - \alpha S)^{-1}\,\| \leq\ \|f\|_\Gamma\ \leq 1.$$

Thus $(2-\alpha S)^{*-1}(2\alpha^2 P - \alpha S)^*(2\alpha^2 P - \alpha S)(2-\alpha S)^{-1} \leq I$. This happens if and only if $(2-\alpha S)^*(2-\alpha S) \geq (2\alpha^2 P - \alpha S)^*(2\alpha^2 P - \alpha S)$. By definition of $\rho(S,P)$, the last inequality is the same as $\rho(\alpha S,\alpha^2 P) \geq 0$.

By continuity, $\rho(\alpha S,\alpha^2 P) \geq 0$ for all $\alpha \in \overline{\mathbb{D}}$. ■

It is clear from the definition that if (S_1,P_1) and (S_2,P_2) are Γ-unitaries, then so is the direct sum $(S,P) = (S_1 \oplus S_2, P_1 \oplus P_2)$. Indeed, the joint spectrum of (S,P) is the union of the joint spectrum of (S_1,P_1) and the joint spectrum of (S_2,P_2) (see [45]). We begin with an elementary lemma whose proof we skip because it is routine.

Lemma 6.9.5. *Let X be a bounded operator on a Hilbert space $\mathcal{H}$. If Re $\beta X \leq 0$ for all complex numbers β of modulus 1, then $X = 0$.*

The following theorem gives characterizations of Γ-unitaries.

Theorem 6.9.6. *Let (S,P) be a pair of commuting operators defined on a Hilbert space $\mathcal{H}$. Then the following are equivalent:*

(1) *(S,P) is a Γ-unitary ;*
(2) *there exist commuting unitary operators U_1 and U_2 on $\mathcal{H}$ such that*

$$S = U_1 + U_2, \quad P = U_1 U_2\ ;$$

(3) *P is a unitary, $S = S^*P$, and $r(S) \leq 2$, where $r(S)$ is the spectral radius of S.*
(4) *(S,P) is a Γ-contraction and P is a unitary.*
(5) *P is a unitary and $S = U + U^*P$ for some unitary U commuting with P.*

Remark 6.9.7. *We draw attention to the similarity between part(5) of this theorem and part(4) of Theorem 6.8.3.*

Proof. $(1) \Rightarrow (2)$: This proof is the same as the one given by Agler and Young in [5]. We include it for the sake of completeness. Let (S, P) be a Γ-unitary. By the spectral theorem for commuting normal operators there exists a spectral measure say $M(.)$ on $\sigma(S, P)$ such that

$$S = \int_{\sigma(S,P)} p_1(z)M(dz), \quad P = \int_{\sigma(S,P)} p_2(z)M(dz),$$

where $p_1, \ p_2$ are the co-ordinate functions on $\mathbb{C}^2$. Now choose a measurable right inverse β of the restriction of the function π to $\mathbb{T}^2$ so that β maps the distinguished boundary $b\Gamma$ of Γ to $\mathbb{T}^2$. Let $\beta = (\beta_1, \beta_2)$ and

$$U_j = \int_{\sigma(S,P)} \beta_j(z)M(dz), \quad j = 1, 2.$$

Then U_1, U_2 are commuting unitary operators on $\mathcal{H}$ and

$$U_1 + U_2 = \int_{\sigma(S,P)} (\beta_1 + \beta_2)(z)M(dz) = \int_{\sigma(S,P)} p_1(z)M(dz) = S.$$

Similarly $U_1U_2 = P$. Thus $(1) \Rightarrow (2)$.

$(2) \Rightarrow (3)$ is clear.

$(3) \Rightarrow (1)$: We have $P^*P = PP^* = I$ and $S = S^*P$. Therefore, $S^* = P^*S$ and as a consequence

$$SS^* = (S^*P)(P^*S) = S^*S$$

as P is unitary. So (S, P) is a commuting pair of normal operators. So we have $r(S) = \|S\|$.
Let $C^*(S, P)$ be the commutative C^*-algebra generated by them. By general theory of joint spectrum (see p-27, Proposition 1.2 of [45]),

$$\sigma(S, P) = \{(\varphi(S)\ , \varphi(P)) : \varphi \in \mathcal{M}\},$$

where $\mathcal{M}$ is the maximal ideal space of $C^*(S, P)$. Consider an element $(s, p) = (\psi(S), \psi(P))$ of $\sigma(S, P)$, where $\psi \in \mathcal{M}$. Then

$$|p|^2 = \bar{p}p = \overline{\psi(p)}\psi(p) = \psi(P^*)\psi(P) = \psi(P^*P) = \psi(I) = 1$$

and

$$\bar{s}p = \overline{\psi(S)}\psi(P) = \psi(S^*P) = \psi(S) = s.$$

Also $|s| = |\psi(S)| \leq \|S\| = r(S) \leq 2$. Therefore, by Theorem 6.8.3, $(s, p) \in b\Gamma$ i.e, $\sigma(S, P) \subseteq b\Gamma$. So (S, P) is a Γ-unitary. Hence $(3) \Rightarrow (1)$.

Thus (1), (2) and (3) are equivalent.

The implication $(1) \Rightarrow (4)$ is trivial.

(4) ⇒ (3) depends on the fact that if (S, P) is a Γ-contraction, then

$$\rho(\alpha S, \alpha^2 P) \geq 0\,, \qquad \text{for all } \alpha \in \overline{\mathbb{D}}.$$

Therefore, for $\beta \in \mathbb{T}$, we have $\rho(\beta S, \beta^2 P) = 2(I - P^*P) - \beta(S - S^*P) - \overline{\beta}(S^* - P^*S) \geq 0$. Using the fact that $P^*P = I$, we get that Re $\beta(S - S^*P) \leq 0$. By invoking Lemma 6.9.5 now, we get that $S - S^*P = 0$. Also since (S, P) is a Γ-contraction, $r(S) \leq \|S\| \leq 2$. Hence done.

(2) ⇒ (5) follows as $S = U_1 + U_2 = U_1 + U_1^*P$ and $U_1P = U_1U_1U_2 = U_1U_2U_1 = PU_1$.

(5) ⇒ (2) follows by taking $U_1 = U$ and $U_2 = U^*P$.

■

Corollary 6.9.8. *The pair (S, I) can be a Γ-contraction only by being a Γ-unitary. It is so if and only if S is a self-adjoint operator of spectral radius not bigger than* 2.

During the course of the proof, we used something which we segregate as a separate result because it will be used later too.

Observation 6.9.9. *If P is a unitary, S commutes with P and $S = S^*P$, then S is normal.*

The structure of Γ-isometries can be deciphered using numerical radius. We recall the definitions of *numerical range* and *numerical radius* and discuss some of their properties which will be useful. The *numerical range* of an operator T on a Hilbert space $\mathcal{H}$ is defined to be

$$\Omega(T) = \{\langle Tx, x\rangle \ :\ \|x\|_{\mathcal{H}} = 1\}.$$

The *numerical radius* of T is defined as

$$\omega(T) = \sup\{|\langle Tx, x\rangle| \ :\ \|x\|_{\mathcal{H}} = 1\}.$$

It is well known that $r(T) \leq \omega(T) \leq \|T\|$ for a bounded operator T. An elementary fact will be used more than once, and hence we state it as a lemma.

Lemma 6.9.10. *The numerical radius of an operator X is not greater than one if and only if Re $\beta X \leq I$ for all complex numbers β of modulus* 1.

Proof. It is obvious that $\omega(X) \leq 1$ implies that Re $\beta X \leq I$ for all $\beta \in \mathbb{T}$. We prove the other way. By hypothesis, $\langle \text{Re } \beta X h, h\rangle \leq 1$ for all $h \in \mathcal{H}$ with $\|h\| \leq 1$ and for all $\beta \in \mathbb{T}$. Note that $\langle \text{Re } \beta X h, h\rangle = \text{Re } \beta\langle Xh, h\rangle$. Write $\langle Xh, h\rangle = e^{i\varphi_h}|\langle Xh, h\rangle|$ for some $\varphi_h \in \mathbb{R}$, and then choose $\beta = e^{-i\varphi_h}$. Then we get $|\langle Xh, h\rangle| \leq 1$ and this holds for each $h \in \mathcal{H}$ with $\|h\| \leq 1$. Hence done. ■

Another interesting fact about the numerical radius is the following.

Lemma 6.9.11. *Suppose the numerical radius of an operator X on a Hilbert space $\mathcal{H}$ is not greater than one. Then the numerical radius of $M_z \otimes X$ on $H^2_{\mathbb{D}} \otimes \mathcal{H}$ is also not greater than one.*

Proof. As we have seen many times by now, the space $H^2_{\mathbb{D}} \otimes \mathcal{H}$ along with the operator $M_z \otimes X$ is unitarily equivalent to the space $H^2_{\mathbb{D}}(\mathcal{H})$ with the operator $M^{\mathcal{H}}_{zX}$. Now, for any unit vector f in $H^2_{\mathbb{D}}(\mathcal{H})$,

$$\begin{aligned} |\langle M^{\mathcal{H}}_{zX} f, f\rangle| &= |\frac{1}{2\pi}\int_0^{2\pi} \langle \exp(i\theta) X f(\exp(i\theta)), f(\exp(i\theta))\rangle d\theta| \\ &\leq \frac{1}{2\pi}\int_0^{2\pi} |\langle \exp^{i\theta} X f((i\theta)), f((i\theta))\rangle| d\theta \leq 1 \end{aligned}$$

because $\omega(X) \leq 1$. That completes the proof that $\omega(M^{\mathcal{H}}_{zX}) \leq 1$. ■

Definition 6.9.12. *A bounded operator X acting on a Hilbert space $\mathcal{H}$ is said to be hyponormal if $X^*X \geq XX^*$.*

Note that this is the same as saying that $\|X^*h\| \leq \|Xh\|$ for all h in $\mathcal{H}$.

Proposition 6.9.13. *[Stampfli] If X is hyponormal, then $\|X^n\| = \|X\|^n$ for all $n \geq 1, 2, \ldots$ and so $\|X\| = r(X)$.*

Proof. For h in $\mathcal{H}$ with $\|h\| = 1$, we have

$$\begin{aligned} \|X^n h\|^2 = \langle X^n h, X^n h\rangle &= \langle X^* X^n h, X^{n-1} h\rangle \\ &\leq \|X^* X^n h\| \|X^{n-1} h\| \\ &\leq \|X^{n+1} h\| \|X^{n-1} h\|. \end{aligned}$$

Hence, $\|X^n\|^2 \leq \|X^{n+1}\| \|X^{n-1}\|$. Now, we apply an induction process. Suppose $\|X^k\| = \|X\|^k$ for $k = 1, 2, \ldots, n$. Then,

$$\|X\|^{2n} = \|X^n\|^2 \leq \|X^{n+1}\| \|X^{n-1}\| = \|X^{n+1}\| \|X\|^{n-1}.$$

Hence, $\|X\|^{n+1} \leq \|X^{n+1}\| \leq \|X\|^{n+1}$. We are done. ■

Lemma 6.9.14. *Let U and V be a unitary and a pure isometry on Hilbert Spaces $\mathcal{H}_1$ and $\mathcal{H}_2$ respectively, and let $X : \mathcal{H}_1 \to \mathcal{H}_2$ be such that $XU = VX$. Then $X = 0$.*

Proof. We have , for any positive integer n, $XU^n = V^n X$ by iteration. Therefore, $U^{*n} X^* = X^* V^{*n}$. Thus X^* vanishes on $\ker V^{*n}$, and since $\bigcup_n \ker V^{*n}$ is dense in $\mathcal{H}_2$ we have $X^* = 0$ $i.e,$ $X = 0$. ■

Theorem 6.9.15. *Let S , P be commuting operators on a Hilbert space $\mathcal{H}$. The following statements are all equivalent:*

(1) *(S, P) is a Γ-isometry,*
(2) *(S, P) is a Γ-contraction and P is isometry,*
(3) *P is an isometry , $S = S^*P$ and $r(S) \leq 2$,*
(4) *$r(S) \leq 2$ and $\rho(\beta S, \beta^2 P) = 0$ for all $\beta \in \mathbb{T}$.*
Moreover if the spectral radius $r(S)$ of S is less than 2 then (1),(2),(3) and (4) are equivalent to:
(5) *$(2\beta P - S)(2 - \beta S)^{-1}$ is an isometry, for all $\beta \in \mathbb{T}$.*

Proof. (1)$\Rightarrow$(2) is obvious.

(2)$\Rightarrow$(3): The fact that (S, P) is a Γ-contraction implies that $\|S\| \leq 2$, whence $r(S) \leq 2$. It also implies that $\rho(\alpha S, \alpha^2 P) \geq 0$ for all α in the closed disk, in particular on the circle. In view of P being an isometry, this means that

$$\text{Re } \beta(S - S^*P) \leq 0$$

for all β of modulus 1. By using Lemma 6.9.5, we get that $S = S^*P$.

(3)$\Rightarrow$(4): This is obvious.

(4)$\Rightarrow$(1): We have

$$\rho(\beta S, \beta^2 P) = 2(I - P^*P) - \beta(S - S^*P) - \overline{\beta}(S^* - P^*S) = 0 \text{ for all } \beta \in \mathbb{T}.$$

Putting $\beta = 1$ and $\beta = -1$, we get $P^*P = I$ from which it follows by the same argument as above that $S = S^*P$. We shall now show that (S, P) is a Γ-isometry by exhibiting a Γ-unitary extension.

Wold decomposition of the isometry P breaks the whole space $\mathcal{H}$ into the direct sum of two reducing subspaces $\mathcal{H}_1$ and $\mathcal{H}_2$ so that P has the form

$$P = \begin{pmatrix} P_1 & 0 \\ 0 & P_2 \end{pmatrix} \text{ on } \mathcal{H}_1 \oplus \mathcal{H}_2 = \mathcal{H},$$

where P_1 is a unitary and P_2 is a pure isometry (a shift of some multiplicity). With respect to this decomposition of $\mathcal{H}$, we write

$$S = \begin{pmatrix} S_{11} & S_{12} \\ S_{21} & S_{22} \end{pmatrix}.$$

By commutativity of S and P and applying Lemma 6.9.14, we see that S takes the form

$$\begin{pmatrix} S_1 & 0 \\ 0 & S_2 \end{pmatrix} \quad \text{on } \mathcal{H}_1 \oplus \mathcal{H}_2 \quad \text{and} \quad S_1 P_1 = P_1 S_1 \ , \ S_2 P_2 = P_2 S_2.$$

Also by $S = S^*P$ and $P^*P = I$ we get $S_i = S_i^*P_i$ and $P_i^*P_i = I$ respectively for $i = 1, 2$.
The pair (S_1, P_1) is Γ-unitary by part (4) of Theorem 6.9.6 because P_1 is unitary and restriction of a Γ-contraction to an invariant subspace is a Γ-contraction.

The pair (S_2, P_2) is a Γ-contraction since (S, P) is so. Since P_2 is a pure isometry it can be identified with the multiplication operator M_z^E on $H^2(E)$ for some Hilbert space E. Again since S_2 commutes with $P_2(\equiv M_z^E)$, it can be identified with the multiplication operator M_φ^E for some $\varphi \in H^\infty(\mathcal{B}(E))$.

Also because P_2 is isometry, $I - P_2P_2^* \geq 0$ and we have

$$S_2^*(I - P_2P_2^*)S_2 \geq 0 \Rightarrow S_2^*S_2 \geq (S_2^*P_2)(P_2^*S_2) = S_2S_2^*, \text{ since } S_2 = S_2^*P_2.$$

Thus S_2 is hyponormal and by the result of Stampfli (Theorem 6.9.13), $r(S_2) = \|S_2\|$ and hence $\|\varphi\| = \|M_\varphi^E\| = \|S_2\| \leq 2$. Since $S_2 = {S_2}^*P_2$, or equivalently $M_\varphi^E = M_\varphi^{E^*}M_z^E$, we have

$$\varphi(z) = \varphi^*(z)z \text{ for all } z \in \mathbb{T}.$$

Consider on $L^2(E)$, the multiplication operators U_φ^E and U_z^E, multiplication by $\varphi(z)$ and z respectively. Obviously U_z^E is a unitary operator on $L^2(E)$. Since $\varphi(z) = \varphi^*(z)z$ we have $U_\varphi^E = U_\varphi^{E^*}U_z^E$, i.e, $U_\varphi^{E^*} = U_z^{E^*}U_\varphi^E$ and hence

$$U_\varphi^E U_\varphi^{E^*} = (U_\varphi^{E^*}U_z^E)(U_z^{E^*}{U_\varphi}^E) = U_\varphi^{E^*}U_\varphi^E$$

and thus U_φ^E is normal. So we have a pair of commuting normal operators (U_φ^E, U_z^E) on $L^2(E))$ such that $r(U_\varphi^E) = \|U_\varphi^E\| = \|\varphi\| \leq 2$, $U_\varphi^E = U_\varphi^{E^*}U_z^E$ and U_z^E is unitary. Therefore by part (3) of Theorem 6.9.6, the pair (U_φ^E, U_z^E) is a Γ-unitary. The restriction to $H^2(E)$ of this Γ-unitary is (M_φ^E, M_z^E). In other words (U_φ^E, U_z^E) is a Γ-unitary extension of the pair (M_φ^E, M_z^E).

Taking $\tilde{S} = S_1 \oplus U_\varphi^E$ and $\tilde{P} = P_1 \oplus U_z^E$ on $\mathcal{H}_1 \oplus L^2(E)$, we see that $(\tilde{S}, \tilde{P})$ is a Γ-unitary extension of (S, P). Hence (S, P) is a Γ-isometry.

Thus (1) through (4) are equivalent.

(4) $\Leftrightarrow$ (5): By hypothesis,

$$\rho(\beta S, \beta^2 P) = \frac{1}{2}\{(2 - \beta S)^*(2 - \beta S) \; - \; (2\beta^2 P - \beta S)^*(2\beta^2 P - \beta S)\} = 0.$$

$$\Rightarrow \; (2 - \beta S)^*(2 - \beta S) = (2\beta^2 P - \beta S)^*(2\beta^2 P - \beta S).$$

Since $r(S) < 2$, the operator $2 - \beta S$ is invertible. Therefore, we have

$$((2-\beta S)^{-1})^*(2\beta^2 P - \beta S)^*(2\beta^2 P - \beta S)(2-\beta S)^{-1} = I.$$

Therefore $(2\beta^2 P - \beta S)(2-\beta S)^{-1}$ and hence $(2\beta P - S)(2-\beta S)^{-1}$ is an isometry for all $\beta \in \mathbb{T}$.

Conversely, let (5) hold. Then $(2\beta^2 P - \beta S)(2-\beta S)^{-1}$ is an isometry for every $\beta \in \mathbb{T}$. Therefore,

$$\begin{aligned} & ((2-\beta S)^{-1})^*(2\beta^2 P - \beta S)^*(2\beta^2 P - \beta S)(2-\beta S)^{-1} = I \\ \text{or} \quad & (2-\beta S)^*(2-\beta S) - (2\beta^2 P - \beta S)^*(2\beta^2 P - \beta S) = 0 \\ \text{or} \quad & \rho(\beta S, \beta^2 P) = 0\,, \quad \forall\, \beta \in \mathbb{T}. \end{aligned}$$

Hence done. ∎

Note 6.9.16. *The* Γ*-isometry* (S_2, P_2) *in the proof above is pure.*

Corollary 6.9.17. *If* (S, P) *is a* Γ*-isometry (respectively a* Γ*-unitary), then* (rS, P) *is also a* Γ*-isometry (respectively a* Γ*-unitary) for* $0 \le r \le 1$.

The following two results are remarkable in their simplicity to characterize Γ-isometries.

Lemma 6.9.18. *A pair of bounded operators* (S, P) *defined on* $\mathcal{H}$ *is a pure* Γ*-isometry if and only if* $S = C + C^*P$ *for some pure isometry* P *and a bounded operator* C *which commutes with* P *and* P^* *and has numerical radius not greater than* 1.

Proof. Let (S, P) be a pure Γ-isometry. Then by Theorem 2.4 of [5], S and P can be identified with M_φ^E and M_z^E respectively on $H^2(E)$ for some separable Hilbert space E, where $\varphi(z) = G + G^*z$ for an operator G defined on E such that $\omega(G) \le 1$. Clearly

$$M_\varphi^E = M_{G+G^*z}^E = M_G^E + M_{G^*}^E M_z^E \equiv (I \otimes G) + (I \otimes G^*)(M_z \otimes I)$$

on $H^2(\mathbb{D}) \otimes E \equiv H^2(E)$. Therefore $S \equiv C + C^*P$ where $P = M_z \otimes I$ and $C = I \otimes G$. Obviously P commutes with C, C^* and $\omega(C) \le 1$. Conversely, let $S = C + C^*P$ where $\omega(C) \le 1$ and P is a pure isometry which commutes with C and C^*. Since P is a pure isometry, $P \equiv M_z^E$ on $H^2(E)$ and hence $C \equiv M_\varphi^E$ on $H^2(E)$ for some $\varphi \in \mathcal{H}^\infty(\mathcal{B}(E))$, by the commutativity of C and P.

Also, since both of M_φ^E and M_φ^{E*} commute with M_z^E, the function φ is a constant say equal to G_1. Clearly,

$$M_\varphi^E \equiv (I \otimes G_1) \ \text{ and } \ M_z^E \equiv M_z \otimes I \text{ on } H^2(\mathbb{D}) \otimes E.$$

By the commutativity of C and P, we have

$$S^*P = (C^* + P^*C)P = S.$$

Now

$$\begin{aligned}\omega(S) &= \omega(C + C^*P)\\ &\leq \omega(C) + \omega(C^*P) = \omega(I \otimes G_1) + \omega(M_z \otimes G_1^*) \leq 1 + \omega(M_z \otimes G_1^*).\end{aligned}$$

Since $\omega(G_1^*) \leq 1$, by Lemma 6.9.11, $\omega(M_z \otimes G_1^*) \leq 1$. Thus, we have $\omega(S) \leq 2$. Therefore by Theorem 6.9.15, part (3), (S, P) is a Γ-isometry where P is a pure isometry i.e, (S, P) is a pure Γ-isometry.

■

Theorem 6.9.19. *A pair of bounded operators (S, P) defined on $\mathcal{H}$ is a Γ-isometry if and only if $S = C + C^*P$ for some isometry P and a bounded operator C which commutes with P and P^* and has numerical radius not greater than* 1.

Proof. By Theorem 6.9.15, (S, P) is a Γ-isometry if and only if $S = S_1 \oplus S_2$ and $P = P_1 \oplus P_2$ where (S_1, P_1) and (S_2, P_2) are Γ-unitary and pure Γ-isometry respectively.

Therefore $S_2 = C + C^*P_2$ where $\omega(C) \leq 1$ and P_2 is a pure isometry which commutes with C and C^*. Also by Theorem 6.9.6, $S_1 = U + U^*P_1$ where U is a unitary which commutes with P_1.
Choosing $C_1 = U \oplus C$ we get

$$S = S_1 \oplus S_2 = C_1 + C_1^*(P_1 \oplus P_2) = C_1 + C_1^*P,$$

where P commutes with C_1, C_1^* and obviously $\omega(C_1) \leq 1$. ■

Observation 6.9.20. *Let (S, P) be a Γ-contraction where P is a projection. Then S and P have the operator matrices*

$$S = \begin{pmatrix} S_1 & 0 \\ 0 & S_2 \end{pmatrix} \quad P = \begin{pmatrix} I & 0 \\ 0 & 0 \end{pmatrix},$$

with respect to the decomposition $\mathcal{H} = \operatorname{ran}(P) \oplus \ker(P)$.

Proof. Clearly P has the stated form as P is a projection. Let $S = [S_{ij}]_{i,j=1}^2$ with respect to the decomposition $\mathcal{H} = \operatorname{ran}(P) \oplus \ker(P)$. By the commutativity of S and P it follows that $S_{12} = S_{21} = 0$. ■

Observation 6.9.21. *If (S, P) is a Γ-contraction where P is a partial isometry then $S - S^*P = \begin{pmatrix} 0 & 0 \\ * & S \end{pmatrix}$ with respect to the decomposition $\mathcal{H} = \overline{\operatorname{ran}}\, P^* \oplus \ker(P)$.*

Proof. Since (S, P) is a Γ-contraction, by Proposition 6.9.4 we have $\rho(\alpha S, \alpha^2 P) \geq 0$ for all α in $\mathbb{T}$ which implies that

$$(I - P^*P) - \text{Re}\, \alpha(S - S^*P) \geq 0.$$

Since P is a partial isometry, P^*P is a projection onto $\overline{\operatorname{ran}}\, P^* = \ker(P)^\perp$. Therefore $I - P^*P$ is a projection onto $\ker(P)$. So we have $P_{\ker(P)} - \text{Re}\, \alpha(S - S^*P) \geq 0$ for all α in $\mathbb{T}$. Therefore for $x \in \ker(P)^\perp = \overline{\operatorname{ran}}\, P^*$ we have $P_{\ker(P)}(x) = 0$ and hence

$$\langle \text{Re}\, \alpha(S - S^*P)x, x\rangle \leq 0, \text{ for all } \alpha \text{ in } \mathbb{T} \text{ and } x \text{ in } \overline{\operatorname{ran}}\, P^*.$$

Therefore, by an argument similar to the proof of Lemma 6.9.5, we have $P|_{\overline{\operatorname{ran}}\, P^*}(S - S^*P)|_{\overline{\operatorname{ran}}\, P^*} = 0$. Hence, $S - S^*P = \begin{pmatrix} 0 & * \\ * & * \end{pmatrix}$ with respect to the decomposition $\mathcal{H} = \overline{\operatorname{ran}}\, P^* \oplus \ker(P)$. Now, the commutativity of S and P shows that $S - S^*P = \begin{pmatrix} 0 & 0 \\ * & S \end{pmatrix}$. ■

A canonical way of constructing a Γ-isometry is to consider the Hardy space $H^2(\mathbb{D}^2)$ of the bidisc with the reproducing kernel $\frac{1}{(1-z_1\overline{w}_1)(1-z_2\overline{w}_2)}$. If M_{z_1} and M_{z_2} are multiplications by the independent variables z_1 and z_2 respectively, then $(M_{z_1} + M_{z_2}, M_{z_1}M_{z_2})$ is a Γ-isometry.

6.10. Γ-contractions - examples

Dilating a contraction operator to an isometry is well studied in the history of dilation theory (see [113]). For the class of examples of Γ-contractions contained in this section, we produce their Γ-isometric dilations. The contents of this section are from [35].

Definition 6.10.1. *Let (S, P) be a Γ-contraction on $\mathcal{H}$. A commuting pair of operators (T, V) acting on a Hilbert space $\mathcal{N}$ containing $\mathcal{H}$ as a subspace is said to be a Γ-isometric dilation of (S, P) if (T, V) is a Γ-isometry and*

$$T^*|_{\mathcal{H}} = S^* \text{ and } V^*|_{\mathcal{H}} = P^*.$$

Thus (T, V) is a Γ-isometric dilation of a Γ-contraction (S, P) is same as saying that (T^*, V^*) is a Γ-co-isometric extension of (S^*, P^*). Moreover, the dilation will be called minimal if

$$\mathcal{N} = \overline{\text{span}}\{V^n h : h \in \mathcal{H} \text{ and } n = 0, 1, 2, \dots\}.$$

We shall see the existence and uniqueness of minimal Γ-isometric dilation in Theorem 6.11.3.

In this section we exhibit a new class of examples of Γ-contractions and using a recent theorem of Douglas, Misra and Sarkar ([52]), find Γ-isometric dilations of some of them. The main result of this section is Theorem 6.10.9.

Lemma 6.10.2. *Let T_1 and T_2 be to commuting contractions defined on $\mathcal{H}$ and let $\mathcal{M} \subseteq \mathcal{H}$ be a subspace invariant under $T_1 + T_2$ and T_1T_2. Then $((T_1 + T_2)|_{\mathcal{M}}, T_1T_2|_{\mathcal{M}})$ is a Γ-contraction.*

Proof. We have to show that Γ is a spectral set for the pair $((T_1 + T_2)|_{\mathcal{M}}, T_1T_2|_{\mathcal{M}})$, that is, for any polynomial p of two variables,

$$\|p((T_1 + T_2)|_{\mathcal{M}}, T_1T_2|_{\mathcal{M}})\| \le \|p(z_1, z_2)\|_{\infty,\Gamma}.$$

Let $\pi : \mathbb{C}^2 \to \mathbb{C}^2$ be defined as

$$\pi(z_1, z_2) = (z_1 + z_2, z_1 z_2).$$

Then by *von Neumann's* inequality in the bidisc $\mathbb{D}^2$, we have,

$$\|p(\pi(T_1, T_2))\| \le \|p \circ \pi\|_{\infty,\mathbb{D}^2}$$
$$\text{or } \|p(T_1 + T_2, T_1T_2)\| \le \|p\|_{\infty,\Gamma}$$

Certainly,

$$\|p((T_1 + T_2)|_{\mathcal{M}}, T_1T_2|_{\mathcal{M}})\| \le \|p(T_1 + T_2, T_1T_2)\|.$$

Hence done. ■

Let us see a particular example of this theorem. For $\lambda, \mu > 1$ we define the weighted Bergman spaces

(6.10.2)

$$\mathbb{A}^{(\lambda,\mu)}(\mathbb{D}^2) = \{f : \mathbb{D}^2 \to \mathbb{C} \ : f \text{ is holomorphic and}$$
$$\int_{\mathbb{D}^2} |f(z_1, z_2)|^2 (1 - |z_1|^2)^{\lambda-2} (1 - |z_2|^2)^{\mu-2} dm^{(\lambda,\mu)}(z_1, z_2) < \infty\},$$

where $m^{(\lambda,\mu)}$ is $\frac{(\lambda-1)(\mu-1)}{\pi^2}$ times the Lebesgue measure on $\mathbb{D}^2$. For $f, g \in \mathbb{A}^{(\lambda,\mu)}(\mathbb{D}^2)$, define the inner product $\langle f, g\rangle_{\mathbb{A}^{(\lambda,\mu)}(\mathbb{D}^2)}$ by

$$\int_{\mathbb{D}^2} f(z_1, z_2)\overline{g(z_1, z_2)}(1 - |z_1|^2)^{\lambda-2}(1 - |z_2|^2)^{\mu-2} dm^{(\lambda,\mu)}(z_1, z_2).$$

It is easy to verify that $\mathbb{A}^{(\lambda,\mu)}(\mathbb{D}^2)$ is a Hilbert space.

Let Γ^0 denote the interior of Γ. Define the Hilbert space

(6.10.3)
$$\mathbb{A}^{(\lambda,\mu)}(\Gamma^0)$$
(6.10.4)
$$= \{f : \Gamma^0 \to \mathbb{C} : f \text{ is holomorphic and } (f \circ \pi) \det J_\pi \in \mathbb{A}^{(\lambda,\mu)}(\mathbb{D}^2)\},$$

with

$$\langle f, g\rangle_{\mathbb{A}^{(\lambda,\mu)}(\Gamma^0)} = \langle (f \circ \pi) \det J_\pi, (g \circ \pi) \det J_\pi\rangle_{\mathbb{A}^{(\lambda,\mu)}(\mathbb{D}^2)},$$

where $J_\pi = \begin{pmatrix} 1 & 1 \\ z_2 & z_1 \end{pmatrix}$ is the Jacobian of the map

$$\pi(z_1, z_2) = (z_1 + z_2, z_1 z_2)$$

so that $\det J_\pi = (z_1 - z_2)$. Let $M_s^{(\lambda,\mu)}$ and $M_p^{(\lambda,\mu)}$ be the multiplication operators on $\mathbb{A}^{(\lambda,\mu)}(\Gamma^0)$ by the co-ordinate functions s and p, respectively, where $(s, p) \in \Gamma^0$. For $\lambda = \mu$, we denote $\mathbb{A}^{(\lambda,\lambda)}(\mathbb{D}^2)$ by $\mathbb{A}^{(\lambda)}(\mathbb{D}^2)$, $\mathbb{A}^{(\lambda,\lambda)}(\Gamma^0)$ by $\mathbb{A}^{(\lambda)}(\Gamma^0)$, $M_s^{(\lambda,\lambda)}$ by $M_s^{(\lambda)}$ and $M_p^{(\lambda,\lambda)}$ by $M_p^{(\lambda)}$. The following lemma serves the purpose of showing that the operator pair $(M_s^{(\lambda,\mu)}, M_p^{(\lambda,\mu)})$, which is obviously a commuting pair, is a Γ-contraction.

Lemma 6.10.3. *For integers $m, n \geq 0$, let $\widetilde{e_{mn}}$ and $\widetilde{f_{mn}}$ be the functions defined on $\mathbb{D}^2$ by*

$$\widetilde{e_{mn}}(z_1, z_2) = z_1^m z_2^n - z_1^n z_2^m \text{ and } \widetilde{f_{mn}}(z_1, z_2) = z_1^m z_2^n + z_1^n z_2^m.$$

Let $\mathbb{A}_{\mathrm{a}}^{(\lambda,\mu)}(\mathbb{D}^2) := \overline{\operatorname{span}}\{\widetilde{e_{mn}} : m > n \geq 0\}$ and $\mathbb{A}_{\mathrm{s}}^{(\lambda,\mu)}(\mathbb{D}^2) := \overline{\operatorname{span}}\{\widetilde{f_{mn}} : m \geq n \geq 0\}$ be subspaces of $\mathbb{A}^{(\lambda,\mu)}(\mathbb{D}^2)$. Then

(1) *both $\mathbb{A}_{\mathrm{a}}^{(\lambda,\mu)}(\mathbb{D}^2)$ and $\mathbb{A}_{\mathrm{s}}^{(\lambda,\mu)}(\mathbb{D}^2)$ are invariant subspaces of $\mathbb{A}^{(\lambda,\mu)}(\mathbb{D}^2)$ under $M_{z_1+z_2}$ and $M_{z_1 z_2}$;*
(2) *the restrictions of the pair $(M_{z_1+z_2}, M_{z_1 z_2})$ to the invariant subspaces $\mathbb{A}_{\mathrm{a}}^{(\lambda,\mu)}(\mathbb{D}^2)$ and $\mathbb{A}_{\mathrm{s}}^{(\lambda,\mu)}(\mathbb{D}^2)$ are Γ-contractions, call them $(S_a^{(\lambda,\mu)}, P_a^{(\lambda,\mu)})$ and $(S_s^{(\lambda,\mu)} P_s^{(\lambda,\mu)})$ respectively. As usual, for $\lambda = \mu$ we use just one index.*
(3) *there is an isometry U from $\mathbb{A}^{(\lambda,\mu)}(\Gamma^0)$ onto $\mathbb{A}_{\mathrm{a}}^{(\lambda,\mu)}(\mathbb{D}^2)$ such that $UM_s^{(\lambda,\mu)}U^* = S_a^{(\lambda,\mu)}$ and $UM_p^{(\lambda,\mu)}U^* = P_a^{(\lambda,\mu)}$.*

Proof. Note that

$$\begin{aligned}(z_1+z_2)(z_1^m z_2^n - z_1^n z_2^m) &= z_1^{m+1} z_2^n - z_1^{n+1} z_2^m + z_1^m z_2^{n+1} - z_1^n z_2^{m+1}\\ &= (z_1^{m+1} z_2^n - z_1^n z_2^{m+1}) + (z_1^m z_2^{n+1} - z_1^{n+1} z_2^m)\end{aligned}$$

Again

$$z_1 z_2 (z_1^m z_2^n - z_1^n z_2^m) = (z_1^{m+1} z_2^{n+1} - z_1^{n+1} z_2^{m+1}).$$

So $\mathbb{A}_{\rm a}^{(\lambda,\mu)}(\mathbb{D}^2)$ is invariant under both of the multiplication operators $M_{z_1+z_2}$ and $M_{z_1 z_2}$. We can show similarly that $\mathbb{A}_{\rm s}^{(\lambda,\mu)}(\mathbb{D}^2)$ is invariant under both the operators $M_{z_1+z_2}$ and $M_{z_1 z_2}$. Hence by Lemma 6.10.2, (1) and (2) above are proved. To prove (3), define

$$U : \mathbb{A}^{(\lambda,\mu)}(\Gamma^0) \longrightarrow \mathbb{A}^{(\lambda,\mu)}(\mathbb{D}^2)$$

by

$$Uf = (f \circ \pi) \det J_\pi.$$

That U is an isometry follows from the definitions of norms on the corresponding spaces. It is easy to check by direct computation that U intertwines $M_s^{(\lambda,\mu)}$ with $S_a^{(\lambda,\mu)}$ and $M_p^{(\lambda,\mu)}$ with $P_a^{(\lambda,\mu)}$. ■

Remark 6.10.4. We observe that $\langle \widetilde{e_{mn}}, \widetilde{f_{mn}} \rangle_{\mathbb{A}^{(\lambda,\mu)}(\mathbb{D}^2)} = \frac{m!n!}{(\lambda)_m(\mu)_n} - \frac{m!n!}{(\mu)_m(\lambda)_n}$, for $m > n \geq 0$, where $(\lambda)_m = \frac{\lambda(\lambda-1)(\lambda-2)\ldots(\lambda-m+1)}{m!}$. Therefore the subspaces $\mathbb{A}_{\rm a}^{(\lambda,\mu)}(\mathbb{D}^2)$ and $\mathbb{A}_{\rm s}^{(\lambda,\mu)}(\mathbb{D}^2)$ of $\mathbb{A}^{(\lambda,\mu)}(\mathbb{D}^2)$ are mutually orthogonal if and only if $\lambda = \mu$.

Consider the weighted Bergman space $\mathbb{A}^{(\lambda)}(\mathbb{D}^2)$, as defined in (6.10.2), on the bidisc for $\lambda > 1$ and its subspaces

$$\mathbb{A}_{\rm a}^{(\lambda)}(\mathbb{D}^2) := \overline{\rm span}\{z_1^m z_2^n - z_1^n z_2^m : m > n \geq 0, (z_1, z_2) \in \mathbb{D}^2\}$$

and

$$\mathbb{A}_{\rm s}^{(\lambda)}(\mathbb{D}^2) := \overline{\rm span}\{z_1^m z_2^n + z_1^n z_2^m : m \geq n \geq 0, (z_1, z_2) \in \mathbb{D}^2\}.$$

They are mutually orthogonal and $\mathbb{A}^{(\lambda)}(\mathbb{D}^2) = \mathbb{A}_{\rm a}^{(\lambda)}(\mathbb{D}^2) \oplus \mathbb{A}_{\rm s}^{(\lambda)}(\mathbb{D}^2)$. Let

$$f_{mn}(z_1, z_2) = \begin{cases} \sqrt{\frac{(\lambda)_m(\lambda)_n}{2(m!n!)}}(z_1^m z_2^n + z_1^n z_2^m) & \text{for } m > n \geq 0; \\ \sqrt{\frac{(\lambda)_n}{n!}}(z_1 z_2)^n & \text{for } m = n \geq 0. \end{cases} \tag{6.10.5}$$

Clearly, $\{f_{mn} : m \geq n \geq 0\}$ is an orthonormal basis for the Hilbert space $\mathbb{A}_{\rm s}^{(\lambda)}(\mathbb{D}^2)$.

Proposition 6.10.5. *The Hilbert space $\mathbb{A}_{\mathrm{s}}^{(\lambda)}(\mathbb{D}^2)$ is a reproducing kernel Hilbert space with its reproducing kernel $K_{\mathrm{s}}^{(\lambda)}$ given by the formula:*

(6.10.6)

$$K_{\mathrm{s}}^{(\lambda)}(\boldsymbol{z},\boldsymbol{w}) = \frac{1}{2}(1-z_1\bar{w}_1)^{-\lambda}(1-z_1\bar{w}_2)^{-\lambda} + \frac{1}{2}(1-z_1\bar{w}_2)^{-\lambda}(1-z_2\bar{w}_1)^{-\lambda},$$

where $\boldsymbol{z}=(z_1,z_2)$ and $\boldsymbol{w}=(w_1,w_2)$ are in $\mathbb{D}^2$

Proof. We shall prove this by expanding the right hand side of the formula (6.10.6) in terms of the basis elements f_{mn}. For $\boldsymbol{z},\boldsymbol{w}\in\mathbb{D}^2$, we have

$$\begin{aligned}
K_{\mathrm{s}}^{(\lambda)}(\boldsymbol{z},\boldsymbol{w}) &= \sum_{m\ge n\ge 0} f_{mn}(z_1,z_2)\overline{f_{mn}(w_1,w_2)}\\
&= \sum_{m>n\ge 0} f_{mn}(z_1,z_2)\overline{f_{mn}(w_1,w_2)} + \sum_{n\ge 0} f_{nn}(z_1,z_2)\overline{f_{nn}(w_1,w_2)}\\
&= \frac{1}{2}\sum_{\substack{m,n\ge 0\\ m\ne n}} f_{mn}(z_1,z_2)\overline{f_{mn}(w_1,w_2)} + \sum_{n\ge 0} f_{nn}(z_1,z_2)\overline{f_{nn}(w_1,w_2)}\\
&= \frac{1}{4}\sum_{\substack{m,n\ge 0\\ m\ne n}} \frac{(\lambda)_m(\lambda)_n}{m!n!}\Big((z_1\bar{w}_1)^m(z_2\bar{w}_2)^n + (z_1\bar{w}_2)^m(z_2\bar{w}_1)^n\\
&+ (z_2\bar{w}_1)^m(z_1\bar{w}_2)^n + (z_2\bar{w}_2)^m(z_1\bar{w}_1)^n\Big) + \sum_{n\ge 0}\frac{(\lambda)_n^2}{(n!)^2}(z_1\bar{w}_1)^n(z_2\bar{w}_2)^n\\
&= \frac{1}{2}(1-z_1\bar{w}_1)^{-\lambda}(1-z_1\bar{w}_2)^{-\lambda} + \frac{1}{2}(1-z_1\bar{w}_2)^{-\lambda}(1-z_2\bar{w}_1)^{-\lambda}
\end{aligned}$$

■

Edigarian and Zwonek found the Bergman kernel for symmetrized polydisc, see [62]. We shall need explicit formulae for the reproducing kernels of the weighted Bergman spaces $\mathbb{A}^{(\lambda)}(\Gamma^0)$, as defined in (6.10.3). These have been extensively studied in [78]. We recall only some relevant facts here. For $\lambda>1$, the reproducing kernel for the weighted Bergman space $\mathbb{A}^{(\lambda)}(\Gamma^0)$ on the interior of the symmetrized bidisc Γ^0 is given by

$$\frac{1}{\lambda}\frac{\{(1-z_1\bar{w}_1)^{-\lambda}(1-z_2\bar{w}_2)^{-\lambda}-(1-z_1\bar{w}_2)^{-\lambda}(1-z_2\bar{w}_1)^{-\lambda}\}}{(z_1-z_2)(\bar{w}_1-\bar{w}_2)} \tag{6.10.7}$$

for $\boldsymbol{z},\boldsymbol{w}\in\mathbb{D}^2$. We shall denote this kernel by $\mathbf{B}_{\Gamma^0}^{(\lambda)}(\pi(\boldsymbol{z}),\pi(\boldsymbol{w}))$.

The kernel above remains a positive definite kernel for $\lambda = 1$. This prompted the authors of [78] to define the Hardy space $H^2(\Gamma^0)$ of the symmetrized bidisc to be the reproducing kernel Hilbert space whose kernel is

$$\frac{(1-z_1\bar{w}_1)^{-1}(1-z_2\bar{w}_2)^{-1}-(1-z_1\bar{w}_2)^{-1}(1-z_2\bar{w}_1)^{-1}}{(z_1-z_2)(\bar{w}_1-\bar{w}_2)}, \tag{6.10.8}$$

for $\boldsymbol{z}, \boldsymbol{w} \in \mathbb{D}^2$. They denoted it by $\mathbb{S}_{\Gamma^0}(\pi(\boldsymbol{z}), \pi(\boldsymbol{w}))$. Before we progress further, we need a lemma. To motivate the lemma, we want to recall the well-known fact that for $n = 1, 2, \ldots$, the kernels

$$k_n(z,w) = \frac{1}{(1-z\bar{w})^n}$$

on the unit disc define reproducing kernel Hilbert spaces, $\mathcal{H}_n$, say (so the Hardy space which is usually denoted by H^2 is $\mathcal{H}_1$ in this notation). If $M_z^{(n)}$ is the co-ordinate multiplication on $\mathcal{H}_n$, then every $M_z^{(n)}$ for $n = 2, 3, \ldots$ dilates to $M_z^{(1)} \otimes I_{\mathcal{E}}$ on $H_1 \otimes \mathcal{E}$ for some $\mathcal{E}$ depending on $M_z^{(n)}$.

Lemma 6.10.6. *Let Ω be an open connected bounded domain in $\mathbb{C}^n$ and let $\mathcal{H}$ and $\mathcal{M}$ be two reproducing kernel Hilbert spaces of holomorphic functions on Ω with their reproducing kernels $k_{\mathcal{H}}$ and $k_{\mathcal{L}}$ being related by $k_{\mathcal{H}}(z,w) = k_{\mathcal{L}}(z,w)k(z,w)$ for some positive definite kernel k which is holomorphic in the first variable and anti-holomorphic in the second. Then there is a Hilbert space $\mathcal{E}$ such that the commuting tuple of operators $(M_{z_1}^{\mathcal{L}} \otimes I_{\mathcal{E}}, M_{z_2}^{\mathcal{L}} \otimes I_{\mathcal{E}}, \ldots, M_{z_n}^{\mathcal{L}} \otimes I_{\mathcal{E}})$ on $\mathcal{L} \otimes \mathcal{E}$ is a commuting simultaneous dilation of the commuting tuple $(M_{z_1}^{\mathcal{H}}, M_{z_2}^{\mathcal{H}}, \ldots, M_{z_n}^{\mathcal{H}})$ on $\mathcal{H}$.*

Proof. We shall sketch the proof. Take $\mathcal{E}$ to be the Hilbert space corresponding to the kernel k. Define $V : \mathcal{H} \to \mathcal{L} \otimes \mathcal{E}$ by defining $Vk_{\mathcal{H}}(\cdot, w) = k_{\mathcal{L}}(\cdot, w) \otimes k(\cdot, w)$ and then extending it linearly. This is an isometry because of the way $k_{\mathcal{H}}$ factorizes. It is easy to see that the range of the isometry is co-invariant under $M_{z_i}^{\mathcal{L}} \otimes I_{\mathcal{E}}$ for each i and the compression of $M_{z_i}^{\mathcal{L}} \otimes I_{\mathcal{E}}$ to the range of V agrees with $VM_{z_i}^{\mathcal{H}}V^*$. ■

For more along the same theme as the lemma above, see Theorem 2 of [52].

Lemma 6.10.7. *The ratio $\mathbb{S}_{\Gamma^0}^{-1}\mathbf{B}_{\Gamma^0}^{(n)}$ of the reproducing kernel of the weighted Bergman space $\mathbb{A}^{(n)}(\Gamma^0)$ with the reproducing kernel of the Hardy space $H^2(\Gamma^0)$ is a positive definite kernel for all positive integers n.*

Proof. For $\boldsymbol{z}, \boldsymbol{w} \in \mathbb{D}^2$, from (6.10.7) and (6.10.8), we have

$$\begin{aligned}&\mathbb{S}_{\Gamma^0}^{-1}\mathbf{B}_{\Gamma^0}^{(n)}(\pi(\boldsymbol{z}),\pi(\boldsymbol{w}))\\ =&\frac{1}{\lambda}\frac{(1-z_1\bar{w}_1)^{-n}(1-z_2\bar{w}_2)^{-n}-(1-z_1\bar{w}_2)^{-n}(1-z_2\bar{w}_1)^{-n}}{(1-z_1\bar{w}_1)^{-1}(1-z_2\bar{w}_2)^{-1}-(1-z_1\bar{w}_2)^{-1}(1-z_2\bar{w}_1)^{-1}}\\ =&\frac{1}{\lambda}\sum_{k=0}^{n-1}a^{n-1-k}b^k,\end{aligned}$$

where $a=(1-z_1\bar{w}_1)^{-1}(1-z_2\bar{w}_2)^{-1}$ and $b=(1-z_1\bar{w}_2)^{-1}(1-z_2\bar{w}_1)^{-1}$. Clearly, the last expression can be expressed as a polynomial in ab and a^k+b^k for $k=1,\ldots,n-1$. Since $ab=\mathbb{S}_{\Gamma^0}(\pi(\boldsymbol{z}),\pi(\boldsymbol{w}))$ is the reproducing kernel for the Hilbert spaces $H^2(\Gamma^0)$ and $a^k+b^k=K_{\mathrm{s}}^{(k)}(\boldsymbol{z},\boldsymbol{w})$ is the reproducing kernel for the Hilbert space $\mathbb{A}_{\mathrm{s}}^{(k)}(\mathbb{D}^2)$, they both are positive definite. Recalling that pointwise product and sum of two positive definite kernels are again positive definite kernels we conclude that $\mathbb{S}_{\Gamma^0}^{-1}\mathbf{B}_{\Gamma^0}^{(n)}$ is a positive definite kernel. ∎

By $H^2(\mathbb{D}^2)$, we shall denote the Hardy space of the bidisc. For convenience of notation, we shall also call it $\mathbb{A}^{(1)}(\mathbb{D}^2)$. This will enable us to talk about the operator pairs $(S_a^{(1)}, P_a^{(1)})$ and $(S_s^{(1)}, P_s^{(1)})$.

Lemma 6.10.8. *The pair (M_s^H, M_p^H) of multiplication operators on $H^2(\Gamma^0)$ by the co-ordinate functions is a Γ-isometry.*

Proof. Let $H^2_{\mathrm{a}}(\mathbb{D}^2) := \overline{\mathrm{span}}\{z_1^m z_2^n - z_1^n z_2^m : m \geq n \geq 0, (z_1,z_2)\in\mathbb{D}^2\}$ and $H^2_{\mathrm{s}}(\mathbb{D}^2) := \overline{\mathrm{span}}\{z_1^m z_2^n + z_1^n z_2^m : m \geq n \geq 0, (z_1,z_2)\in\mathbb{D}^2\}$. Clearly, $H^2(\mathbb{D}^2)=H^2_{\mathrm{a}}(\mathbb{D}^2)\oplus H^2_{\mathrm{s}}(\mathbb{D}^2)$. For $\lambda=\mu=1$, analogous arguments as in Lemma 6.10.3, shows that

(i) the subspaces $H^2_{\mathrm{a}}(\mathbb{D}^2)$ and $H^2_{\mathrm{s}}(\mathbb{D}^2)$ are invariant subspaces of $H^2(\mathbb{D}^2)$ under $M_{z_1+z_2}$ and $M_{z_1z_2}$;
(ii) there is an isometry U from $H^2(\Gamma^0)$ onto $H^2_{\mathrm{a}}(\mathbb{D}^2)$ such that $UM_s^HU^*=S_a^{(1)}$ and $UM_p^HU^*=P_a^{(1)}$.

By Theorem 6.9.6–(2), the pair $(M_{z_1+z_2}, M_{z_1z_2})$ is a Γ-unitary on $L^2(\mathbb{T}^2)$. Moreover, $S_a^{(1)}=M_{z_1+z_2}|_{H^2_{\mathrm{a}}(\mathbb{D}^2)}$ and $P_a^{(1)}=M_{z_1z_2}|_{H^2_{\mathrm{a}}(\mathbb{D}^2)}$. So $(S_a^{(1)}, P_a^{(1)})$ is a Γ-isometry. Noting that U is a unitary, it follows from (ii) that (M_s^H, M_p^H) is a Γ-isometry. ∎

Recall that the pair $(M_s^{(\lambda)}, M_p^{(\lambda)})$ denotes the commuting pair of multiplication operators by the coordinate functions s and p, respectively, on the Hilbert space $\mathbb{A}^{(\lambda)}(\Gamma^0)$ for $\lambda > 1$. For $\lambda = 1$, this space is the Hardy space $H^2(\Gamma^0)$ and the operator pair is (M_s^H, M_p^H). Thus, by Lemma 6.10.7 and Lemma 6.10.6, we have proved the following theorem which is the main result of this section.

Theorem 6.10.9. *For every positive integer n, the Γ-contraction $(M_s^{(n)}, M_p^{(n)})$ acting on $\mathbb{A}^{(n)}(\Gamma^0)$ can be dilated to the Γ-isometry $(M_s^H \otimes I_{\mathcal{L}}, M_p^H \otimes I_{\mathcal{L}})$ on $H^2(\Gamma^0) \otimes \mathcal{L}$ for some Hilbert space $\mathcal{L}$.*

Recalling the notations from Lemma 6.10.3, we have the following corollary.

Corollary 6.10.10. *For every positive integer n, the commuting pair of operators $(S_a^{(n)}, P_a^{(n)})$ acting on the Hilbert space $\mathbb{A}_\mathrm{a}^{(n)}(\mathbb{D}^2)$ has a Γ-isometric dilation to the commuting pair of operators $(S_a^{(1)}, P_a^{(1)})$ on the Hilbert space $H_\mathrm{a}^2(\mathbb{D}^2) \otimes \mathcal{L}$ for some Hilbert space $\mathcal{L}$.*

Proof. Observing that the isometry U in part 3 of Lemma 6.10.3 is actually a unitary the proof follows from Theorem 6.10.9. ■

Lemma 6.10.11. *The Γ-isometric dilation $(S_a^{(1)}, P_a^{(1)})$ on the Hilbert space $H_\mathrm{a}^2(\mathbb{D}^2) \otimes \mathcal{L}$ of the commuting pair of operators $(S_a^{(n)}, P_a^{(n)})$ on the Hilbert space $\mathbb{A}_\mathrm{a}^{(n)}(\mathbb{D}^2)$ is minimal.*

Proof. To prove minimality, we need to show that $\overline{\text{span}}\{P_1^k h : h \in H_\mathrm{a}^2(\mathbb{D}^2), k \geq 0\} = H_\mathrm{a}^2(\mathbb{D}^2)$. Recalling that $\widetilde{e_{mn}}(z_1, z_2) = z_1^m z_2^n - z_1^n z_2^m$ and $H_\mathrm{a}^2(\mathbb{D}^2) = \overline{\text{span}}\{\widetilde{e_{mn}} : m > n \geq 0\}$, it suffices to show that $\overline{\text{span}}\{P_1^k(\widetilde{e_{mn}}) : m > n \geq 0, k \geq 0\} = \overline{\text{span}}\{\widetilde{e_{mn}} : m > n \geq 0\}$. Since $P_1^k(\widetilde{e_{mn}}) = \widetilde{e_{m+k,n+k}}$, we have $\overline{\text{span}}\{P_1^k(\widetilde{e_{mn}}) : m > n \geq 0, k \geq 0\} = \overline{\text{span}}\{\widetilde{e_{m+k,n+k}} : m > n \geq 0, k \geq 0\} = \overline{\text{span}}\{\widetilde{e_{mn}} : m > n \geq 0\}$. Hence the proof is complete. ■

We have a corollary of the above lemma.

Corollary 6.10.12. *The dilation $(M_s^H \otimes I_{\mathcal{L}}, M_p^H \otimes I_{\mathcal{L}})$ on the Hilbert space $H^2(\Gamma^0) \otimes \mathcal{L}$ of the commuting pair of operators $(M_s^{(n)}, M_p^{(n)})$ on the Hilbert space $\mathbb{A}^{(n)}(\Gamma^0)$ is minimal.*

Proof. Set

$$\epsilon_{mn}(s, p) = \epsilon \circ \pi(z_1, z_2) = \frac{z_1^m z_2^n - z_1^n z_2^m}{z_1 - z_2}$$

for $(s,p) \in \Gamma^0, (z_1,z_2) \in \mathbb{D}^2$. So $H^2(\Gamma^0) = \overline{\text{span}}\{\epsilon_{mn} : m > n \geq 0\}$ and $M_p\epsilon_{mn} = \epsilon_{m+1,n+1}$. Now analogous arguments as in the previous corollary shows that

$$\overline{\text{span}}\{M_p^k\epsilon_{mn} : m > n \geq 0, k \geq 0\} = \overline{\text{span}}\{\epsilon_{mn} : m > n \geq 0,\} = H^2(\Gamma^0).$$

This proves minimality of $(M_s^H \otimes I_{\mathcal{L}}, M_p^H \otimes I_{\mathcal{L}})$ on the Hilbert space $H^2(\Gamma^0) \otimes \mathcal{L}$. ∎

We move on to general discussion of dilation in the next section.

6.11. Dilation

As in many occasions in operator theory, in our case too, finding a solution to an operator equation turns out to be of utmost importance. As is clear by now, a crucial role in deciphering the structure of a Γ-contraction (S,P) is played by the operator $S - S^*P$. For a pair (S,P) of commuting bounded operators with $\|P\| \leq 1$, we shall denote from now on by Σ and Σ_*, the operators $S - S^*P$ and $S^* - SP^*$ respectively. We denote by D_P and $\mathcal{D}_P$ the operator $(I - P^*P)^{\frac{1}{2}}$ and its range closure respectively. For a pair of commuting bounded operators (S,P) with $\|P\| \leq 1$, the *fundamental equation* is defined to be

$$\Sigma = D_P X D_P, \text{ where } X \in \mathcal{B}(\mathcal{D}_P) \tag{6.11.9}$$

and the same for the pair (S^*, P^*) is

$$\Sigma_* = D_{P^*} Y D_{P^*}, \text{ where } Y \in \mathcal{B}(\mathcal{D}_{P^*}). \tag{6.11.10}$$

We start with a pivotal theorem which guarantees the existence and uniqueness of solutions of such equations for Γ-contractions. The proof of the theorem needs the following lemma and the proof of the lemma given here is from a private communication with Michael A. Dritschel.

Lemma 6.11.1. *Let Σ and D be two bounded operators on $\mathcal{H}$. Then*

$$DD^* \geq \text{Re}\,(e^{i\theta}\Sigma) \quad \textit{for all } \theta \in [0, 2\pi)$$

if and only if there is $F \in \mathcal{B}(\mathcal{D}_*)$ with numerical radius of F not greater than one such that $\Sigma = DFD^*$, where $\mathcal{D}_* = \overline{\text{ran}}\, D^*$.

The proof of this result needs the operator Fejer-Riesz factorization theorem, that we studied in an earlier section of this chapter, along with Douglas's lemma (Lemma 2.1 of [54]) and the familiar result that an operator X has numerical radius not greater than one if and only if Re $\beta X \leq I$ for all complex numbers β of modulus 1 (Lemma 6.9.11).

Proof of Lemma 6.11.1. Let there be an operator $F \in \mathcal{B}(\mathcal{H})$ with numerical radius not bigger than one such that $\Sigma = DFD^*$. Since $I - \text{Re}\,(e^{i\theta}F) \geq 0$, for all $\theta \in [0, 2\pi)$, we have

$$D(I - \text{Re}\, e^{i\theta}F)D^* \geq 0, \quad \text{for all } \theta.$$

So we have

$$DD^* \geq D\text{Re}\,(e^{i\theta}F)D^* = \text{Re}\,(e^{i\theta}\ DFD^*) = \text{Re}\,(e^{i\theta}\Sigma)$$

for all $\theta \in [0, 2\pi)$.

The nontrivial part of this lemma, however, is the converse of the above. Suppose that $DD^* \geq \text{Re}\,(e^{i\theta}\Sigma)$ for all $\theta \in [0, 2\pi)$. This means that the Laurent polynomial

$$DD^* - \frac{1}{2}(z\Sigma + \bar{z}\Sigma^*)$$

is non-negative for z on the unit circle. By the operator Fejér-Riesz theorem, we thus have a factorization

$$DD^* - \frac{1}{2}(z\Sigma + \bar{z}\Sigma^*) = (X - zY)(X^* - \bar{z}Y^*), \quad |z| = 1,$$

with $X, Y \in \mathcal{B}(\mathcal{H})$. Thus $DD^* = XX^* + YY^*$ and $\Sigma = 2YX^*$. Since $DD^* \geq XX^*$ and $DD^* \geq YY^*$, Douglas's lemma tells us that there exist contractions Q and R such that $X = DQ$ and $Y = DR$. Thus $\Sigma = DFD^*$ for

$$F = P_{\mathcal{D}_*}2RQ^*|_{\mathcal{D}_*}.$$

To show that the numerical radius of F is not greater than one, note that $DD^* \geq \text{Re}\,(e^{i\theta}\Sigma) = \text{Re}\,(e^{i\theta}DFD^*)$ for all $\theta \in [0, 2\pi)$ which implies that

$$D(I_{\mathcal{D}_*} - \text{Re}\,(e^{i\theta}F))D^* \geq 0, \quad \text{for all } \theta \in [0, 2\pi).$$

Hence

$$\langle (I_{\mathcal{D}_*} - \text{Re}\,(e^{i\theta}F))D^*h, D^*h\rangle = \langle D(I_{\mathcal{D}_*} - \text{Re}\,(e^{i\theta}F))D^*h, h\rangle \geq 0$$

for all $\theta \in [0, 2\pi)$ and as a consequence, the numerical radius of A is no bigger than one. ■

Now here is the theorem which guarantees the existence and uniqueness of solution to the fundamental equation of a Γ-contraction.

Theorem 6.11.2. *Let (S, P) be a Γ-contraction. Then there is a unique solution A to its fundamental equation*

$$S - S^*P = D_P X D_P.$$

Moreover, A has numerical radius less than or equal to one.

Proof. Since (S,P) is a Γ-contraction, by Proposition 6.9.4, we have

$$\rho(\alpha S, \alpha^2 P) \geq 0 \quad \text{for all } \alpha \in \overline{\mathbb{D}}.$$

So in particular for all β with modulus 1, we have $D_P^2 - \text{Re}\,\beta(S - S^*P) \geq 0$. Therefore by Lemma 6.11.1, there exists an operator $A \in \mathcal{B}(\mathcal{D}_P)$ with numerical radius not greater than one such that $S - S^*P = D_P A D_P$.

For uniqueness let there be two such solutions A_1 and A_2. Then

$$D_P \tilde{A} D_P = 0, \quad \text{where } \tilde{A} = A_1 - A_2 \in \mathcal{B}(\mathcal{D}_P).$$

Then

$$\langle \tilde{A} D_P h, D_P h' \rangle = \langle D_P \tilde{A} D_P h, h' \rangle = 0$$

which shows that $\tilde{A} = 0$ and hence $A_1 = A_2$. ∎

This theorem allows us to construct an explicit Γ-isometric dilation of a Γ-contraction, which is one of our main results and is shown in the following theorem.

Theorem 6.11.3. *Let (S,P) be a Γ-contraction on a Hilbert space $\mathcal{H}$. Let A be the unique solution of the fundamental equation (6.11.9) and let $\mathcal{K}_0 = \mathcal{H} \oplus \mathcal{D}_P \oplus \mathcal{D}_P \oplus \mathcal{D}_P \oplus \cdots = \mathcal{H} \oplus l^2(\mathcal{D}_P)$. Consider the operators T_A, V_0 defined on $\mathcal{K}_0$ by*

$$T_A(h_0, h_1, h_2, \dots) = (Sh_0, A^* D_P h_0 + A h_1, A^* h_1 + A h_2, A^* h_2 + A h_3, \dots)$$
$$V_0(h_0, h_1, h_2, \dots) = (Ph_0, D_P h_0, h_1, h_2, \dots).$$

Then

(1) *(T_A, V_0) is a Γ-isometric dilation of (S,P).*
(2) *If $(\widehat{T}, V_0)$ on $\mathcal{K}_0$ is a Γ-isometric dilation of (S,P), then $\widehat{T} = T_A$.*
(3) *If (T,V) is a Γ-isometric dilation of (S,P) where V is a minimal isometric dilation of P, then (T,V) is unitarily equivalent to (T_A, V_0).*

Thus (2) and (3) guarantee the uniqueness of Γ-isometric dilation (T,V) of (S,P) where V is minimal isometric dilation of P.

Proof. (1): It is evident from the definition that V_0 on $\mathcal{K}_0$ is the minimal isometric dilation of P. Obviously T_A^* and V_0^* are defined on $\mathcal{K}_0$ as

$$T_A^*(h_0, h_1, h_2, \dots) = (S^* h_0 + D_P A h_1, A^* h_1 + A h_2, A^* h_2 + A h_3, \dots)$$
$$V_0^*(h_0, h_1, h_2, \dots) = (P^* h_0 + D_P h_1, h_2, h_3, \dots).$$

The space $\mathcal{H}$ can be embedded inside $\mathcal{K}_0$ by the map $h \mapsto (h, 0, 0, \dots)$. It is clear that $\mathcal{H}$, considered as a subspace of $\mathcal{K}_0$ is co-invariant under T_A and V_0 and $T_A^*|_{\mathcal{H}} = S^*, V_0^*|_{\mathcal{H}} = P^*$. Since V_0 is an isometry, in order to show that (T_A, V_0) is a Γ-*isometric dilation* of (S, P) one has to justify (by virtue of Theorem 6.9.15-(3)) the following:

(a) $T_A V_0 = V_0 T_A$
(b) $T_A = T_A^* V_0$
(c) $r(T_A) \leq 2$.

We start with (a). Note that

$$T_A V_0(h_0, h_1, h_2, \dots) = T_A(Ph_0, D_P h_0, h_1, h_2, \dots)$$

which is the same as

$$(SPh_0, A^* D_P h_0 + AD_P h_0, A^* D_P h_0 + Ah_1, A^* h_1 + Ah_2, A^* h_2 + Ah_3, \dots)$$

and

$$\begin{aligned} & V_0 T_A(h_0, h_1, h_2, \dots) \\ =& V_0(Sh_0, A^* D_P h_0 + Ah_1, A^* h_1 + Ah_2, A^* h_2 + Ah_3, \dots) \\ =& (PSh_0, D_P Sh_0, A^* D_P h_0 + Ah_1, A^* h_1 + Ah_2, A^* h_2 + Ah_3, \dots). \end{aligned}$$

Let $G = A^* D_P P + AD_P - D_P S$. Then G is defined from $\mathcal{H} \to \mathcal{D}_P$. Since A is a solution of the equation (6.11.9), we have

$$\begin{aligned} D_P G &= D_P A^* D_P P + D_P A D_P - D_P^2 S \\ &= (S^* - P^* S)P + (S - S^* P) - (I - P^* P)S = 0. \end{aligned}$$

Now $\langle Gh, D_P h' \rangle = \langle D_P G h, h' \rangle = 0$ for all $h, h' \in \mathcal{H}$. This shows that $G = 0$ and hence $A^* D_P P + AD_P = D_P S$. Therefore $T_A V_0 = V_0 T_A$. Now for (b),

$$\begin{aligned} & T_A^* V_0(h_0, h_1, h_2, \dots) \\ &= T_A^*(Ph_0, D_P h_0, h_1, h_2, \dots) \\ =& (S^* Ph_0 + D_P A D_P h_0, A^* D_P h_0 + Ah_1, A^* h_1 + Ah_2, A^* h_2 + Ah_3, \dots). \end{aligned}$$

Since A is a solution of (6.11.9), we have $S^* P + D_p A D_p = S$. Therefore we have $T_A^* V_0 = T_A$.

We now show that $r(T_A) \leq 2$ which completes the proof. The numerical radius of A is not greater than 1 by Theorem 6.11.2. It is clear from the definition that T_A has the matrix form

$$T_A = \begin{pmatrix} S & 0 & 0 & 0 & \dots \\ A^* D_p & A & 0 & 0 & \dots \\ 0 & A^* & A & 0 & \dots \\ 0 & 0 & A^* & A & \dots \\ \dots & \dots & \dots & \dots & \dots \end{pmatrix},$$

with respect to the decomposition $\mathcal{H}\oplus\mathcal{D}_P\oplus\mathcal{D}_P\oplus\mathcal{D}_P\oplus\cdots$ of $\mathcal{K}_0$. Again since $T_A = \begin{pmatrix} S & 0 \\ C & D \end{pmatrix}$ on $\mathcal{H}\oplus l^2(\mathcal{D}_P) = \mathcal{K}_0$, where $C = \begin{pmatrix} A^*D_P \\ 0 \\ 0 \\ \vdots \end{pmatrix}$ and $D = \begin{pmatrix} A & 0 & 0 & \dots \\ A^* & A & 0 & \dots \\ 0 & A^* & A & \dots \\ \dots & \dots & \dots & \dots \end{pmatrix}$, we have by Lemma 1 of [61] that $\sigma(T_A) \subseteq \sigma(S) \cup \sigma(D)$. We shall be done if we show that $r(S)$ and $r(D)$ are not greater than 2. We show that $\|D\| \leq 2$. Let us define

$$\varphi : \mathbb{D} \to \mathcal{B}(\mathcal{D}_P)$$
$$z \to A + A^*z.$$

Clearly φ is holomorphic, bounded and continuous on the boundary $\partial\mathbb{D} = \mathbb{T}$ of the disc. For $z = e^{-2i\theta} \in \mathbb{T}$ we have

$$\begin{aligned}
&\|A + A^*z\| \\
=&\|A + e^{-2i\theta}A^*\| \\
=&\|e^{i\theta}A + e^{-i\theta}A^*\| \\
=&\sup_{\|x\|\leq 1} |\langle (e^{i\theta}A + e^{-i\theta}A^*)x, x\rangle| \quad [\text{since } \ e^{i\theta}A + e^{-i\theta}A^* \text{ is self adjoint}] \\
\leq&\omega(A) + \omega(A^*) \leq 2.
\end{aligned}$$

Therefore by *Maximum Modulus Principle*, $\|A + A^*z\| \leq 2$ for all $z \in \overline{\mathbb{D}}$ and

$$\|\varphi\| = \sup_{z\in\overline{\mathbb{D}}} \|A + A^*z\| \leq 2.$$

By noting that D is just

$$M_{A+A^*z} = T_{A+A^*z}$$

in the notation of Section 6.6 and by applying the inequality (6.6.2), we get that $\|D\| \leq 2$. Again since (S, P) is a Γ-contraction, $r(S) \leq \|S\| \leq 2$. Since both of $r(S), r(D)$ are not greater than 2, $r(T_A) \leq 2$. Hence done.

(2): Obviously $V_0 = \begin{pmatrix} P & 0 \\ C_1 & D_1 \end{pmatrix}$ with respect to the decomposition $\mathcal{H} \oplus l^2(\mathcal{D}_P)$ of $\mathcal{K}_0$, where

$$C_1 = \begin{pmatrix} D_P \\ 0 \\ 0 \\ \vdots \end{pmatrix} \text{ from } \mathcal{H} \to \mathcal{D}_P \oplus \mathcal{D}_P \oplus \mathcal{D}_P \oplus \dots \text{ and}$$

$$D_1 = \begin{pmatrix} 0 & 0 & 0 & \dots \\ I & 0 & 0 & \dots \\ 0 & I & 0 & \dots \\ \dots & \dots & \dots & \dots \end{pmatrix} \text{ on } \mathcal{D}_P \oplus \mathcal{D}_P \oplus \mathcal{D}_P \oplus \dots .$$

Since $(\widehat{T}, V_0)$ on $\mathcal{K}_0$ is a Γ-isometric dilation of (S, P), we have $\widehat{T}^*|_{\mathcal{H}} = S^*$ and $V_0^*|_{\mathcal{H}} = P^*$. Therefore $\widehat{T}$ on $\mathcal{H} \oplus l^2(\mathcal{D}_P)$ has matrix form $\widehat{T} = \begin{pmatrix} S & 0 \\ E & F \end{pmatrix}$. Let us define

$$\begin{aligned} U_1 : & H^2(\mathcal{D}_P) \to \mathcal{D}_P \oplus \mathcal{D}_P \oplus \mathcal{D}_P \oplus \dots \\ & z^n \mapsto (\underbrace{0, 0, \dots, 0}_{n}, 1, 0, 0, \dots). \end{aligned}$$

The action of U_1 on an arbitrary vector is clear from its action on the basis $\{1, z, z^2, \dots\}$ of $H^2(\mathcal{D}_P)$. Since it maps a basis of $H^2(\mathcal{D}_P)$ to a basis of $\mathcal{D}_P \oplus \mathcal{D}_P \oplus \mathcal{D}_P \oplus \dots$ in a one-to-one fashion, U_1 is a unitary operator. Therefore the spaces $\mathcal{K}_0$ and $\mathcal{H} \oplus H^2(\mathcal{D}_P)$ are isomorphic. Let $U = U_1^*$. Then $\widehat{T}$ and V_0 on $\mathcal{K}_0$ are respectively identified with the operators

$$\widetilde{T} = \begin{pmatrix} S & 0 \\ UE & UFU^* \end{pmatrix} \text{ and } \widetilde{V_0} = \begin{pmatrix} P & 0 \\ UC_1 & UD_1U^* \end{pmatrix} \text{ on } \mathcal{H} \oplus H^2(\mathcal{D}_P).$$

Therefore $(\widetilde{T}, \widetilde{V_0})$ is a Γ-isometric dilation of (S, P). We now show that UD_1U^* is same as the multiplication operator $M_z^{\mathcal{D}_P}$ on $H^2(\mathcal{D}_P)$. For a basis vector z^n of $H^2(\mathcal{D}_P)$ we have

$$\begin{aligned} UD_1U^*(z^n) &= U \begin{pmatrix} 0 & 0 & 0 & 0 & \dots \\ I & 0 & 0 & 0 & \dots \\ 0 & I & 0 & 0 & \dots \\ 0 & 0 & I & 0 & \dots \\ \dots & \dots & \dots & \dots & \dots \end{pmatrix} \begin{pmatrix} 0 \\ \vdots \\ 0 \\ 1 \\ 0 \\ \vdots \end{pmatrix}, \text{1 at } (n+1)\text{th place} \\ &= U \begin{pmatrix} 0 \\ \vdots \\ 0 \\ 1 \\ 0 \\ \vdots \end{pmatrix}, \quad \text{1 at } (n+2)\text{th place} \\ &= z^{n+1} = M_z^{\mathcal{D}_P}(z^n). \end{aligned}$$

Hence $UD_1U^* = M_z^{\mathcal{D}_P}$. By the commutativity of $\widetilde{T}$ and $\widetilde{V}_0$ we have the commutativity of UFU^* and $UD_1U^*(= M_z^{\mathcal{D}_P})$. Therefore $UFU^* =$

$M_\varphi^{\mathcal{D}_P}$ for some $\varphi \in H^\infty(\mathcal{B}(\mathcal{D}_P))$. Thus

$$\widetilde{T} = \begin{pmatrix} S & 0 \\ UE & M_\varphi^{\mathcal{D}_P} \end{pmatrix} \text{ and } \widetilde{V_0} = \begin{pmatrix} P & 0 \\ UC_1 & M_z^{\mathcal{D}_P} \end{pmatrix} \text{ on } \mathcal{H} \oplus H^2(\mathcal{D}_P).$$

By $\widetilde{T} = \widetilde{T}^*\widetilde{V_0}$, we get

$$\begin{pmatrix} S & 0 \\ UE & M_\varphi^{\mathcal{D}_P} \end{pmatrix} = \begin{pmatrix} S^* & E^*U^* \\ 0 & M_\varphi^{\mathcal{D}_P *} \end{pmatrix} \begin{pmatrix} P & 0 \\ UC_1 & M_z^{\mathcal{D}_P} \end{pmatrix}$$
$$= \begin{pmatrix} S^*P + E^*C_1 & E^*U^*M_z^{\mathcal{D}_P} \\ M_\varphi^{\mathcal{D}_P *}UC_1 & M_\varphi^{\mathcal{D}_P *}M_z^{\mathcal{D}_P} \end{pmatrix},$$

which gives

$$\text{(6.11.11)} \qquad \begin{cases} \text{(i) } S - S^*P = E^*C_1 \\ \text{(ii) } UE = M_\varphi^{\mathcal{D}_P *}UC_1 \\ \text{(iii) } M_\varphi^{\mathcal{D}_P} = M_\varphi^{\mathcal{D}_P *}M_z. \end{cases}$$

From (6.11.11)-(iii), it is clear by considering the power series expansion that $\varphi(z) = A_0 + A_0^* z$, for some $A_0 \in \mathcal{B}(\mathcal{D}_P)$. We now show that if $D_0 = \begin{pmatrix} A_0 & 0 & 0 & \dots \\ A_0^* & A_0 & 0 & \dots \\ 0 & A_0^* & A_0 & \dots \\ \dots & \dots & \dots & \dots \end{pmatrix}$ on $\mathcal{D}_P \oplus \mathcal{D}_P \oplus \mathcal{D}_P \oplus \dots$, then $UD_0U^* = M_\varphi^{\mathcal{D}_P}$. For a basis vector z^n of $H^2(\mathcal{D}_P)$ we have

$$UD_0U^*(z^n) = U \begin{pmatrix} A_0 & 0 & 0 & \dots \\ A_0^* & A_0 & 0 & \dots \\ 0 & A_0^* & A_0 & \dots \\ \dots & \dots & \dots & \dots \end{pmatrix} \begin{pmatrix} 0 \\ \vdots \\ 0 \\ 1 \\ 0 \\ \vdots \end{pmatrix}$$

$$= U \begin{pmatrix} 0 \\ \vdots \\ 0 \\ A_0(1) \\ A_0^*(1) \\ 0 \\ \vdots \end{pmatrix}$$

$$= A_0(1)z^n + A_0^*(1)z^{n+1} = M_{A_0 + A_0^* z}^{\mathcal{D}_P}(z^n).$$

Thus $UD_0U^* = M_\varphi^{\mathcal{D}_P}$ and hence $F = D_0$. Combining this with (6.11.11)-(ii), we get $UE = M_\varphi^{\mathcal{D}_P *}UC_1 = UD_0^*U^*UC_1 = UD_0^*C_1$, i.e, $E = D_0^*C_1$.

Therefore

$$\widehat{T} = \begin{pmatrix} S & 0 \\ D_0^* C_1 & D_0 \end{pmatrix} \text{ on } \mathcal{H} \oplus l^2(\mathcal{D}_P).$$

Considering the above stated matrix forms of D_0 and C_1 we get $D_0^*C_1 = \begin{pmatrix} A_0^* D_P \\ 0 \\ 0 \\ \vdots \end{pmatrix}$. Hence with respect to the decomposition $\mathcal{H} \oplus \mathcal{D}_P \oplus \mathcal{D}_P \oplus \dots$ of $\mathcal{K}_0$, we have

$$\widehat{T} = \begin{pmatrix} S & 0 & 0 & 0 & \dots \\ A_0^* D_P & A_0 & 0 & 0 & \dots \\ 0 & A_0^* & A_0 & 0 & \dots \\ 0 & 0 & A_0^* & A_0 & \dots \\ \dots & \dots & \dots & \dots & \dots \end{pmatrix}.$$

Also by (6.11.11)-(i),

$$\begin{aligned} S - S^*P = E^*C_1 &= C_1^* D_0 C_1 \\ &= \begin{pmatrix} D_P & 0 & 0 & \dots \end{pmatrix} \begin{pmatrix} A_0 & 0 & 0 & \dots \\ A_0^* & A_0 & 0 & \dots \\ 0 & A_0^* & A_0 & \dots \\ \dots & \dots & \dots & \dots \end{pmatrix} \begin{pmatrix} D_P \\ 0 \\ 0 \\ \vdots \end{pmatrix} \\ &= D_P A_0 D_P, \end{aligned}$$

which shows that A_0 satisfies the fundamental equation (6.11.9). By uniqueness of solution, $A = A_0$ and hence $\widehat{T} = T_A$.

(3): Let (T, V) defined on $\mathcal{K}$ be a minimal isometric dilation of (S, P), where V is a minimal isometric dilation of P. Since V on $\mathcal{K}$ is a minimal isometric dilation of P, there is a unitary

$$U : \mathcal{K} \to \mathcal{K}_0 (= \mathcal{H} \oplus \mathcal{D}_P \oplus \mathcal{D}_P \oplus \dots)$$

such that $UVU^* = V_0$. Let $T^\flat = UTU^*$. Then $(T^\flat, V_0)$ on $\mathcal{K}_0$ is a Γ-isometry dilation of (S, P). Therefore by part-(2), $T^\flat = T_A$ and consequently (T, V) is unitarily equivalent to (T_A, V_0). ■

As a consequence of the dilation theorem above, we have a new and elegant characterization for Γ-contractions.

Theorem 6.11.4. *Let (S, P) be a commuting pair of operators defined on $\mathcal{H}$. Then (S, P) is a Γ-contraction if and only if spectral radius of S is not greater than 2 and the fundamental equation $S - S^*P = D_P X D_P$ has a solution A with $\omega(A) \le 1$.*

Proof. Let there be a solution A to the fundamental equation $S - S^*P = D_P X D_P$ with $\omega(A) \leq 1$ for such a pair (S, P). Then by the dilation theorem (Theorem 6.11.3), we can construct a Γ-isometry (T_A, V_0) of (S, P). Now clearly (S, P) can be recovered by compressing (T_A, V_0) to the common co-invariant subspace $\mathcal{H}$. So (S, P) is a Γ-contraction.

The converse is just the Theorem 6.11.2. ■

Remark 6.11.5. We call the unique solution A of the operator equation (6.11.9) for a Γ-contraction (S, P), the fundamental operator of (S, P).

We now give another explicit construction of a Γ-isometric dilation of a pure Γ-contraction. This is very convenient to reap some beautiful consequences.

Theorem 6.11.6. *Let (S, P) be a Γ-contraction on a Hilbert space $\mathcal{H}$ where P is in $C_{\cdot 0}$. Let B be the solution of the fundamental equation (6.11.10). Let us consider the operators T, V on $\mathcal{E} = H^2(\mathbb{D}) \otimes \mathcal{D}_{P^*}$ defined as*

$$T = I \otimes B^* + M_z \otimes B \text{ and } V = M_z \otimes I.$$

Then (T, V) is a Γ-isometric dilation of (S, P).

Proof. Since B is the solution of the equation (6.11.10), by Theorem 6.11.2, the numerical radius of B is not greater than one. In order to prove that (T, V) is a Γ-isometric dilation of (S, P) we shall show the following steps:

(1) the pair (T,V) is a Γ-*isometry* on $\mathcal{E}$.
(2) The space $\mathcal{H}$ can be thought of as a subspace of $\mathcal{E}$, i.e, there is an isometric embedding of $\mathcal{H}$ in $\mathcal{E}$.
(3) After identification of $\mathcal{H}$ with this isometric image, $V^*\mathcal{H} \subseteq \mathcal{H}$ and $V^*|_{\mathcal{H}} = P^*$. Also, $T^*\mathcal{H} \subseteq \mathcal{H}$ and $T^*|_{\mathcal{H}} = S^*$.

V is clearly an isometry (it is a shift of some multiplicity) and obviously it commutes with T. Also

$$T = (I \otimes B^*) + (I \otimes B)(M_z \otimes I) = C + C^*V,$$

where $C = I \otimes B^*$. Obviously C and C^* commute with V and $\omega(C) \leq 1$. Therefore by Theorem 6.9.19, (T, V) is a Γ-isometry.

Now we embed $\mathcal{H}$ isometrically inside $H^2 \otimes \mathcal{D}_{P^*}$ by defining $W : \mathcal{H} \to \mathcal{E}$ as $h \mapsto \sum_{n=0}^{\infty} z^n \otimes D_{P^*} P^{*n} h$.

$$\begin{aligned}
\|Wh\|^2 &= \|\sum_{n=0}^{\infty} z^n \otimes D_{P^*} P^{*n} h\|^2 \\
&= \langle \sum_{n=0}^{\infty} z^n \otimes D_{P^*} P^{*n} h \,,\, \sum_{m=0}^{\infty} z^m \otimes D_{P^*} P^{*m} h \rangle \\
&= \sum_{m,n=0}^{\infty} \langle z^n, z^m \rangle \langle D_{P^*} P^{*n} h \,,\, D_{P^*} P^{*m} h \rangle \\
&= \sum_{n=0}^{\infty} \langle P^n D_{P^*}^2 P^{*n} h, h \rangle \\
&= \sum_{n=0}^{\infty} \langle P^n (I - PP^*) P^{*n} h, h \rangle \\
&= \sum_{n=0}^{\infty} \{ \langle P^n P^{*n} h, h \rangle - \langle P^{n+1} P^{*n+1} h, h \rangle \} \\
&= \|h\|^2 - \lim_{n \to \infty} \|P^{*n} h\|^2.
\end{aligned}$$

Since $P \in C_{\cdot 0}$, $\lim_{n \to \infty} \|P^{*n} h\|^2 = 0$ and hence $\|Wh\| = \|h\|$. Therefore W is an isometry. Let $L = W^*$.

For a basis vector $z^n \otimes \xi$ of $\mathcal{E}$ we have

$$\begin{aligned}
\langle L(z^n \otimes \xi), h \rangle &= \langle z^n \otimes \xi, \sum_{k=0}^{\infty} z^k \otimes D_{P^*} P^{*k} h \rangle \\
&= \langle \xi, D_{P^*} P^{*n} h \rangle = \langle P^n D_{P^*} \xi, h \rangle.
\end{aligned}$$

This implies that

$$L(z^n \otimes \xi) = P^n D_{P^*} \xi, \quad for\ n = 0, 1, 2, 3, ...$$

Therefore

$$\begin{aligned}
\langle L(M_z \otimes I)(z^n \otimes \xi), h \rangle &= \langle z^{n+1} \otimes \xi, \sum_{k=0}^{\infty} z^k \otimes D_{P^*} P^{*k} h \rangle \\
&= \langle \xi, D_{P^*} P^{*n+1} h \rangle \\
&= \langle P^{n+1} D_{P^*} \xi, h \rangle.
\end{aligned}$$

Consequently, $LV = PL$ on vectors of the form $z^n \otimes \xi$ which span $H^2 \otimes \mathcal{D}_{P^*}$ and hence

$$LV = PL. \tag{6.11.12}$$

Therefore V^* leaves the range of L^* (isometric copy of $\mathcal{H}$) invariant and $V^*|_{L^*\mathcal{H}} = L^*P^*L$ which is the isometric copy of the operator P^* on range of L^*. For the next step,

$$\begin{aligned} LT(z^n \otimes \xi) &= L(I \otimes B^* + M_z \otimes B)(z^n \otimes \xi) \\ &= L(I \otimes B^*)(z^n \otimes \xi) + L(M_z \otimes B)(z^n \otimes \xi) \\ &= L(z^n \otimes B^*\xi) + L(z^{n+1} \otimes B\xi) \\ &= P^n D_{P^*} B^*\xi + P^{n+1} D_{P^*} B\xi. \end{aligned}$$

Again $SL(z^n \otimes \xi) = SP^n D_{P^*}\xi$. Therefore for showing $LT = SL$ it is enough to show that

$$\begin{aligned} &P^n D_{P^*} B^* + P^{n+1} D_{P^*} B = SP^n D_{P^*} = P^n S D_{P^*} \\ &\text{i.e, } D_{P^*} B^* + P D_{P^*} B = S D_{P^*}. \end{aligned}$$

Let $H = D_{P^*}B^* + PD_{P^*}B - SD_{P^*}$. Then $H = 0$ by an argument similar to the one given in the proof of Theorem 6.11.3 to show that $G = 0$. So we have

$$D_{P^*} B^* + P D_{P^*} B = S D_{P^*}$$

and hence

$$L(I \otimes B^* + M_z \otimes B) = SL \tag{6.11.13}$$

which is similar to the equation (6.11.12). This shows that T^* leaves $L^*(\mathcal{H})$ invariant as well as $T^*|_{L^*(\mathcal{H})} = L^*S^*L$. Hence we are done. ∎

Remark 6.11.7. In particular when $\|P\| < 1$ the unique solutions A of (6.11.9) and B of (6.11.10) coincide with $D_P^{-1}(S - S^*P)D_P^{-1}$ and $D_{P^*}^{-1}(S^* - SP^*)D_{P^*}^{-1}$ respectively.

Corollary 6.11.8. *If (S, P) is a Γ-contraction with $P \in C_{\cdot 0}$, then $S = C + PC^*$ for some C with $\omega(C) \leq 1$.*

Proof. By the previous theorem, if (T, V) is a Γ-isometric dilation of (S, P) from (6.11.13) we have

$$\begin{aligned} &LT = L(I \otimes B^* + M_z \otimes B) = SL \\ &\text{or } L(I \otimes B^* + M_z \otimes B)L^* = S, \text{ since } L^* \text{ is isometry} \\ &\text{or } L(I \otimes B^*)L^* + L(M_z \otimes B)L^* = S \\ &\text{or } L(I \otimes B^*)L^* + L(M_z \otimes I)(I \otimes B)L^* = S \\ &\text{or } L(I \otimes B^*)L^* + PL(I \otimes B)L^* = S, \text{ since } L(M_z \otimes I) = PL. \end{aligned}$$

Taking $C = L(I \otimes B^*)L^*$ we get the stated form of S, and $\omega(C) \leq 1$ is obvious. ∎

Observation 6.11.9. *If (S, P) is a Γ-contraction with $P^* \in C_{\cdot 0}$, then S has the form $S = C_1 + C_1^* P$, where $\omega(C_1) \leq 1$.*

Proof. Clearly (S^*, P^*) is also a Γ-contraction and by the previous result, $S^* = C + P^* C^*$ for some C with $\omega(C) \leq 1$. This implies that $S = C^* + CP = C_1 + C_1^* P$ where $C_1 = C^*$. ■

Observation 6.11.10. *If (S, P) is a Γ-contraction with $\|P\| < 1$, then there is a unique C such that $S = C + C^* P$.*

Proof. Let there be C_1 and C_2 such that $S = C_1 + C_1^* P$ and $S = C_2 + C_2^* P$. Then we have $C + C^* P = 0$, where $C = C_1 - C_2$. Now

$$\|C\| = \| - C^* P\| \leq \|C\| \|P\| < \|C\| \quad \text{as } \|P\| < 1.$$

This shows that $C = 0$ and consequently $C_1 = C_2$. ■

The two main contributions of this chapter have been to construct an explicit dilation for a Γ-contraction and a host of examples.

Work on the symmetrized bidisc in various directions are going on. There have been attempts to generalize part of the theory to symmetrized polydisc. Interesting generalization has also been done to domains invariant under the action of the symmetric group. Both complex analytic and functional analytic work continue to be carried out.

CHAPTER 7

An Abstract Dilation Theory

7.1. Set theoretic dilation

Sz. Nagy dilation allowed us to dilate contractions to isometries or unitaries. Later we saw tuples of contractions getting dilated to isometries with orthogonal ranges. The main idea has been to realize operators which are a priori quite arbitrary or not easy to handle as compressions of better behaved operators such as isometries or unitaries. We saw several instances of the same. This chapter has a different flavor. Here we stand back and look at the edifice of dilation theory of operators from a distance. In the process some details of the theory are lost. But we get a broad brush picture of the subject.

Here we obtain several dilation theory results in a much simpler setting with very little structure. We replace Hilbert spaces by sets and bounded operators by arbitrary functions. Injective maps would play the role of isometries and bijective maps replace unitaries. Direct sums of Hilbert spaces would be replaced by disjoint unions of sets. Finally what we have is not much of operator theory but just set theory. But we believe that this showcases some essential features of dilation theory. Almost all the results of this chapter are borrowed almost verbatim from [29].

7.2. Single variable dilation theory

The most basic dilation theorem for contraction operators on Hilbert spaces is the Halmos Dilation which we recall from the preamble of Section 3 of Chapter 1: Let T be a contraction ($\|T\| \leq 1$) on a Hilbert space $\mathcal{H}$. Define $U : \mathcal{H} \oplus \mathcal{H} \to \mathcal{H} \oplus \mathcal{H}$ by

$$U = \begin{bmatrix} T & D_{T^*} \\ D_T & -T^* \end{bmatrix}$$

where $D_T = (I - T^*T)^{\frac{1}{2}}$ and $D_{T^*} = (I - TT^*)^{\frac{1}{2}}$. Then U is a unitary. In other words, every contraction can be enlarged or dilated to a unitary on a larger space. If we identify $\mathcal{H}$ with $\mathcal{H} \oplus 0$ in $\mathcal{H} \oplus \mathcal{H}$, the action of U

on $\mathcal{H}$ restricted to $\mathcal{H}$ is same as the action by T, that is,

$$T = P_{\mathcal{H}}U|_{\mathcal{H}}.$$

We observe that this dilation has three components to it. First, there is the embedding of the original space in a larger space. Then there is a 'good' map in the larger space, which is usually called as the dilation. This is followed by a compression (or projection) to the range of the original space in the larger space. The last map is usually an idempotent. This is true of most dilation theory results. In probability theory contexts this map is a conditional expectation map. In a very concrete way Markov chains or Markov processes can be thought of as dilation of stochastic matrices or semigroups of stochastic maps.

Here is a simple Halmos type dilation for functions. Let $h : A \to A$ be a function. Now we want a bijection (analogue of unitary) on two copies of A, that is, on $B = A\times\{0,1\}$, which could be called as dilation. Indeed define $i : A \to B$ by $i(a) = (a, 0)$ and $u : B \to B$ by $u(a, m) = (a, 1 - m)$, $(a, m) \in B$. Then i is injective and u is bijective. Further define $p : B \to B$ by

$$p(a,m) = \begin{cases} (a,0) & \text{if } m = 0; \\ (h(a),0) & \text{if } m = 1. \end{cases}$$

Then p is an idempotent ($p^2 = p$) and the range of p is $i(A)$. The quadruple (B, i, u, p) has the property

$$i(h(a)) = p(u(i(a))) \; \forall a \in A.$$

In other words, we are suggesting that this quadruple can be considered as Halmos type dilation for functions (self-maps) on sets. Below we explore this idea further.

Let us recall the Sz. Nagy dilation of contractions which is perhaps the most famous dilation of all. Here we are considering the 'isometric dilation'. The isometry can be further dilated to a unitary and that we will come to later. This Theorem has been discussed at length with different constructions of the dilation in Sections 3-5 of Chapter 1. Now in the abstract setting here is the basic definition. Here and elsewhere for any function h, h^0 is taken as the identity function.

Definition 7.2.1. *Let A be a non-empty set and let $h : A \to A$ be a function. An* injective power dilation *(or simply a dilation) of h, is a quadruple (B, i, v, p) where B is a set, $i : A \to B$, $v : B \to B$ are injective and $p : B \to B$ is an idempotent with $p(B) = i(A)$. Further,*

$$i(h^n(a)) = pv^n(i(a)) \; \forall a \in A, n \in \mathbb{Z}_+.$$

Any such dilation is said to be minimal *if* $B = \bigcup_{n=0}^{\infty} v^n(i(A))$. *Two dilations* (B, i, v, p) *and* (B', i', v', p') *are said to be bijectively equivalent if there exists a bijection* $u : B \to B'$ *such that* $i' = ui$, $v' = uvu^{-1}$, *and* $p' = upu^{-1}$.

Theorem 7.2.2. *Every function* $h : A \to A$ *admits a minimal injective power dilation.*

Proof. The construction is simple. Take

$$\begin{aligned} B &= A \times \mathbb{Z}_+; \\ i(a) &= (a, 0),\ a \in A; \\ v(a, m) &= (a, m+1),\ (a, m) \in B; \\ p(a, m) &= (h^m(a), 0),\ (a, m) \in B. \end{aligned}$$

It is easily verified that (B, i, v, p) is an injective power dilation of h and it is minimal. ■

We will call the dilation constructed in this theorem as the *standard dilation* of h. Here there is no uniqueness statement. Although the construction gives technically a minimal dilation, it may not be optimal, for instance when h is injective there is actually no need to enlarge the space. To mitigate this to some extent we make the following definition.

Definition 7.2.3. *Let* $h : A \to A$ *be a function. A subset D of A is said to be a* defect space *for* h *if* $h|_{D^c}$ *is injective (Here* $D^c = A \backslash D$ *is the complement of D). A defect space D is said to be a minimal defect space if there is no proper subset of D which is a defect space for* h.

This Definition is of course motivated by the notion of defect operators we have extensively used in earlier chapters. The defect space measures as to how far the map is from being injective. In this Definition, it is important to note that h need not leave D^c invariant. We observe that if a family $\{D_j : j \in J\}$ of defect spaces of h, forms a chain under inclusion, that is, for any $j, k \in J$ either $D_j \subseteq D_k$ or $D_k \subseteq D_j$, then $\bigcap_{j \in J} D_j$ is also a defect space. Now it follows by Zorn's lemma that every function admits a minimal defect space. A minimal defect space is empty if and only if the function is injective.

Let $h : A \to A$ be a function and let D be a defect space of h. Define (B_D, i_D, v_D, p_D) by taking

$$\begin{aligned} B_D &= (D^c \times \{0\}) \cup (D \times \mathbb{Z}_+); \\ i_D(a) &= (a, 0),\ a \in A; \\ v_D(a,m) &= \begin{cases} (h(a), 0), & (a,m) \in D^c \times \{0\}; \\ (a, m+1), & (a,m) \in D \times \mathbb{Z}_+; \end{cases} \\ p_D(a,m) &= (h^m(a), 0),\ (a,m) \in B_D. \end{aligned}$$

Then it is seen that (B_D, i_D, v_D, p_D) is a minimal injective dilation. This is same as the standard dilation when $D = A$.

The following definition is motivated by the fact that every minimal isometric dilation V of a contraction T on a Hilbert space $\mathcal{H}$ leaves $\mathcal{H}^\perp$ invariant. It is well-known that such a property is important while considering dilations of tuples of operators and also in dilation theory of completely positive maps. This property is called 'coinvariance' in [14], 'regular' in [25] and co-increasing in [111]. We follow Arveson here, as his terminology looks most appropriate in the present situation.

Definition 7.2.4. *An injective dilation* (B, i, v, p) *of a function* $h : A \to A$ *is said to be* co-invariant *if* $v(i(A)^c) \subseteq i(A)^c$.

It may be noted that the dilation (B_D, i_D, v_D, p_D) we constructed above using a defect space D is co-invariant. Here is an example of a dilation which is not co-invariant.

Example 7.2.5. *Let* $A = \{1\}$ *and consider the function* $h = Id_A$. *Let* $B = \{1, 2\}$ *and define* $v : B \to B$ by $v(1) = 2, v(2) = 1$. *Let* $i : A \to B$ *be the embedding given by* $i(1) = 2$. *Define* $p : B \to B$ by $p(1) = p(2) = 2$. *Then clearly* p *is an idempotent with* $p(B) = \{2\} = i(A)$. *Further note that* $pv^n i(1) = 2 = i(h^n(1))$ *for all* $n \in \mathbb{Z}_+$. *Thus,* (B, i, v, p) *is an injective dilation for* h *which, obviously, is minimal. But this dilation is not co-invariant as* $v(i(A)^c) = v(\{1\}) = \{2\} = i(A)$.

Theorem 7.2.6. *Let* (B, i, v, p) *be a co-invariant injective dilation of a function* $h : A \to A$. *Then* $D = D(B, i, v, p) = \{a \in A : v(i(a)) \notin i(A)\}$ *is a defect space for* h. *Moreover,* (B, i, v, p) *is in bijective correspondence with* (B_D, i_D, v_D, p_D).

Proof. Let $a, b \in D^c$ be such that $h(a) = h(b)$. Then $v(i(a)), v(i(b)) \in i(A)$ and hence, $v(i(a)) = pv(i(a)) = i(h(a)) = i(h(b)) = pv(i(b)) = v(i(b))$ and since, i, v are injective, it follows that $a = b$.

Define a map $\psi : B_D \to B$ by $\psi((a, m)) = v^m(i(a))$ for $(a, m) \in B_D$. We first show that ψ is a bijection. Let $(a, m), (b, n) \in B_D$ be such that

$v^m(i(a)) = v^n(i(b))$. Note that if both $a, b \in D^c$, then we must have that $m = n = 0$. If one of a, b belongs to D while the other one in D^c, say $a \in D^c$ and $b \in D$, then, of course, $m = 0$ so that $i(A) \ni i(a) = \psi((a, 0)) = \psi((b, n)) = v^n(i(b))$. Now if $n > 0$, the facts that $b \in D$ and the dilation is co-invariant together imply that $v^n(i(b)) \notin i(A)$ which leads to a contradiction and thus, in this case, we have that $m = n = 0$. Finally, if a, b both belong to D and if we assume that $m \neq n$, say, without loss of generality $m > n$, then $v^m(i(a)) = v^n(i(b))$ would imply that $v^{m-n}(i(a)) = i(b) \in i(A)$. But, as before, it follows from the facts $a \in D$, co-invariance of the dilation and $m - n > 0$ that $v^{m-n}(i(a)) \notin i(A)$ and thus, we arrive at a contradiction. Thus, from $\psi((a, m)) = \psi((b, n))$ we obtain that $m = n$, from which, using injectivity of v and i, it follows at once that $a = b$, proving injectivity of the map ψ.

We now show that ψ is surjective. Let $x \in B$. If $x \in i(A)$, say, $x = i(a)$ for some $a \in A$, then clearly $x = \psi((a, 0))$. If $x \notin i(A)$, we assert that there is $(a, m) \in D \times \mathbb{N}$ such that $\psi((a, m)) = x$. Since, by minimality of the dilation, $B = \cup_{n=0}^{\infty} v^n(i(A))$, $x = v^n(i(a))$ for some $n > 0, a \in A$. If $a \in D$, then $x = \psi((a, n))$ and we are done. If $a \notin D$, we must have that $n > 1$ and let k be the largest integer, $1 \leq k < n$, such that $v^k(i(a)) \in i(A)$ and $v^{k+1}(i(a)) \notin i(A)$. Let $v^k(i(a)) = i(b)$ where $b \in A$. Clearly, $b \in D$ and $\psi((b, n-k)) = v^{n-k}(i(b)) = v^n(i(a)) = x$. Therefore, ψ is surjective and hence, bijective. It is easy to verify that

$$v\psi = \psi v_D, \ p\psi = \psi p_D, \text{ and } \psi i_D = i,$$

and we leave these verifications to the reader. Thus (B, i, v, p) is in bijective correspondence with (B_D, i_D, v_D, p_D). ■

Now we analyze injective maps on sets a little, which we have treated as analogues of isometries. From the Wold decomposition (Chapter 1, Theorem 1.12.2), an isometry on a Hilbert space decomposes as a direct sum of a pure isometry and a unitary. The pure isometry is just the unilateral shift in $l^2(\mathbb{Z}_+)$ with some multiplicity, whereas the unitary part is understood using the spectral theory. We can carry out a similar analysis for injective functions. Suppose $v : A \to A$ is an injective function. For a, b in A, write $a \sim b$ if either $a = v^n(b)$ or $v^n(a) = b$ for some $n \in \mathbb{Z}_+$. Clearly this defines an equivalence relation on A. The corresponding equivalence classes are known as orbits of v.

We first classify injective maps which have exactly one orbit. In that direction consider the following examples.

(i) *Cyclic permutations :* For $d \in \mathbb{N}$, take $\mathbb{Z}_d = \{0, 1, 2, \ldots, d-1\}$ with addition modulo d. Define $s_d : \mathbb{Z}_d \to \mathbb{Z}_d$ by $s_d(n) = n + 1$ modulo d. (Note: $\mathbb{Z}_1 = \{0\}$ and $s_0(0) = 0.$). Then s_d is bijective and has single orbit.

(ii) *Bilateral translation:* Define $s : \mathbb{Z} \to \mathbb{Z}$ by $s(n) = n + 1$. Then s is bijective and has single orbit. (We refrain from calling this as a shift as $\bigcap_n s^n(\mathbb{Z}) \neq \emptyset$.)

(iii) *Unilateral translation/shift:* Define $s_+ : \mathbb{Z}_+ \to \mathbb{Z}_+$ by $s_+(n) = n + 1$. Then s_+ is a shift with single orbit.

It is not hard to see that any injective function with single orbit is in bijective correspondence with exactly one of these examples.

If X is any set, we define $1_X \times s : X \times \mathbb{Z} \to X \times \mathbb{Z}$ by

$$(1_X \times s)(x, n) = (x, n + 1).$$

This will be called as the bilateral translation with multiplicity X. In a similar way, we define $1_X \times s_d$ (d-cyclic permutation with multiplicity X) on $X \times \mathbb{Z}_d$ and $1 \times s_+$ (unilateral shift with multiplicity X) on $X \times \mathbb{Z}_+$.

Later, to deal with multi-variable theory, we need to understand the structure of commuting bijections. Suppose $v_1, v_2, \ldots, v_d$ is a commuting d-tuple of bijections on a set A. For a, b in A, we write $a \sim b$ if there exists $(m_1, m_2, \ldots, m_d) \in \mathbb{Z}^d$ such that $v_1^{m_1} v_2^{m_2} \cdots v_d^{m_d}(a) = b$. Clearly, this defines an equivalence relation on A. The equivalence classes are known as orbits generated by the tuple $v_1, v_2, \ldots, v_d$. For fixed $a \in A$, the orbit containing a is given by:

$$\{v_1^{m_1} v_2^{m_2} \cdots v_d^{m_d}(a) : (m_1, m_2, \ldots, m_d) \in \mathbb{Z}^d\}.$$

It has an abelian group structure given by $v_1^{m_1} v_2^{m_2} \cdots v_d^{m_d}(a).v_1^{n_1} v_2^{n_2} \cdots v_d^{n_d}(a) = v_1^{m_1+n_1} v_2^{m_2+n_2} \cdots v_d^{m_d+n_d}(a)$, and

$$\mathbb{Z}^d \ni (m_1, m_2, \ldots, m_d) \mapsto v_1^{m_1} v_2^{m_2} \cdots v_d^{m_d}(a),$$

is a group homomorphism. It follows that orbits of the tuple $v_1, v_2, \ldots, v_d$ are naturally isomorphic to quotients of $\mathbb{Z}^d$, and conversely any such quotient can be realized as an orbit.

We analyze the case of $d = 2$ a bit more closely. So let v_1, v_2 be a pair of commuting bijections on a set A. Note that in order to understand the pair (v_1, v_2) on A, it suffices to understand the restriction of v_1, v_2 to orbits generated by both v_1 and v_2.

Fix $a \in A$ and let B denote the orbit of a generated by v_1 and v_2, that is, $B = \{v_1^m v_2^n(a) : m, n \in \mathbb{Z}\}$. Also, let B_1 denote the orbit of a generated by v_1, that is, $B_1 = \{v_1^n(a) : n \in \mathbb{Z}\}$. Clearly, the set B_1 can be identified with the set $\mathbb{W}$ where $\mathbb{W} = \mathbb{Z}$ or $\mathbb{Z}_d$ according as B_1 is infinite or finite with cardinality d for some integer $d \geq 1$. Observe that $\mathbb{W} = \mathbb{Z}_d$ if and only if d is the smallest natural number such that $v_1^d(a) = a$. We use the notation $s_{\mathbb{W}}$ to denote the bilateral translation s on $\mathbb{Z}$ or the cyclic permutation s_d on $\mathbb{Z}_d$ according as $\mathbb{W} = \mathbb{Z}$ or $\mathbb{Z}_d$. It is evident that the restriction of v_1 to B_1 is in bijective correspondence with $s_{\mathbb{W}}$ on $\mathbb{W}$. If there is no natural number n such that $v_2^n(a) \in B_1$, then

$$\mathbb{Z} \times \mathbb{W} \ni (k, l) \mapsto v_2^k v_1^l(a) \in B$$

is a bijection and one can easily see that $\left(v_1|_B, v_2|_B\right)$ is in bijective correspondence with $\left(1 \times s_{\mathbb{W}}, s \times 1\right)$. Now, if there is a natural number n such that $v_2^n(a) \in B_1$ and if m is the smallest natural number such that $v_2^m(a) \in B_1$, say, $v_2^m(a) = v_1^p(a)$, a little thought should convince the reader that, in this case,

$$\mathbb{Z}_m \times \mathbb{W} \ni (k, l) \mapsto v_2^k v_1^l(a) \in B$$

establishes a bijection and $\left(v_1|_B, v_2|_B\right)$ is in bijective correspondence with $\left(1 \times s_{\mathbb{W}}, g\right)$ where $g : \mathbb{Z}_m \times \mathbb{W} \to \mathbb{Z}_m \times \mathbb{W}$ is the map given by

$$g(k, l) = \begin{cases} (k+1, l), & \text{if } 0 \leq k < m-1, \\ (0, l+p), & \text{if } k = m-1. \end{cases}$$

Summarising the foregoing discussions we obtain that:

Lemma 7.2.7. *With notations as above, $\left(v_1|_B, v_2|_B\right)$ is in bijective correspondence with $\left(1 \times s_{\mathbb{W}}, s \times 1\right)$ if $B_1 \cap \{v_2^n(a) : n \in \mathbb{N}\} = \emptyset$ and if $B_1 \cap \{v_2^n(a) : n \in \mathbb{N}\} \neq \emptyset$ with m being the smallest natural number such that $v_2^m(a) \in B_1$, say, $v_2^m(a) = v_1^p(a)$, then $\left(v_1|_B, v_2|_B\right)$ is in bijective correspondence with $\left(1 \times s_{\mathbb{W}}, g\right)$ where $g : \mathbb{Z}_m \times \mathbb{W} \to \mathbb{Z}_m \times \mathbb{W}$ is the map given by*

$$g(k, l) = \begin{cases} (k+1, l), & \textit{if } 0 \leq k < m-1, \\ (0, l+p), & \textit{if } k = m-1. \end{cases}$$

Now suppose $v : A \to A$ is injective. A little bit of thought shows that the action of v on any of the orbits is in bijective correspondence with one of the examples above. Decomposing A into equivalence classes we see that (A, v) is in bijective correspondence with (B, w) where $B = \sqcup_{d \in \mathbb{N}}(X_d \times \mathbb{Z}_d) \sqcup X \times \mathbb{Z} \sqcup X_+ \times \mathbb{Z}_+$, with suitable multiplicity spaces $\{X_d\}_d \in \mathbb{N}, X, X_{\mathbb{Z}_+}$ (some of these terms could be absent) and w is equal to $1 \times s_d$ or $1 \times s$ or $1 \times s_+$ in respective spaces. Cardinalities of these multiplicity spaces are uniquely determined.

Just as we have wandering subspaces for isometries we have wandering subsets for injective maps.

Definition 7.2.8. *Let A be a set and let $v : A \to A$ be an injective function. Then a subset W of A is said to be* wandering *for v if*

$$v^m(W) \bigcap v^n(W) = \emptyset \text{ for } m \neq n.$$

An injective function $v : A \to A$ is said to be a shift *if $\bigcap_{n=0}^{\infty} v^n(A) = \emptyset$.*

Here is the Wold decomposition under the current setting.

Theorem 7.2.9. *Let A be a set and let $v : A \to A$ be an injective function. Then A decomposes uniquely as $A = A_0 \sqcup A_1$ where A_0 and A_1 are left invariant by v and $v|_{A_0}$ is a shift and $v|_{A_1}$ is a bijection.*

Proof. Take $W = A \backslash v(A)$. If $a \in v^m(W) \bigcap v^{m+k}(W)$ for some $m, k \in \mathbb{Z}_+$, we get $a = v^m(w_1) = v^{m+k}(w_2)$ for some $w_1, w_2 \in W$. As v^m is injective, we get $w_1 = v^k(w_2)$. But this is not possible unless $k = 0$ as $w_1 \in A \backslash v(A)$. This shows that W is a wandering subset for v. Now take $A_0 = \sqcup_{n=0}^{\infty} v^n(W)$ and $A_1 = A_0^c$. Then $A = A_0 \sqcup A_1$. Clearly $v|_{A_0}$ is a shift and $v|_{A_1}$ is a bijection.

Suppose that $A = A_0' \sqcup A_1'$ such that v leaves A_0' and A_1' invariant and $v|_{A_0'}$ is a shift and $v|_{A_1'}$ is a bijection. Then $A_0' = \sqcup_{n=0}^{\infty} (v|_{A_0'})^n W'$ where $W' = A_0' \setminus v(A_0')$. Thus we have that

$$\begin{aligned} W &= A_0 \setminus v(A_0) = (A_0 \sqcup A_1) \setminus (v(A_0) \sqcup A_1) = A \setminus v(A) \\ &= (A_0' \sqcup A_1') \setminus (v(A_0') \sqcup A_1') = A_0' \setminus v(A_0') = W'. \end{aligned}$$

Consequently, $A_0 = \sqcup_{n=0}^{\infty} v^n(W) = \sqcup_{n=0}^{\infty} v^n(W') = A_0'$ and hence, $A_1 = A_1'$. This proves uniqueness. ∎

Recall from Section 1.8 that a contraction T on a Hilbert space $\mathcal{H}$ is said to be pure if $(T^*)^n$ converges in strong operator topology to 0. It was seen that minimal isometric dilation of a pure contraction is a shift. We wish to have a similar theorem in the present set up. But we do not seem to have an exact analogue of pureness. An isometry V is pure if and only if it is a shift that is $\bigcap_{n=0}^{\infty} V^n(\mathcal{H}) = \{0\}$. For contractions T we know that $\bigcap_{n=0}^{\infty} T^n(\mathcal{H}) = \{0\}$, then T is pure, but the converse is not true. For instance if $T = tI$ is a scalar with $0 < t < 1$, then T is pure but $\bigcap_{n=0}^{\infty} T^n(\mathcal{H}) = \mathcal{H}$.

Theorem 7.2.10. *Let $h : A \to A$ be a function and let (B, i, v, p) be a co-invariant, minimal injective dilation of h. Let $D = \{a \in A : v(i(a)) \notin i(A)\}$ be the associated defect space. Then v is a shift if and only if $\bigcap_{n=0}^{\infty} h^n(D^c) = \emptyset$. If h is a function such that $\bigcap_{n=0}^{\infty} h^n(A) = \emptyset$, then for every (not necessarily co-invariant) minimal injective dilation (B, i, v, p) of h, v is a shift.*

Proof. By the previous theorem, (B, i, v, p) is in bijective correspondence with (B_D, i_D, v_D, p_D). Now if $a_0 \in \bigcap_{n=0}^{\infty} h^n(D^c)$, as $v(i(a)) = i(h(a))$ for $a \in D^c$, $i(a_0) \in \bigcap_{n=0}^{\infty} v^n i(D^c) \subseteq \bigcap_{n=0}^{\infty} v^n(B)$ Therefore v is not a shift. Conversely suppose $x \in \bigcap_{n=0}^{\infty} v^n(B)$ or equivalently $x \in \bigcap_{n=0}^{\infty} v_D^n(B_D)$. Recall that $B_D = (D^c \times \{0\}) \bigcup (D \times \mathbb{Z}_+)$. If $x = (a, 0)$ with $a \in D^c$, we see that $a \in \bigcap_{n=0}^{\infty} h^n(D^c)$ and we are done. Now, suppose $x = (a, m)$ for some $a \in D$ and $m \in \mathbb{Z}_+$. For $n \geq 1$, as $x \in v^{n+m}(B)$, from looking at the action of v_D, we see that there exist $a_j \in D^c$ for $n \geq j \geq 1$, such that $v_D(a_n, 0) = (a_{n-1}, 0), v_D(a_{n-1}, 0) = (a_{n-2}, 0), \ldots, v(a_2, 0) = (a_1, 0), v_D(a_1, 0) = (a, 0), v_D(a, 0) = (a, 1), v_D(a, 1) = (a, 2), \ldots, v_D(a, m-1) = (a, m)$. Further as h is injective on D^c, $a_j \in D^c$ for $n \geq j \geq 1$ are uniquely determined and $h(a_n) = a_{n-1}, h(a_{n-1}) = a_{n-2}, \ldots h(a_2) = a_1, h(a_1) = a$. In particular $a_1 = h^{n-1}(a_n)$. So we get $a_1 \in h^{n-1}(D^c)$. As this holds for every n, we get $\bigcap_{n=0}^{\infty} h^n(D^c) \neq \emptyset$. This proves the first part.

Now for the second part, assume $\bigcap_{n=0}^{\infty} h^n(A) = \emptyset$. Let $x \in \bigcap_{n=0}^{\infty} v^n(B)$. We know that $p(x) \in i(A)$. As i is injective there exists unique $a \in A$ such that $p(x) = i(a)$. Now for any $n \in \mathbb{Z}_+$, as $x \in v^n(B)$, there exists some $b_n \in B$ such that $x = v^n(b_n)$. By minimality of the dilation, b_n has the form $b_n = v^k(i(a_n))$ for some $k \in \mathbb{Z}_+$ and $a_n \in A$. So $x = v^{n+k}(i(a_n))$. Then by the dilation property, $i(a) = p(x) = i(h^{n+k}(a_n))$. Therefore for every n there exists $k \geq 0$ such that $a \in h^{n+k}(A)$. Hence $a \in \bigcap_{n=0}^{\infty} h^n(A)$. This contradicts purity of h. ■

Here is an example of a dilation which is not co-invariant to illustrate the second part of the previous theorem.

Example 7.2.11. *Let $A = \mathbb{Z}_+$ and let $h : A \to A$ be defined by $h(n) = n+1$. Clearly h is pure. Take $B = \mathbb{Z}_+ \times \{0, 1\}$. Define $i(a) = (a, 0)$, $v(a, 0) = (a, 1), v(a, 1) = (a+2, 0)$ and $p(a, 0) = (a, 0), p(a, 1) = (a+1, 0)$. It is clear that (B, i, v, p) is a minimal injective dilation of h which is not co-invariant. Nevertheless v is a shift with multiplicity 2.*

We recall the intertwining lifting theorem of Sz. Nagy and C. Foias which we have seen in Section 3.2.

Theorem 7.2.12. *[Intertwining lifting theorem] Let T_1, T_2 be contractions on Hilbert spaces $\mathcal{H}_1, \mathcal{H}_2$ respectively. Let V_1, V_2 acting on $\mathcal{K}_1, \mathcal{K}_2$ be minimal isometric dilations of T_1, T_2 respectively. Suppose $S : \mathcal{H}_2 \to \mathcal{H}_1$ is a bounded operator such that $T_1 S = S T_2$. Then there exists a bounded operator $R : \mathcal{K}_2 \to \mathcal{K}_1$ such that $V_1 R = R V_2$, $P_{\mathcal{H}_1} R|_{\mathcal{H}_2^\perp} = 0$, $P_{\mathcal{H}_1} R|_{\mathcal{H}_2} = S$ and $\|R\| = \|S\|$. Conversely if $R : \mathcal{K}_2 \to \mathcal{K}_1$ is a bounded operator such that $V_1 R = R V_2$, and $P_{\mathcal{H}_1} R|_{\mathcal{H}_2^\perp} = 0$, then $S : \mathcal{H}_2 \to \mathcal{H}_1$ defined by $S = P_{\mathcal{H}_1} R|_{\mathcal{H}_2}$ satisfies $T_1 S = S T_2$.*

Here is an analogue of this theorem.

Theorem 7.2.13. *Let $h_1 : A_1 \to A_1$ and $h_2 : A_2 \to A_2$ be two functions. Let (B_j, i_j, v_j, p_j) be standard dilation of h_j for $j = 1, 2$. Suppose $s : A_2 \to A_1$ is a function such that $sh_2 = h_1 s$. Then there exists a function $r : B_2 \to B_1$ such that $rv_2 = v_1 r$, $rp_2 = p_1 r$ and $ri_2 = i_1 s$. Conversely if $r : B_2 \to B_1$ is a function satisfying $rv_2 = v_1 r$ and $rp_2 = p_1 r$, then there exists unique function $s : A_2 \to A_1$ satisfying $ri_2 = i_1 s$ and $sh_2 = h_1 s$.*

Proof. Define $r : B_2 \to B_1$ by $r(a, m) = (s(a), m)$ for $(a, m) \in B_2$. Recall $v_j(a, m) = (a, m + 1)$ for $(a, m) \in B_j$ and $i_j(a) = (a, 0)$, for $j = 1, 2$. Clearly $rv_1(a, m) = r(a, m + 1) = (s(a), m + 1) = v_2 r(a, m)$. Similarly

$$\begin{aligned} rp_2(a, m) &= r(h_2^m(a), 0) = (sh_2^m(a), 0) = (h_1^m s(a), 0) = p_1(s(a), m) \\ &= p_1 r(a, m), \end{aligned}$$

and $ri_2(a) = r(a, 0) = (s(a), 0) = i_1 s(a)$.

For the converse, consider any $a \in A_2$. As $rp_2 = p_1 r$, $rp_2(a, 0) \in i_1(A_1)$. As i_1 is injective there exists unique $s(a) \in A_1$ such that $rp_2(a, 0) = (s(a), 0)$. This defines $s : A_2 \to A_1$ satisfying $ri_2 = i_1 s$. Now $rp_2 v_2 = p_1 r v_2 = p_1 v_1 r$ and $rp_2 v_2(a, 0) = rp_2(a, 1) = r(h_2(a), 0) = (sh_2(a), 0)$ and $p_1 v_1 r(a, 0) = p_1 v_1(s(a), 0) = p_1(s(a), 1) = (h_1(s(a)), 0)$, and hence $sh_2 = h_1 s$. ∎

We can generalize this intertwining lifting for dilations taking defect spaces into consideration as follows.

Theorem 7.2.14. *For each $j = 1, 2$, let $h_j : A_j \to A_j$ be a function and let D_j be a defect space for h_j and consider the injective minimal dilation $(B_{D_j}, i_{D_j}, v_{D_j}, p_{D_j})$ of h_j. Suppose that $s : A_2 \to A_1$ be a function such that $s(D_2^c) \subseteq D_1^c$, $s(D_2) \subseteq D_1$, and $sh_2 = h_1 s$. Then there exists a function $r : B_{D_2} \to B_{D_1}$ such that $rv_{D_2} = v_{D_1} r$, $rp_{D_2} = p_{D_1} r$ and $ri_{D_2} = i_{D_1} s$. Conversely, if $r : B_{D_2} \to B_{D_1}$ is a function satisfying $rv_{D_2} = v_{D_1} r$, $rp_{D_2} = p_{D_1} r$, then there exists a unique function $s : A_2 \to A_1$ satisfying $ri_{D_2} = i_{D_1} s$ and $sh_2 = h_1 s$. Moreover, $s(D_2^c) \subseteq D_1^c$.*

Proof. Similar to the proof of previous Theorem. ∎

Here is a special case of the famous Sarason's Lemma [108]. We present a basic version suitable for our purposes. For a more complete presentation see [110].

Theorem 7.2.15. *[Sarason's Lemma [108]] Let $\mathcal{H}$ be a closed subspace of a Hilbert space $\mathcal{K}$. Let $V : \mathcal{K} \to \mathcal{K}$ be a bounded operator. Define $T : \mathcal{H} \to \mathcal{H}$ by*

$$T = P_{\mathcal{H}} V|_{\mathcal{H}}.$$

Then V is a power dilation of T, that is, $T^n = P_{\mathcal{H}} V^n|_{\mathcal{H}}$ for all $n \in \mathbb{Z}_+$ if and only if there exist two closed subspaces $\mathcal{M} \subseteq \mathcal{N} \subseteq \mathcal{K}$, invariant under V such that

$$\mathcal{H} = \mathcal{N} \ominus \mathcal{M}.$$

Now we present a similar result for injective maps.

Theorem 7.2.16. *Let $v : B \to B$ be an injective function and let $A \subset B$. Suppose $h : A \to A$ is a function such that $h(a) = v(a)$ for $a \in A$ with $v(a) \in A$. Suppose $A = A_2 \backslash A_1$ where A_1, A_2 are invariant under v. Then there exists $p : B \to B$ such that $p^2 = p$, $p(B) = A$ and*

$$pv^n(a) = h^n(a)\ \forall n \in \mathbb{Z}_+, a \in A. \tag{7.2.14}$$

Proof. All we need to show is that p defined by equation (7.2.14) on $\bigcup_{n \in \mathbb{Z}_+} v^n(A)$ is well-defined. Then it can be arbitrarily extended to some idempotent on B. Suppose $v^n(a_1) = v^{n+m}(a_2)$ for some a_1, a_2 in A and $n, m \in \mathbb{Z}_+$. Then as v is injective, $a_1 = v^m(a_2)$. Note that $A \bigcap A_1 = \emptyset$. As A_1 is left invariant under v, if $v^k(a_2) \in A_1$ for some k, $1 \leq k \leq m$, then we get a contradiction as $a_1 \in A$. Consequently $v^k(a_2) \in A$, and so $v^k(a_2) = h^k(a_2)$ for $1 \leq k \leq m$. Hence $a_1 = h^m(a_2)$ and $h^n(a_1) = h^{n+m}(a_2)$. ∎

The converse of this theorem is false. Here is a simple example to show this.

Example 7.2.17. *Let $B = \mathbb{Z}_+$, the set of non-negative integers, and let $v : B \to B$ be the injective function defined by $v(0) = 1, v(1) = 0$ and $v(n) = n, n \geq 2$. Let $A = \mathbb{N}$ and let $h : A \to A$ be the identity function. Clearly, $h(n) = v(n)$ for $n \in A$ with $v(n) \in A$. Further, if we define $p : B \to B$ by $p(0) = 1$ and $p(n) = n$ for $n \in \mathbb{N}$, then clearly p is an idempotent with $p(B) = A$ and obviously, $pv^m(n) = h^m(n)$ for every $m \geq 0$ and $n \in A$. Suppose $A = A_2 \setminus A_1$ for some subsets A_1, A_2 of B. Note that there are only two choices, namely, $A_2 = B, A_1 = \{0\}$ and $A_2 = A, A_1 = \emptyset$. But clearly neither A nor $\{0\}$ is v-invariant.*

We know that contractions on Hilbert spaces dilate not only to isometries but also to unitaries. This is the well-known Sz. Nagy dilation and we have studied it in detail in Section 12 of Chapter1. Keeping this in mind we can try to dilate arbitrary functions to bijective functions. We get the following result.

Theorem 7.2.18. *Let $h : A \to A$ be a function. Then there exists (C, i, u, p) where C is a set, $i : A \to C$ is injective, $u : C \to C$ is a bijection, $p : C \to C$ is an idempotent with $p(C) = p(A)$ such that*

$$i(h^n(a)) = pu^n(i(a)) \ a \in A, n \in \mathbb{Z}_+,$$

and $C = \bigcup_{n\in\mathbb{Z}} u^n((i(A))$.

Proof. Take $C = A \times \mathbb{Z}, i(a) = (a, 0), u(a, n) = (a, n+1), p(a, n) = (h^n(a), 0)$ for $n \in \mathbb{Z}_+$ and $p(a, n) = (a, 0)$ for $n < 0$. ■

This result is not quite satisfactory as the action of p on (a, n) with $n < 0$ is rather arbitrary and there is no uniqueness.

7.3. Multivariable dilation theory

One would like to dilate commuting contractions on a Hilbert space to commuting isometries and not arbitrary isometries. In this context Ando's dilation theorem [9] is well-known. We have presented it with a detailed proof in Section 1 of Chapter 3. There we have also noted that this theorem can not be extended to triples or to general d-tuples with $d \geq 3$. In contrast, we have the following theorem for functions.

Theorem 7.3.1. *Let J be an index set and let $\{h_j : j \in J\}$ be a commuting family of functions on a set A. Then there exists a quadruple $(B, i, (v_j)_{j\in J}, p)$ where B is a set, $i : A \to B$ is an injective map, $v_j : A \to B$, $j \in J$ is a commuting family of injective maps, $p : B \to B$ is an idempotent, such that*

$$i(h_{j_1}h_{j_2}\ldots h_{j_k}(a)) = p(v_{j_1}v_{j_2}\ldots v_{j_k})(i(a)), \ j_1, j_2, \ldots, j_k \in J, a \in A$$

and $B = \{v_{j_1}v_{j_2}\ldots v_{j_k}(i(a)) : j_1, j_2, \ldots, j_k \in J, a \in A\}$.

Proof. Take $B = A \times \mathbb{Z}^J_{+,0}$, where $\mathbb{Z}^J_{+,0}$ denotes the space of functions from J to $\mathbb{Z}_+$ taking 0 as their values except for finitely many points. Then using multi-index notation, for $\alpha \in \mathbb{Z}^J_{+,0}$, h^α is defined as $h^\alpha = \Pi_{j\in J} h_j^{\alpha(j)}$. Here the product means composition of powers of h_j's and since they form a commuting family, the order of composition does not matter. Define $i : A \to B$ by $i(a) = (a, 0)$ (Here 0 is the zero function). For $j \in J$, let δ_j be the function $\delta_j(k) = \delta_{jk}$. Define $v_j : B \to B$ by $v(a, \alpha) = (a, \alpha + \delta_j)$ and $p(a, \alpha) = (h^\alpha(a), 0)$.

It is not difficult to verify that $(B, i, (v_j)_{j\in J}, p)$ is a minimal commuting dilation of $\{h_j : j \in J\}$. ■

In the previous theorem also we have not bothered about optimality.

Definition 7.3.2. *A commuting injective dilation* $(B, i, (v_j)_{j\in J}, p)$ *of a commuting family of functions* $\{h_j : j \in J\}$ *on some set* A *is said to be* co-invariant *if every* v_j *leaves* $i(A)^c$ *invariant.*

To ensure commutativity of the dilation, we are forced to assume in the following theorem that D^c is left invariant by all $h_j, j \in J$. This was not the case in one variable. Moreover, now we do not have any uniqueness result.

Theorem 7.3.3. *Let* $\{h_j : j \in J\}$ *be a commuting family of functions on a set A. Let D be a defect space for each of* h_j*, that is,* $h_j|_{D^c}$ *is injective for every* j *and suppose* D^c *is left invariant by every* $h_j, j \in J$*. Define* $(B_D, i_D, (v_{j,D})_{j\in J}, p_D)$ *by*

$$
\begin{aligned}
B_D &= D^c \times \{0\} \bigcup D \times \mathbb{Z}^J_{+,0};\\
i_D(a) &= (a, 0),\ a \in A;\\
v_{j,D}(a, \alpha) &= \begin{cases} (h_j(a), 0), & (a, \alpha) \in D^c \times \{0\};\\ (a, \alpha + \delta_j), & (a, \alpha) \in D \times \mathbb{Z}^J_{+,0};\end{cases}\\
p(a, \alpha) &= (h^\alpha(a), 0).
\end{aligned}
$$

Then $(B_D, i_D, (v_{j,D})_{j\in J}, p_D)$ *is a co-invariant, minimal, commuting, and injective dilation.*

Proof. Clearly, $(v_{j,D})_{j\in J}$ is a family of injective maps on B_D. Now we show that this is a commuting family. Let $(a, 0) \in D^c \times \{0\}$. Then, $v_{i,D}v_{j,D}((a, 0)) = (h_i h_j(a), 0) = (h_j h_i(a), 0) = v_{j,D}v_{i,D}(a, 0)$. Let $(a, \alpha) \in D \times \mathbb{Z}^J_{+,0}$. Then $v_{i,D}v_{j,D}((a, \alpha)) = (a, \alpha + \delta_j + \delta_i) = v_{j,D}v_{i,D}((a, \alpha))$. Finally, note that if $a \in D^c$, then

$$
\begin{aligned}
p_D v_{j_1,D} v_{j_2,D} \cdots v_{j_n,D}(i_D(a)) &= p_D v_{j_1,D} v_{j_2,D} \cdots v_{j_n,D}((a, 0))\\
&= p_D(h_{j_1} h_{j_2} \cdots h_{j_n}(a), 0)\\
&= (h_{j_1} h_{j_2} \cdots h_{j_n}(a), 0)\\
&= i_D(h_{j_1} h_{j_2} \cdots h_{j_n}(a)),
\end{aligned}
$$

and if $a \in D$, then

$$
\begin{aligned}
p_D v_{j_1,D} \cdots v_{j_n,D}(i_D(a)) &= p_D v_{j_1,D} v_{j_2,D} \cdots v_{j_n,D}((a, 0))\\
&= p_D(a, \delta_{j_1} + \delta_{j_2} + \cdots + \delta_{j_n})\\
&= (h_{j_1} \cdots h_{j_n}(a), 0)\\
&= i_D(h_{j_1} h_{j_2} \cdots h_{j_n}(a)).
\end{aligned}
$$

Thus, $(B_D, i_D, (v_{j,D})_{j\in J}, p_D)$ is a commuting injective dilation of the family $(h_j)_{j\in J}$. Obviously this dilation is minimal. Observe that $i(A)^c = D \times \mathbb{Z}^J_{+,0} \setminus \{0\}$ and given $(a, \alpha) \in i(A)^c$ (so that $a \in D$), $v_{j,D}(a, \alpha) = (a, \alpha + \delta_j) \in i(A)^c$. Thus, $(B_D, i_D, (v_{j,D})_{j\in J}, p_D)$ is a co-invariant, minimal, commuting injective dilation of $(h_j)_{j\in J}$. ∎

For non-commuting operator tuples forming a row contraction, one natural dilation is to have isometries with mutually orthogonal ranges. We studied such a construction under the name 'standard noncommuting dilation', in Chapter 4. To have it in our abstract framework we make the following Definition.

Definition 7.3.4. *Let J be an index set and let $\{h_j : j \in J\}$ be a family of functions on a set A. A quadruple $(B, i, (v_j)_{j\in J}, p)$, is an* injective non-commutative dilation *of $\{h_j : j \in J\}$ if B is a set, $i : A \to B$ is an injective function, for every j, $v_j : B \to B$ is injective such that $v_j(B) \bigcap v_k(B) = \emptyset$ for $j \neq k$, $p : B \to B$ is an idempotent with $p(B) = i(A)$ and*

$$i(h_{j_1} h_{j_2} \ldots h_{j_k}(a)) = p(v_{j_1} v_{j_2} \ldots v_{j_k})(i(a))$$

for $j_1, j_2, \ldots, j_k$ in J, $k \in \mathbb{N}$ and $a \in A$. Such a dilation is said to be minimal *if*

$$B = \{v_{j_1} v_{j_2} \ldots v_{j_k}(i(a)) : j_1, j_2, \ldots, j_k \in J, k \in \mathbb{Z}_+, a \in A\}.$$

Theorem 7.3.5. *Every family of functions $\{h_j : j \in J\}$ on a set A admits a minimal injective non-commuting dilation.*

Proof. Define

$$\Gamma^J = \{(j_1, j_2, \ldots, j_k) : j_1, j_2, \ldots, j_k \in J, k \in \mathbb{N}\} \bigcup \{\omega\}$$

where ω is the empty tuple or 'vacuum'. Take

$$\begin{aligned} B &= A \times \Gamma^J; \\ i(a) &= (a, \omega), a \in A; \\ v_j(a, \omega) &= (a, (j)); \\ v_j(a, (j_1, j_2, \ldots, j_k)) &= (a, (j, j_1, j_2, \ldots, j_k)); \\ p(a, \omega) &= (a, \omega); \\ p(a, (j_1, j_2, \ldots, j_k)) &= (h_{j_1} h_{j_2} \ldots h_{j_k}(a), \omega). \end{aligned}$$

Now the verification of claims made above is easy. ∎

Definition 7.3.6. *Let $\{h_j : j \in J\}$ be a family of functions on a set A. Then a subset D of A is said to be a joint defect space for $\{h_j : j \in J\}$ if $h_j|_{D^c}$ are injective and*

$$h_j(D^c) \bigcap h_k(D^c) = \emptyset \textit{ for } j \neq k.$$

The condition in the previous definition can also be stated as $H : J \times A \to A$ defined by $H(j,a) = h_j(a)$ is injective on $J \times D^c$.

Now suppose D is a joint defect space for $\{h_j : j \in J\}$. Consider $(B_D, i_D, (v_{j,D})_{j\in J}, p)$, where

$$\begin{aligned}
B_D &= D^c \times \{\omega\} \bigcup D \times \Gamma^J, \\
i_D(a) &= (a, \omega), \\
v_{j,D}(a,\omega) &= (h_j(a), \omega) \text{ if } a \in D^c, \\
v_{j,D}(a,\omega) &= (a, (j)) \ \text{ if } a \in D \\
v_{j,D}(a, (j_1, j_2, \ldots, j_k)) &= (a, (j, j_1, j_2, \ldots, j_k)), \\
p(a,\omega) &= (a, \omega), \\
p(a, (j_1, j_2, \ldots, j_k)) &= (h_{j_1} h_{j_2} \ldots h_{j_k}(a), \omega).
\end{aligned}$$

Then $(B_D, i_D, (v_{j,D})_{j\in J}, p)$ is a non-commutative injective dilation of $\{h_j : j \in J\}$. We also observe that if $v_{j,D}(i_D(a)) \notin i_D(A)$ then $v_{k,D}(i_D(a)) \notin i_D(A)$ for every $k \in J$. This is a property crucial for the next theorem.

Theorem 7.3.7. Let $\{h_j : j \in J\}$ be a family of functions on a set A. Let $(B, i, (v_j)_{j\in J}, p)$ be an injective, non-commutative, minimal, and co-invariant dilation of $\{h_j : j \in J\}$ and has the property that if for some $a \in A$ and $j \in J$, $v_j(i(a)) \notin i(A)$, then $v_k(i(a)) \notin i(A)$ for every $k \in J$. Set

$$D = \{a \in A : v_j(i(a)) \notin i(A) \text{ for some } j \in J\}.$$

Then D is a joint defect space for $\{h_j : j \in J\}$ and $(B_D, i_D, (v_{j,D})_{j\in J}, p_D)$ is bijectively isomorphic to (B, i, v, p).

Proof. Observe that given any $a \in A$, either $v_j(i(a)) \in i(A)$ for every $j \in J$ or $v_j(i(a)) \notin i(A)$ for all $j \in J$. We first show that D is a joint defect space for $\{h_j : j \in J\}$. Let $j \in J$ and let $a, b \in D^c$ be such that $h_j(a) = h_j(b)$. As $a, b \in D^c$, both $v_j(i(a))$ and $v_j(i(b))$ belong to $i(A)$. Thus, $v_j(i(a)) = pv_j(i(a)) = i(h_j(a)) = i(h_j(b)) = pv_j(i(b)) = v_j(i(b))$ and hence, $a = b$. Also for $j \neq k$, $h_j(D^c) \cap h_k(D^c) = \emptyset$ for if $h_j(D^c) \cap h_k(D^c) \neq \emptyset$, say $x \in h_j(D^c) \cap h_k(D^c)$, then $x = h_j(a) = h_k(b)$ for some $a, b \in D^c$ which would imply that $v_j(i(a)) = v_k(i(b))$ and consequently, $v_j(B) \cap v_k(B) \neq \emptyset$, a contradiction. Hence, $h_j(D^c) \cap h_k(D^c) = \emptyset$ for $j, k \in J$ with $j \neq k$.

Consider the map $\psi : B_D \to B$ defined by $\psi(a,\omega) = i(a)$ and $\psi(a,\alpha) = v^\alpha(i(a))$ for $(a,\alpha) \in B_D$ with $\alpha \neq \omega$. We show that ψ is injective. Let $(a,\alpha), (b,\beta) \in B_D$ be such that $\psi(a,\alpha) = \psi(b,\beta)$. If one of a, b, say a, belongs to D and the other one, that is, b belongs to D^c, then we must have that $\beta = \omega$ and $\alpha \neq \omega$ and hence, $v^{\alpha}(i(a)) = i(b)$. Let $\alpha = (j_1, j_2, \cdots, j_n)$ where $n \geq 1$. As $a \in D$, $v_j(i(a)) \notin i(A)$ for every $j \in J$ and so, in particular, $v_{j_n}(i(a)) \notin i(A)$. Now the co-invariance of the dilation allows us to conclude that $v^{\alpha}(i(a)) \notin i(A)$, leading to a contradiction. Thus, either $a, b \in D$ or $a, b \in D^c$. If both a and b belong to D^c, then clearly, $\alpha = \beta = \omega$ from which it follows that $i(a) = i(b)$ and so, $a = b$. Let $a, b \in D$. If $\alpha = \beta = \omega$, then obviously $a = b$. We assert that it can not happen that one of α, β is ω and the other one is different from ω for if, say $\beta = \omega$ and $\alpha \neq \omega$, then $v^{\alpha}(i(a)) = i(b)$ which is a contradiction since similar argument as before establishes that $v^{\alpha}(i(a)) \notin i(A)$. Let $\alpha = (j_1, j_2, \cdots, j_n)$ and $\beta = (k_1, k_2, \cdots, k_m)$ where $m, n \geq 1$. Then $v_{j_1} \cdots v_{j_n}(i(a)) = v_{k_1} \cdots v_{k_m}(i(b))$. If possible let $n > m$. It follows from $v_j(B) \cap v_k(B) = \emptyset$ for $j \neq k$ that $j_1 = k_1, \cdots, j_m = k_m$ and thus, $i(b) = v_{j_{m+1}} \cdots v_{j_n}(i(a))$, a contradiction. Thus $m = n$ and once again applying the fact that $v_j(B) \cap v_k(B) = \emptyset$ for $j \neq k$ we obtain that $i(a) = i(b)$ so that $a = b$. Thus ψ is injective.

Next we show that ψ is surjective. Let $x \in B$. If $x \in i(A)$, say, $x = i(a)$ for some $a \in A$, then $\psi(a, \omega) = i(a) = x$. Assume that $x \notin i(A)$. It follows from from minimality of the dilation that $x = v_{j_1} v_{j_2} \cdots v_{j_n}(i(a))$ where $n \geq 1, a \in A$ and $j_1, \cdots, j_n \in J$. Clearly, if $a \in D$, then $\psi((a, (j_1, j_2, \cdots, j_n))) = x$. If $a \in D^c$, $v_{j_n}(i(a)) \in i(A)$ and let k be the smallest positive integer, $1 < k \leq n$, such that $v_{j_k} v_{j_{k+1}} \cdots v_{j_n}(i(a)) \in i(A)$ and $v_{j_{k-1}} v_{j_k} \cdots v_{j_n}(i(a)) \notin i(A)$. Let $v_{j_k} v_{j_{k+1}} \cdots v_{j_n}(i(a)) = i(b)$ where $b \in A$. As $v_{j_{k-1}}(i(b)) \notin i(A), b \in D$ and $\psi((b, (j_1, \cdots, j_{k-1}))) = v_{j_1} \cdots v_{j_{k-1}}(i(b))$ $= x$. Therefore, ψ is surjective and hence, bijective.

One can easily verify that

$$\psi v_{j,D} = v_j \psi, \ \psi p_D = p\psi, \text{ and } \psi i_D = i, \text{ for every } j \in J$$

and consequently, (B, i, v, p) is bijectively isomorphic to the tuple $(B_D, i_D, (v_{j,D})_{j \in J}, p_D)$.

■

7.4. Berger, Coburn and Lebow Theorem

Recall from Chapter 1, Section 1.13 that the theorem of Berger, Coburn and Lebow ([23]) describes the structure of two commuting isometries. In order to make the comparison with the current set up clearer, we rewrite the statement of this theorem using slightly different notation.

Let V_1, V_2 be two commuting isometries on a Hilbert space $\mathcal{H}$ and $V = V_1 V_2$. Then by Wold decomposition of V, $\mathcal{H} = \mathcal{H}_0 \oplus \mathcal{H}_1$ decomposing V

as $V = V|_{\mathcal{H}_0} \oplus V|_{\mathcal{H}_1}$, where $V|_{\mathcal{H}_1}$ is a unitary and $V|_{\mathcal{H}_0}$ is a shift with some multiplicity. So up to unitary isomorphism $\mathcal{H}_0 = \mathcal{H}^2 \otimes \mathcal{M}$, and $V|_{\mathcal{H}_0} = M_z \otimes I_{\mathcal{D}}$, where M_z is the standard shift isometry on the Hardy space and $\mathcal{M}$ is a multiplicity space.

It is not hard to see that $\mathcal{H}_0$ and $\mathcal{H}_1$ reduce V_1, V_2 and so they decompose say as $V_1 = V_{10} \oplus V_{11}$ and $V_2 = V_{20} \oplus V_{21}$. Of course, this may not be Wold decompositions of V_1 or V_2. However V_{11} and V_{21} are commuting unitaries. Further, V_{10} and V_{20} are commuting isometries related by a simple relation as described below.

Under the set up as above, the Berger, Coburn and Lebow theorem [23] tells us that there exists a projection P on $\mathcal{D}$ and a unitary U on $\mathcal{D}$ such that $V_{10} = S \otimes U^*P + I \otimes U^*(I-P)$ and $V_{20} = I \otimes PU + S \otimes (I-P)U$. Conversely, any pair of a projection P and a unitary U on $\mathcal{D}$ would give a commuting pair of isometries by this formula.

Now we look at a set theoretic version of this result. Let A be a non-empty set and let $v : A \to A$ be an injective map. Suppose v factorizes as $v = v_1v_2 = v_2v_1$ where $v_i : A \to A$ are injective for $i = 1, 2$. Consider the Wold type decomposition of v. So take $A_1 = \bigcap_{n=0}^{\infty} v^n(A)$ and $A_0 = A_1^c$. Using commutativity of v_1, v_2, if $a = v^n(a_1)$, then $v_1(a) = v^n(a_2)$ where $a_2 = v_1(a_1)$. This shows that A_1 is invariant under v_1. Now if $v_1(a_1) = v^{n+1}(a_2)$ then $a_1 = v^n(a_2)$ where $a_2 = v_2(a_1)$. This shows that if $v_1(a_1) \in A_1$, then $a_1 \in A_1$. So A_0 is also invariant under v_1. Similarly A_0 and A_1 are also invariant under v_2. Now $v|_{A_1}$ is a bijection. So $A_1 = v(A_1) = v_1v_2(A_1) \subseteq v_1(A_1) \subseteq A_1$. Hence v_1, v_2 are surjective on A_1. Consequently, v_1, v_2 on A_1 are commuting bijections. By virtue of Lemma 7.2.7, the structure of the pair $(v_1|_{A_1}, v_2|_{A_1})$ is quite transparent. Thus, in order to understand the structure of v, it suffices to consider $v|_{A_0}$. Hence, without loss of generality, hereafter we assume that v is a shift.

Take $W = A \backslash v(A)$. As observed before W is a wandering subset for v. Similarly take $W_i = A \backslash v_i(A)$ for $i = 1, 2$. Now if $a \in W$, then either $a \in W_1$ or $a = v_1(a_1)$ for some a_1. But then a_1 has to be in W_2, otherwise we would have $a \in v_1v_2(A)$. Arguing this way from the commutativity of v_1, v_2 we get

$$W = W_1 \sqcup v_1(W_2) = v_2(W_1) \sqcup W_2.$$

Define $u : W \to W$, that is, from $W_1 \sqcup v_1(W_2)$ to $v_2(W_1) \sqcup W_2$, by $u(w_1) = v_2(w_1)$ and $u(v_1(w_2)) = w_2$ for $w_1 \in W_1, w_2 \in W_2$. It is easily seen that u is a bijection with inverse u^{-1} given by $u^{-1}(v_2(w_1)) = w_1$ and $u^{-1}(w_2) = v_1(w_2)$.

From Wold type decomposition, as v is a shift, $A = \sqcup_{n=0}^{\infty} v^n(W) = W \sqcup v(W) \sqcup v^2(W) \sqcup \cdots$. For $n \geq 1$ we extend the definition of u to $v^n(W)$ by taking $u(v^n(w)) = v^n(u(w))$ for any $w \in W$. Then $u : A \to A$ is a bijection commuting with v. Take $C_0 = \sqcup_{n=0}^{\infty} v^n(W_2)$ and $C_1 = \sqcup_{n=0}^{\infty} v^n(v_2(W_1))$, so that $A = C_0 \sqcup C_1$

Moreover, $v_1 : A \to A$ is given by

$$v_1(x) = \begin{cases} u^{-1}(x) & \text{if } x \in C_0 \\ u^{-1}(v(x)) & \text{if } x \in C_1. \end{cases}$$

Similarly, $v_2 : A \to A$ is given by

$$v_2(x) = \begin{cases} u(x) & \text{if } x \in u^{-1}(C_1) \\ u(v(x)) & \text{if } x \in u^{-1}(C_0). \end{cases}$$

This can be written using unilateral shift as follows. Take $B = W \times \mathbb{Z}_+$. Let $1 \times s_+ : B \to B$ be the canonical unilateral shift with wandering space W (identified with $W \times 0$). Let $g : A \to B$ defined by $g(v^n(w)) = (w, n)$ for $w \in W$ and $n \in \mathbb{Z}_+$ is a bijection such that $v = g^{-1}(1 \times s_+)g$. Take $s_i = g v_i g^{-1}$ for $i = 1, 2$. Then

$$s_1(w,n) = \begin{cases} (u^{-1}(w), n) & (w,n) \in W_2 \times \mathbb{Z}_+; \\ (u^{-1}(w), n+1) & (w,n) \in v_2(W_1) \times \mathbb{Z}_+; \end{cases}$$

and

$$s_2(w,n) = \begin{cases} (u(w), n) & (w,n) \in W_1 \times \mathbb{Z}_+; \\ (u(w), n+1) & (w,n) \in v_1(W_2) \times \mathbb{Z}_+. \end{cases}$$

Recall that W decomposes as $W = W_2 \sqcup v_2(W_1)$, and $u : W \to W$ is a bijection such that $u(W_1) = v_2(W_1) = W_2^c$. In other words we have the following theorem.

Theorem 7.4.1. *With notation as above, (A, v, v_1, v_2) is in bijective correspondence with $(W \times \mathbb{Z}_+, 1_W \times s_+, s_1, s_2)$ where,*

$$s_1 = (u^{-1} \times \textit{ id. })|_{W_2 \times \mathbb{Z}_+} + (u^{-1} \times s_+)|_{W_2^c \times \mathbb{Z}_+};$$

$$s_2 = (u \times \textit{ id. })|_{u^{-1}(W_2^c) \times \mathbb{Z}_+} + (u \times s)|_{u^{-1}(W_2) \times \mathbb{Z}_+}.$$

Now we extend our BCL type theorem to families of maps. Let $\{v_j : j \in \{1, 2, \ldots n\}\}$ be a commuting family of injective maps on a set A. Assume that $v = v_1 v_2 \ldots v_n$ is a shift. Take $W = A \setminus v(A)$ and $W_j = A \setminus v_j(A)$ for $j = 1, 2, \ldots, n$. Let S_n denote the group of permutations of $\{1, 2, \ldots, n\}$. Now for any $\sigma \in S_n$, W has a decomposition given by $W =$

$$W_{\sigma(1)} \sqcup v_{\sigma(1)}(W_{\sigma(2)}) \sqcup v_{\sigma(1)} v_{\sigma(2)}(W_{\sigma(3)}) \sqcup \cdots \sqcup v_{\sigma(1)} v_{\sigma(2)} \cdots v_{\sigma(n-1)}(W_{\sigma(n)}),$$

and we use the notation $W(\sigma)$ to denote this decomposition of W. For $\tau, \sigma \in S_n$, let u_τ^σ denote the bijection from $W(\sigma)$ to $W(\tau)$ defined as follows:

$$u_\tau^\sigma(v_{\sigma(1)}v_{\sigma(2)}\cdots v_{\sigma(k-1)}(w)) = v_{\tau(1)}v_{\tau(2)}\cdots v_{\tau(r-1)}(w)$$

for $w \in W_{\sigma(k)}$, with r being chosen such that $\tau(r) = \sigma(k)$. For $k = 0, 1, \cdots, n-1$, let $\sigma_k \in S_n$ be given by

$$\sigma_k(j) = \begin{cases} k+j, & \text{for } 1 \leq j \leq n-k, \\ k-n+j, & \text{for } n-k+1 \leq j \leq n. \end{cases}$$

and we consider the family $\{u_{\sigma_{k-1}}^{\sigma_k} : 1 \leq k \leq n\}$ of bijections of W where the notation σ_n stands for σ_0. Using these bijections, we can describe the maps v_k $(1 \leq k \leq n)$ on W as follows:

$$v_k = \begin{cases} u_{\sigma_{k-1}}^{\sigma_k}, & \text{on } W \setminus \prod_{i=1, i\neq k}^n v_i(W_k), \\ vu_{\sigma_{k-1}}^{\sigma_k}, & \text{on } \prod_{i=1, i\neq k}^n v_i(W_k). \end{cases}$$

Now, as before, using the fact that $A = \sqcup_{m=0}^\infty v^m(W)$, we may extend the bijections $u_{\sigma_{k-1}}^{\sigma_k}$ of W to bijections of A by setting $u_{\sigma_{k-1}}^{\sigma_k}(v^m w) = v^m(u_{\sigma_{k-1}}^{\sigma_k} w)$ for $w \in W$ and $m \geq 0$. Consequently,

$$v_k = \begin{cases} u_{\sigma_{k-1}}^{\sigma_k}, & \text{on } \sqcup_{m=0}^\infty v^m(W \setminus \prod_{i=1, i\neq j}^n v_i(W_k)), \\ vu_{\sigma_{k-1}}^{\sigma_k}, & \text{on } \sqcup_{m=0}^\infty v^m(\prod_{i=1, i\neq j}^n v_i(W_k)). \end{cases}$$

Note that as before, $g^{-1}(1 \times s_+)g = v$. Set $s_k = gv_kg^{-1}$ for $k = 1, 2, \cdots, n$. We then have the following result.

Theorem 7.4.2. *With the same notations as in Theorem 7.4.1,* $(A, v, v_1, v_2, \cdots, v_n)$ *is in bijective correspondence with* $(W \times \mathbb{Z}_+, 1_W \times s_+, s_1, s_2, \cdots, s_n)$ *where*

$$s_k = (u_{\sigma_{k-1}}^{\sigma_k} \times id.)|_{W_k'' \times \mathbb{Z}_+} + (u_{\sigma_{k-1}}^{\sigma_k} \times s_+)|_{W_k' \times \mathbb{Z}_+}, \ 1 \leq k \leq n;$$

and $W_k' = \prod_{i=1, i\neq k}^n v_i(W_k) = W \setminus (u_{\sigma_k}^{\sigma_{k-1}}(W(\sigma_{k-1}) \setminus W_k)), W_k'' = W \setminus W_k'$.

7.5. Generalizations and variations

We considered dilations of $\{h^n : n \in \mathbb{Z}_+\}$ for a map $h : A \to A$. Now instead of $\mathbb{Z}_+$ we can consider $\mathbb{R}_+$ or more general monoids. Dilation theory on general monoids have been considered by many authors, see for instance [111]. Recall that a monoid is a set S with an associative binary operation (say '.') and an identity element (say 1). The associative operation need not be abelian. It seems eminently feasible to extend most of what we did in previous sections to this setting. Here is the standard dilation in this setting.

Theorem 7.5.1. *Let S be a left cancellative monoid. Suppose A is a set and $\{h_s : s \in S\}$ is a family of functions such that $h_{s.t} = h_s \circ h_t$ for all $s, t \in S$ and $h_1 =$ id.. Then there exists a quadruple $(B, i, \{v_s : s \in S\}, p)$ where B is a set, $i : A \to B$, $v_s : B \to B$ are injective functions, $v_{s.t} = v_s \circ v_t$ for all s, t in S, $v_1 =$ id., $p : B \to B$ is idempotent with $p(B) = i(A)$ such that*

$$p(v_s(i(a))) = i(h_s(a))\ \forall s \in S, a \in A$$

and $B = \bigcup_{s\in S}\{v_s(i(a)) : s \in S, a \in A\}$.

Proof. Take $B = A \times S$. Define $i : A \to B$ by $i(a) = (a, 1)$, $v_s(a, t) = (a, s.t)$ and $p(a, s) = (h_s(a), 1)$. The left cancellative property of the monoid S ensures that v_s are isometries. ∎

Instead of working with sets and functions, we can try to develop the dilation theory working with vector spaces and linear maps. Now bounded operators on Hilbert spaces gets replaced by linear maps, isometries by injective linear maps and unitaries by bijective linear maps. Direct sums of Hilbert spaces gets replaced by direct sums of vector spaces. Here is a formal definition and a sample result.

Definition 7.5.2. *Let A be a vector space and let $h : A \to A$ be a linear map. A quadruple (B, i, v, p) is said to be a minimal injective linear dilation of h if B is a vector space, $i : A \to B$ is an injective linear map, $v : B \to B$ is an injective linear map, $p : B \to B$ is an idempotent linear map with $p(B) = i(A)$, satisfying*

$$i(h^n(a)) = p(v^n(i(a))\ \forall n \in \mathbb{Z}_+, a \in A.$$

Such a dilation is said to be minimal if $B =$ span $\{v^n(i(a)) : n \in \mathbb{Z}_+, a \in A\}$.

Theorem 7.5.3. *Every linear map $h : A \to A$ on a vector space admits a minimal injective linear dilation.*

Proof. Take $B = A_0^{\mathbb{Z}_+}$, the space of functions from $\mathbb{Z}_+$ to A which take value 0 at all but finitely many points. It is a vector space under natural linear operations. Define $i : A \to B$ by $i(a)(0) = a$ and $i(a)(n) = 0$ for $n \neq 0$. Define $v : B \to B$ by $v(b)(0) = 0$ and $v(b)(n) = b(n-1)$ for $n \geq 1$. Finally define $p : B \to B$ by $p(b) = \sum_{n=0}^{\infty} ih^n(b(n))$. This map p is well defined as $b(n) = 0$ for all but finitely many n. It is easy to see that (B, i, v, p) is a minimal injective linear dilation of h.

∎

We observe that the construction of dilation here is similar to the standard dilation of functions to injective maps, however there are certain differences. Now the addition operation of vector spaces plays a non-trivial role. This has been explored further by [74].

It is conceivable that the dilation theory can be developed in other settings.

Bibliography

[1] J. Agler. "Rational dilation on an annulus". In: *Ann. of Math. (2)* 121.3 (1985), pp. 537 –563.

[2] J. Agler. "The Arveson extension theorem and coanalytic models". In: *Integral Equations and Operator Theory* 5 (1982), pp. 608–631.

[3] J. Agler and J. E. McCarthy. "Distinguished varieties". In: *Acta Math* 194.2 (2005), 133–153.

[4] J. Agler and N. J. Young. "A commutant lifting theorem for a domain in C^2 and spectral interpolation". In: *J. Funct. Anal.* 161.2 (1999), pp. 452 –477.

[5] J. Agler and N. J. Young. "A model theory for Γ-contractions". In: *J. Oper. Th.* 49.1 (2003), pp. 45 –60.

[6] J. Agler and N. J. Young. "The hyperbolic geometry of the symmetrized bidisc". In: *J. Geom. Anal.* 14.3 (2004), pp. 375 –403.

[7] H. Alexander and J. Wermer. *Several complex variables and Banach algebras*. Third edition. Graduate Texts in Mathematics 35. Springer-Verlag, New York, 1998., 2002.

[8] D. Alpay, T. Bhattacharyya, A. Jindal, and P. Kumar. "A dilation theoretic approach to approximation by inner functions". In: *arXiv article https://arxiv.org/abs/2203.10936* ().

[9] T. Ando. "On a pair of commuting contractions". In: *Acta Sci. Math.* 24 (1963), pp. 88 –90.

[10] A. Arias and G. Popescu. "Noncommutative interpolation and Poisson transforms". In: *Israel J. Math.* 115 (2000), pp. 205 –234.

[11] A. Arias and G. Popescu. "Noncommutative interpolation and Poisson transforms II". In: *Houston J. Math.* 25 (1999), pp. 79 – 98.

[12] N. Aronszajn. "Theory of reproducing kernels". In: *Trans. Amer. Math. Soc.* 68 (1950), 337–404.

[13] W. Arveson. "Dilation theory yesterday and today". In: *Axler, S., Rosenthal, P., Sarason, D. (eds) A Glimpse at Hilbert Space Operators. Operator Theory Advances and Applications* 207 (2010).

[14] W. Arveson. *Noncommutative dynamics and E–semigroups*. Monographs in Mathematics. Springer, 2003.

[15] W. Arveson. "On the index and dilations of completely positive semigroups". In: *Internat. J. Math.* 10 (1999), pp. 791 –823.

[16] W. Arveson. "Subalgebras of C^*-algebras II". In: *Acta Math.* 128 (1972), pp. 271 –308.

[17] W. Arveson. "Subalgebras of C*-algebras III: multivariable operator theory". In: *Acta Math.* 181 (1998), pp. 159 –228.

[18] W. Arveson. "The curvature invariant of a Hilbert module over $\mathbb{C}[z_1, \ldots, z_d]$". In: *J. Reine Angew. Math.* 522 (2000), pp. 173 –236.

[19] W. Arveson. "The index of a quantum dynamical semigroup". In: *J. Funct. Anal.* 146 (1997), pp. 557 –588.

[20] A. Athavale. "Model theory on the unit ball in $\mathbb{C}^m$". In: *J. Operator Theory* 27 (1992), pp. 347 –358.

[21] A. Athavale. "On the intertwining of joint isometries". In: *J. Operator Theory* 23 (1990), pp. 339 –350.

[22] M. Bakonyi and T. Constantinescu. *Schur's algorithm and several applications*. Vol. 261. Pitman Research Notes in Mathematics Series. Harlow: Longman Scientific & Technical, 1992.

[23] C. A. Berger, L. A. Coburn, and A. Lebow. "Representation and index theory for C*-algebras generated by commuting isometries". In: *J. Funct. Anal.* 27.1 (1978), pp. 51 –99.

[24] B. V. R. Bhat. "An index theory for quantum dynamical semigroups". In: *Trans. Amer. Math. Soc.* 348 (1996), pp. 561 –583.

[25] B. V. R. Bhat. "Atomic dilations". In: *Advances in quantum dynamics (G.L. Price, B .M. Baker, P.E.T. Jorgensen, and P.S. Muhly, eds.), Contemporary Mathematics* 335 (2003), pp. 99 –107.

[26] B. V. R. Bhat. "Minimal dilations of quantum dynamical semigroups to semigroups of endomorphisms of C^*-algebras". In: *J. Ramanujan Math. Soc.* 14 (1999), pp. 109 –124.

[27] B. V. R. Bhat and T. Bhattacharyya. "A model theory for q-commuting contractive tuples". In: *J. Operator Theory* 47 (2002), pp. 97 –116.

[28] B. V. R. Bhat, T. Bhattacharyya, and S. Dey. "Standard non-commuting and commuting dilations for commuting tuples". In: *Trans. Amer. Math. Soc.* 356 (2004), pp. 1551 –1568.

[29] B. V. R. Bhat, S. De, and N. Rakshit. "A caricature of dilation theory". In: *Advances in Operator theory* 6 (2021).

[30] B. V. R. Bhat, R. Hillier, N. Mallick, and U. Vijayakumar. "Roots of completely positive maps". In: *Linear Algebra Appl.* 587 (2020), pp. 143 –165.

[31] R. Bhatia. *Matrix Analysis*. Graduate Texts in Mathematics. Springer-Verlag New York, 1997.

[32] T. Bhattacharyya, J. Eschmeier, and J. Sarkar. "Characteristic function of a pure commuting contractive tuple". In: *Integral Equations Operator Theory* 53 (2005), pp. 23 –32.

[33] T. Bhattacharyya, J. Eschmeier, and J. Sarkar. “On CNC commuting contractive tuples”. In: *Proc. Indian Acad. Sci. Math. Sci.* 116 (2006), pp. 299 –316.

[34] T. Bhattacharyya, P. Kumar, and H. Sau. “Distinguished varieties through the Berger–Coburn–Lebow Theorem”. In: *Anal. PDE* 15.2 (2022), pp. 477–506.

[35] T. Bhattacharyya, S. Pal, and S Shyam Roy. “Dilation of Γ-contractions by solving operator equations”. In: *Adv. Math.* 230.2 (2012), pp. 577 –606.

[36] T. Bhattacharyya and H. Sau. “Holomorphic functions on the symmetrized bidisk - Realization, interpolation and extension”. In: *J. Funct. Anal.* 274 (2018), pp. 504 –524.

[37] J. Bunce. “The joint spectrum of commuting nonnormal operators”. In: *Proc. Amer. Math. Soc.* 29 (1971), 499–505.

[38] J. W. Bunce. “Models for n-tuples of noncommuting operators”. In: *J. Funct. Anal.* 57 (1984), pp. 21 –30.

[39] P. G. Casazza. “Every frame is a sum of three (but not two) orthonormal bases–and other frame representations”. In: *J. Fourier Anal. Appl.* 4 (1998), pp. 727 –732.

[40] M. D. Choi and Ch. Davis. “The spectral mapping theorem for joint approximate point spectrum”. In: *Bull. Amer. Math. Soc.* 80 (1974), pp. 317 –321.

[41] J. B. Conway. *A course in functional analysis. Second edition.* Vol. 96. Graduate Texts in Mathematics. Springer-Verlag, New York, 1990.

[42] J. B. Conway. *Functions of one complex variable. Second edition.* Vol. 11. Graduate Texts in Mathematics. Springer-Verlag, New York-Berlin, 1978.

[43] C. Costara. “The symmetrized bidisc and Lempert’s theorem”. In: *Bull. London Math. Soc.* 36.5 (2004), 656 – 662.

[44] M. Crabb and A. Davie. “von Nemann’s inequality for Hilbert space operators”. In: *Bull. London Math. Soc.* 7 (1975), pp. 49 – 50.

[45] R. E. Curto. “Applications of several complex variables to multiparameter spectral theory”. In: *Surveys of some recent results in operator theory.* Vol. II. Pitman Res. Notes Math. Ser., 192. Longman Sci. Tech., Harlow, 1988, pp. 25–90.

[46] B. Krishna Das and J. Sarkar. “Ando dilations, von Neumann inequality, and distinguished varieties”. In: *J. Funct. Anal.* 272.5 (2017), pp. 2114 –2131.

[47] B. Krishna Das, J. Sarkar, and S. Sarkar. “Factorizations of contractions”. In: *Adv. Math.* 322 (2017), 186–200.

[48] K. R. Davidson. *C^*-algebras By Example.* American Mathematical Society, 1996.

[49] S. Dey. “Standard dilations of q-commuting tuples”. In: *Colloq. Math.* 107 (2007), pp. 141 –165.

[50] R. G. Douglas. *Banach algebra techniques in operator theory. Second edition, Graduate Texts in Mathematics, 179.* Springer-Verlag, New York, 1998.

[51] R. G. Douglas and G. Misra. “Geometric invariants for resolutions of Hilbert modules”. In: *Nonselfadjoint operator algebras, operator theory, and related topics, Oper. Theory Adv. Appl.* 104 (1998), pp. 83 –112.

[52] R. G. Douglas, G. Misra, and J. Sarkar. “Contractive Hilbert modules and their dilations”. In: *Israel J. Math.* 187 (2012), pp. 141 –165.

[53] R. G. Douglas, G. Misra, and C. Varughese. “On quotient modules - the case of arbitrary multiplicity”. In: *J. Funct. Anal.* 174 (2000), pp. 364 –398.

[54] R. G. Douglas, P. S. Muhly, and C. Pearcy. “Lifting commuting operators”. In: *Michigan Math. J.* 15 (1968), pp. 385 –395.

[55] R. G. Douglas and V. I. Paulsen. *Hilbert modules over function algebras*. Longman Scientific & Technical, Harlow, 1989.

[56] M. A. Dritschel. “Factoring Non-negative Operator Valued Trigonometric Polynomials in Two Variables”. In: *arXiv:1811.06005* (2018).

[57] M. A. Dritschel and S. McCullough. “The failure of rational dilation on a triply connected domain”. In: *J. Amer. Math. Soc.* 18 (2005), pp. 873 –918.

[58] M. A. Dritschel and J. Rovnyak. “The operator Fejér-Riesz theorem”. In: *A glimpse at Hilbert space operators*. Oper. Theory Adv. Appl. 207 (2010), 223 – 254.

[59] S. Drury. “A generalization of von Neumann’s inequality to the complex ball”. In: *Proc. Amer. Math. Soc.* 68 (1978), pp. 300 –304.

[60] S. Drury. “Remarks on von Neumann’s inequality”. In: *Banach Spaces, Harmonic Analysis and Probability Theory.* Ed. by R. Blei and S. Sydney. Vol. 995. Springer Lecture Notes. Springer Verlag, 1983, pp. 14 –32.

[61] H. K. Du and P. Jin. “Perturbation of spectrums of 2×2 operator matrices”. In: *Proc. Amer. Math. Soc.* 121 (1994), pp. 761 –766.

[62] A. Edigarian and W. Zwonek. “Geometry of the symmetrized polydisc”. In: *Arch. Math. (Basel)* 84 (2005), pp. 364 –374.

[63] A. E. Frazho. “Models for non-commuting operators”. In: *J. Funct. Anal.* 48 (1982), pp. 1 –11.

[64] P. M. Gauthier. *Lectures on Several Complex Variables*. Springer International Publishing, Switzerland, 2014.

[65] P. R. Halmos. *A Hilbert Space Problem Book.* Second Edition. Graduate Texts in Mathematics No. 19. Springer-Verlag, New York-Berlin, 1982.

[66] P. R. Halmos. "Normal dilations and extensions of operators". In: *Summa Brasil. Math.* 2 (1950), 125 – 134.

[67] P. R. Halmos. "Shifts on Hilbert spaces". In: *J. Reine Angew Math.* 208 (1961), pp. 102 –112.

[68] J. Harland J. Agler and B. J. Raphael. *Classical function theory, operator dilation theory, and machine computation on multiply-connected domains*. Memoirs of the American mathematical Society no. 892, 2008.

[69] R. V. Kadison and G. K. Pedersen. "Means and convex combinations of unitary operators". In: *Math. Scand.* 57 (1985), pp. 249 –266.

[70] R. V. Kadison and J. R. Ringrose. *Fundamentals of the theory of operator algebras. Vol. I. Elementary theory. Reprint of the 1983 original*. Vol. 15. Graduate Studies in Mathematics. American Mathematical Society, Providence, RI, 1997.

[71] R. V. Kadison and J. R. Ringrose. *Fundamentals of the theory of operator algebras. Vol. II. Advanced theory. Corrected reprint of the 1986 original*. Vol. 16. Graduate Studies in Mathematics. American Mathematical Society, Providence, RI, 1997.

[72] L. Kosiński and W. Zwonek. "Nevanlinna-Pick problem and uniqueness of left inverses in convex domains, symmetrized bidisc and tetrablock". In: *J. Geom. Anal.* 26 (2016), pp. 1863 –1890.

[73] S. G. Krantz. *Function Theory of Several Complex Variables*. AMS Chelsea Publishing, 2001.

[74] K. Mahesh Krishna and P. Sam Johnson. "Dilations of linear maps on vector spaces". In: *Oper. Matrices* 16.2 (2022), 465–477.

[75] E. Levy and O.M. Shalit. "Dilation theory in finite dimensions: the possible, the impossible and the unknown". In: *Rocky Mountain J. Math.* 44 (2014), 203 – 221.

[76] R. A. Martínez-Avendaño and P. Rosenthal. Vol. 237. Graduate Texts in Mathematics. Springer, New York.

[77] G. Misra and V. Pati. "Contractive and completely contractive modules, matricial tangent vectors and distance decreasing metrics". In: *J. Operator Theory* 30 (1993), pp. 353 –380.

[78] G. Misra, S. Shyam Roy, and G. Zhang. "Reproducing kernel for a class of weighted Bergman spaces on the symmetrized polydisc". In: *Proc. Amer. Math. Soc.* 141.7 (2013), 2361 – 2370.

[79] G. Misra and N. S. N. Sastry. "Bounded modules, extremal problems, and a curvature inequality". In: *J. Funct. Anal.* 88 (1990), pp. 118 –134.

[80] G. Misra and N. S. N. Sastry. "Completely bounded modules and associated extremal problems". In: *J. Funct. Anal.* 91 (1990), pp. 213 –220.

[81] M. Sal Moslehian, M. Kian, and Q. Xu. "Positivity of 2×2 block matrices of operators". In: *Banach J. Math. Anal.* 13.3 (2019), pp. 726–743.

[82] V. Müller and F.-H. Vasilescu. "Standard models for some commuting multioperators". In: *Proc. Amer. Math. Soc.* 117 (1993), pp. 979 –989.

[83] Gerard J. Murphy. *C^*-algebras and Operator Theory*. Academic Press, London, 1990.

[84] J. von Neumann. "Eine spektraltheorie fur allgemeine operatoren eines unitaren raumes". In: *Math. Nachr.* 4 (1951), pp. 258 –281.

[85] O. Nevanlinna. "Matrix valued versions of a result of von Neumann with an application to time discretization". In: *J. Comput. Appl. Math.* 12&13 (1985), pp. 475 –489.

[86] S. Parrott. "Unitary dilations for commuting contractions". In: *Pacific J. Math.* 34 (1970), pp. 481 –490.

[87] K. R. Parthasarathy. *An Introduction to Quantum Stochastic Calculus*. Birkhauser, Basel, 1992.

[88] V. I. Paulsen. "Every completely polynomially bounded operator is similar to a conraction". In: *J. Funct. Anal.* 55 (1984), pp. 1 –17.

[89] V. I. Paulsen. "Representations of function algebras, abstract operator spaces, and Banach space geometry". In: *J. Funct. Anal.* 109 (1992), pp. 113 –129.

[90] V. I. Paulsen and M. Raghupathi. *An Introduction to the Theory of Reproducing Kernel Hilbert Spaces*. Cambridge Studies in Advanced Mathematics. Cambridge University Press, 2016.

[91] G. Pisier. *Introduction to operator space theory*. Vol. 294. London Mathematical Society Lecture Series. Cambridge University Press, 2002.

[92] G. Pisier. *Similarity Problems and Completely Bounded Maps*. 2nd. Lecture Notes in Mathematics 1618. Springer-Verlag, 2001.

[93] G. Polya and G. Szego. *Problems and theorems in analysis. I. Series, integral calculus, theory of functions*. Springer-Verlag, Berlin, 1998, 1998.

[94] G. Popescu. "Curvature invariant for Hilbert modules over free semigroup algebras". In: *Advances in Mathematics* 158 (2001), pp. 264 –309.

[95] G. Popescu. "Isometric dilations for infinite sequences of noncommuting operators". In: *Trans. Amer. Math. Soc.* 316 (1989), pp. 523 –536.

[96] G. Popescu. "Models for infinite sequences of noncommuting operators". In: *Acta Sci. Math. (Szeged)* 53 (1989), pp. 355 –368.

[97] G. Popescu. "Poisson transforms on some C^*-algebras generated by isometries". In: *J. Funct. Anal.* 161 (1999), pp. 27 –61.

[98] V. Ptak and N. J. Young. "A generalization of the zero location theorem of Schur and Cohn". In: *IEEE Trans. Automat. Control* 25.5 (1980), pp. 978 –990.

[99] M. Putinar. "Uniqueness of Taylor's Functional Calculus". In: *Proc. Amer. Math. Soc.* 89.4 (1983), pp. 647 –650.

[100] C. E. Rickart. "An elementary proof of a fundamental theorem in the theory of Banach algebras". In: *Michigan Math. J.* 5 (1958), pp. 75 –78.

[101] F. Riesz and B. Sz.-Nagy. *Functional analysis. Translated from the second French edition by Leo F. Boron. Reprint of the 1955 original.* Dover Books on Advanced Mathematics. Dover Publications, Inc., New York, 1990.

[102] M. Rosenblum. "Vectorial Toeplitz operators and the Fejér-Riesz theorem". In: *J. Math. Anal. Appl.* 23 (1968), 139–147.

[103] M. Rosenblum and J. Rovnyak. *Hardy classes and operator theory. Corrected reprint of the 1985 original.* Dover Publications, Inc., Mineola, NY, 1997.

[104] W. Rudin. *Real and complex analysis*. McGraw-Hill Book Co., New York, 1987.

[105] B. Russo and H. A. Dye. "A note on unitary operators in C^*-algebras". In: *Duke Math. J.* 33 (1966), pp. 413 –416.

[106] R.V.Kadison. "Isometries of operator algebras". In: *Annals of Math.* 54 (1951), pp. 325–338.

[107] D. Sarason. "Generalized Interpolation in H^∞". In: *Trans. Amer. Math. Society* 127 (1967), pp. 179 –203.

[108] D. Sarason. "On spectral sets having connected complement". In: *Acta Sci. Math. (Szeged)* 26 (1965), pp. 289–299.

[109] J. J. Schäffer. "On unitary dilations of contractions". In: *Proc. Amer. Math. Soc.* 6 (1955), p. 322.

[110] O. M. shalit. "Dilation theory: a guided tour". In: *Bastos, M.A., Castro, L., Karlovich, A.Y. (eds) Operator Theory, Functional Analysis and Applications. Operator Theory: Advances and Applications* 282 (2021).

[111] O. M. Shalit and M. Skeide. "CP-semigroups and dilations, subproduct systems and superproduct systems: The multi-parameter case and beyond". In: *arXiv:2003.05166* ().

[112] V. S. Sunder. *Functional analysis. Spectral theory.* Texts and Readings in Mathematics 13. Hindustan Book Agency, 1997.

[113] B. Sz.-Nagy and C. Foias. *Harmonic Analysis of Operators on Hilbert Space*. North-Holland, 1970.

[114] F. H. Szafraniec. "Reproducing kernel Hilbert Spaces". In: *Operator Theory, Vol.1, Edited by Daniel Alpay* (2015), pp. 1–58.

[115] J. L. Taylor. "A joint spectrum for several commuting operators". In: *J. Funct. Anal.* 6 (1970), pp. 172–191.

[116] J. L. Taylor. "The analytic functional calculus for several commuting operators". In: *Acta Math.* 125 (1970), pp. 1–38.

[117] N. Varopoulos. "On an inequality of von Neumann and an application of the metric theory of tensor products to operator theory". In: *J. Funct. Anal.* 16 (1974), pp. 83 –100.

Index

Γ- unitary, 186
Γ-contraction, 171
Γ-isometric dilation, 195
Γ-isometry, 186
$\mathcal{A}$-morphism, 117
n-positive map, 80

Ando dilation, 91
Ando's dilation theorem, 95
Ando's inequality, 95

Banach-algebras, 47
Bilateral translation, 220

C*-algebras, 58
Character, 55
Characteristic function, 32, 151
Characterizations of Γ-unitaries, 187
Co-invariant injective dilation, 218
Co-ordinate multipliers, 144
Coincidence, 159
Commutant lifting theorem, 97
Commuting d-shift, 115
Complete unitary invariant, 37, 159
Completely positive map, 78, 80
Complex homomorphism, 55
Cone, 11
Continuity of the GNS construction, 5
Continuous functional calculus, 63
Contractive tuple, 109
Creation operator tuple, 112
Cyclic permutation, 220
Cyclic representation, 69, 72
Cyclic vector, 69

Defect space, 217
Dilation, 16
Dilation of a commuting contractive tuple, 122
Dilation of a contractive tuple, 127
Dilation of an operator tuple, 91
Direct sum, 9
Distinguished boundary, 184
Drury-Arveson space, 141, 142

Fejér-Riesz Theorem, 173
Full FOck space, 110
Full Fock space, 111
Functional model, 36, 154
Fundamental equation, 203

Gelfand map, 57
Gelfand's theorem for unital commutative C^*-algebras, 61
Gelfand-Mazur Theorem, 55
GNS construction, 5, 7, 9, 10, 69
GNS pair, 5, 9
GNS space, 5
Gram matrix, 2

Halmos dilation, 16
Hardy space, 8
Hardy space norm, 28
Hyponormal operator, 190

Ideal, 55
Injective power dilation, 216
Intertwining lifting theorem, 98
Isometric dilation, 18

Joint spectrum, 148

Kadison - Schwarz inequality, 87
Kolmogorov construction, 5

Lowdenslager's Criterion, 178

Matrix-valued rational inner function, 106
Maximal commuting piece, 133
Maximal ideal, 55

Minimal isometric dilation, 19
Multi-index, 113
Multiplier, 143
Multiplier algebra, 28

n-positivity, 81
N-step unitary dilation, 17
Neumann series, 49
New characterization for
  Γ-contractions, 210
Non-commutative disk algebra, 125
Normal element in a C^*-algebra, 59
Numerical radius, 189
Numerical range, 189

Oka-Weil Theorem, 173
Operator Fejér-Riesz Theorem, 177
Outer operator, 177
Outer polynomial, 177

Parametrized coordinate functions,
  182
Poisson transform, 124
Polynomial convexity, 172
Polynomial spectral mapping
  theorem, 51
Positive element in a C^*-algebra, 59
Positive kernel, 1, 2
Positive linear functional, 67
Positive map, 80
Positive operators, 11
Projection in a C^*-algebra, 59
Pure Γ-isometry, 186
Pure commuting contractive tuple,
  155
Pure isometric dilation, 106

Realization formula, 99
Representation, 69
Reproducing Kernel, 3
Reproducing kernel Hilbert space, 7
Resolvent set, 50
Riesz's theorem, 8
RKHS, 7, 8

Schäffer construction, 24
Schur map, 81
Schur product, 10
Self-adjoint element in a C^*-algebra,
  59
Set theoretic dilation, 215
Spectral radius, 50
Spectral radius formula, 51
Spectral sets, 169
Spectrum, 50
Spherical unitary, 137
Standard commuting dilation, 121
Standard dilation, 217
Standard non-commuting dilation,
  124
State, 67
Stinespring dilation, 85
Symmetric Fock space, 110, 111
Symmetric tensor product, 110
Symmetrized bidisc, 171
Szegő kernel, 8

Tensor product, 9
Toeplitz C^*-algebra, 116
Transfer function, 104

Unilateral shift, 25
Unilateral translation, 220
Unitary dilation, 18, 40
Unitary element in a C^*-algebra, 59

Vacuum space, 111
Vacuum vector, 111
von Neumann's inequality, 17, 121,
  126

Wandering subset, 222
Wold decomposition, 40

Zaremba formula, 8

Texts and Readings in Mathematics

1. R. B. Bapat: Linear Algebra and Linear Models (3/E)
2. Rajendra Bhatia: Fourier Series (2/E)
3. C.Musili: Representations of Finite Groups
4. Henry Helson: Linear Algebra (2/E)
5. Donald Sarason: Complex Function Theory (2/E)
6. M. G. Nadkarni: Basic Ergodic Theory (3/E)
7. Henry Helson: Harmonic Analysis (2/E)
8. K. Chandrasekharan: A Course on Integration Theory
9. K. Chandrasekharan: A Course on Topological Groups
10. Rajendra Bhatia(ed.): Analysis, Geometry and Probability
11. K. R. Davidson: C* -Algebras by Example
12. Meenaxi Bhattacharjee et al.: Notes on Infinite Permutation Groups
13. V. S. Sunder: Functional Analysis - Spectral Theory
14. V. S. Varadarajan: Algebra in Ancient and Modern Times
15. M. G. Nadkarni: Spectral Theory of Dynamical Systems (2/E)
16. A. Borel: Semi-Simple Groups and Symmetric Spaces
17. Matilde Marcoli: Seiberg Witten Gauge Theory
18. Albrecht Bottcher:Toeplitz Matrices, Asymptotic Linear Algebra and Functional Analysis
19. A. Ramachandra Rao and P Bhimasankaram: Linear Algebra (2/E)
20. C. Musili: Algebraic Geomtery for Beginners
21. A. R. Rajwade: Convex Polyhedra with Regularity Conditions and Hilbert's Third Problem
22. S. Kumaresen: A Course in Differential Geometry and Lie Groups
23. Stef Tijs: Introduction to Game Theory
24. B. Sury: The Congruence Subgroup Problem - An Elementary Approach Aimed at Applications
25. Rajendra Bhatia (ed.): Connected at Infinity - A Selection of Mathematics by Indians
26. Kalyan Mukherjea: Differential Calculas in Normed Linear Spaces (2/E)
27. Satya Deo: Algebraic Topology - A Primer (2/E)
28. S. Kesavan: Nonlinear Functional Analysis - A First Course (2/E)
29. Sandor Szabo: Topics in Factorization of Abelian Groups
30. S. Kumaresan and G.Santhanam: An Expedition to Geometry
31. David Mumford: Lectures on Curves on an Algebraic Surface (Reprint)
32. John. W Milnor and James D Stasheff: Characteristic Classes(Reprint)
33. K.R. Parthasarathy: Introduction to Probability and Measure
34. Amiya Mukherjee: Topics in Differential Topology
35. K.R. Parthasarathy: Mathematical Foundation of Quantum Mechanics (Corrected Reprint)
36. K. B. Athreya and S.N.Lahiri: Measure Theory
37. Terence Tao: Analysis - I (4E)
38. Terence Tao: Analysis - II (4E)
39. Wolfram Decker and Christoph Lossen: Computing in Algebraic Geometry
40. A. Goswami and B.V.Rao: A Course in Applied Stochastic Processes
41. K. B. Athreya and S.N.Lahiri: Probability Theory
42. A. R. Rajwade and A.K. Bhandari: Surprises and Counterexamples in Real Function Theory
43. Gene H. Golub and Charles F. Van Loan: Matrix Computations (Reprint of the 4/E)

44. Rajendra Bhatia: Positive Definite Matrices
45. K.R. Parthasarathy: Coding Theorems of Classical and Quantum Information Theory (2/E)
46. C.S. Seshadri: Introduction to the Theory of Standard Monomials (2/E)
47. Alain Connes and Matilde Marcolli: Noncommutative Geometry, Quantum Fields and Motives
48. Vivek S. Borkar: Stochastic Approximation - A Dynamical Systems Viewpoint (2/E)
49. B.J. Venkatachala: Inequalities - An Approach Through Problems (2/E)
50. Rajendra Bhatia: Notes on Functional Analysis (2/E)
51. A. Clebsch: Jacobi's Lectures on Dynamics (2/E)
52. S. Kesavan: Functional Analysis (2/E)
53. V.Lakshmibai and Justin Brown: Flag Varieties - An Interplay of Geometry, Combinatorics and Representation Theory (2/E)
54. S. Ramasubramanian: Lectures on Insurance Models (2/E)
55. Sebastian M. Cioaba and M. Ram Murty: A First Course in Graph Theory and Combinatorics (2/E)
56. Bamdad R. Yahaghi: Iranian Mathematics Competitions 1973-2007
57. Aloke Dey: Incomplete Block Designs
58. R.B. Bapat: Graphs and Matrices (2/E)
59. Hermann Weyl: Algebraic Theory of Numbers(Reprint)
60. C L Siegel: Transcendental Numbers(Reprint)
61. Steven J. Miller and RaminTakloo-Bighash: An Invitation to Modern Number Theory (Reprint)
62. John Milnor: Dynamics in One Complex Variable (3/E)
63. R. P. Pakshirajan: Probability Theory: A Foundational Course
64. Sharad S. Sane: Combinatorial Techniques
65. Hermann Weyl: The Classical Groups-Their Invariants and Representations (Reprint)
66. John Milnor: Morse Theory (Reprint)
67. Rajendra Bhatia(Ed.): Connected at Infinity II- A Selection of Mathematics by Indians
68. Donald Passman: A Course in Ring Theory (Reprint)
69. Amiya Mukherjee: Atiyah-Singer Index Theorem- An Introduction
70. Fumio Hiai and Denes Petz: Introduction to Matrix Analysis and Applications
71. V. S. Sunder: Operators on Hilbert Space
72. Amiya Mukherjee: Differential Topology
73. David Mumford and Tadao Oda: Algebraic Geometry II
74. Kalyan B. Sinha and Sachi Srivastava: Theory of Semigroups and Applications
75. Arup Bose and Snigdhansu Chatterjee: U-Statistics, M_m-Estimators and Resampling
76. Rajeeva L. Karandikar and B. V. Rao: Introduction to Stochastic Calculus
77. S. Kesavan: Measure and Integration
78. Avrim Blum, John Hopcroft and Ravi Kannan: Foundations of Data Science
79. Anima Nagar, Riddhi Shah and Shrihari Sridharan (Eds.): Elements of Dynamical Systems: Lecture Notes from NCM School
80. Hervé Queffélec and Martine Queffélec: Diophantine Approximation and Dirichlet Series (2/E)
81. M. Vidyasagar: An Introduction to Compressed Sensing
82. Apoorva Khare: Matrix Analysis and Entrywise Positivity Preservers
83. Vivek Borkar and Mallikarjuna Rao: Elementary Convexity with Optimization